T0332363

Manufacturing of
Polymer Composites

To Carol and Sara, who each in their own way made this book possible

Manufacturing of Polymer Composites

B. T. Åström
Department of Aeronautics
Royal Institute of Technology
Stockholm
Sweden

Text © B. Tomas Åström 1997

The right of B.Tomas Åström to be identified as author of this work has been asserted by him in accordance with the Copyright, Designs and Patents Act 1988.

All rights reserved. No part of this publication may be reproduced or transmitted in any form or by any means, electronic or mechanical, including photocopy, recording or any information storage and retrieval system, without permission in writing from the publisher or under licence from the Copyright Licensing Agency Limited, of 90 Tottenham Court Road, London W1T 4LP.

Any person who commits any unauthorised act in relation to this publication may be liable to criminal prosecution and civil claims for damages.

First published in 1997 by:
Chapman & Hall

Reprinted in 2002 by:
Nelson Thornes Ltd
Delta Place
27 Bath Road
CHELTENHAM
GL53 7TH
United Kingdom

02 03 04 05 06 / 10 9 8 7 6 5 4 3 2

A catalogue record for this book is available from the British Library

ISBN 13: 978-0-7487-7076-2

Page make-up by GreenGate Publishing Services

CONTENTS

Disclaimer

While every effort has been made to check the accuracy of the information in this book, no responsibility is assumed by Author or Publisher for any damage or injury to or loss of property or persons as a matter of product liability, negligence or otherwise, or from any use of materials, techniques, methods, instructions, or ideas contained herein.

PREFACE

This book has been written in the belief that a designer of structural polymer-matrix composites must have a good understanding of post-design issues to be able to create an efficient design that results in an easily manufacturable component with predictable properties. While much has been written about composite design and analysis without taking post-design issues into account, preciously little comprehensive material is available on constituent materials, manufacturing techniques, secondary processing, and other post-design issues. While there is unquestionably an immense knowledge of such issues within the composite industry, it is rare that designers possess it, which inevitably leads to designs and eventually composites with disappointing properties and unnecessarily high cost. The intention of this book is to provide a conceptual, albeit still fairly detailed, knowledge of post-design issues to enable avoidance of common design pitfalls. The book is written for engineers or individuals with similar technical background, but it is not primarily written for the worker on the shop floor. Although he (or she) may find parts of the book valuable, he is likely to find it too detailed in areas mainly of relevance to the designer. He may also find that it provides too little substance on hands-on issues, since the intention is not to provide information detailed enough for anyone to go into business manufacturing composites without investigating other sources.

The main intended audience of this book is engineering students, particularly final-year undergraduate students or graduate students. To fully appreciate the entire contents of the book, knowledge of composite design is required and it is therefore recommended that this book be studied following, or possibly concurrently with, a course on composite design. The structure of the book is such that it is best suited for sequential reading, but the contents will be considerably more meaningful if complemented by some hands-on composite manufacturing and characterization labs and factory visits to "get to smell the styrene."

While "no one reads the introduction chapter to a textbook," I believe that at least *Section 1.5* is important enough for careful study, since, in addition to providing a range of application examples, it discusses reasons why composites were selected over competitive materials in these cases; an understanding of the pros and cons of composites is fundamental to sensible use of the material concept. The second most important part of the book (after *Chapter 4*) is *Chapter 2*, which provides sufficient understanding of common constituents to get the gist of their processing behaviors and requirements. The treatment of polymers starts from a level that allows mechanical engineering students with little or no recollection of past chemistry classes to follow it, meaning that the first sections may be trivial to some readers. *Chapter 3* gives quantitative substance to the qualitative treatment in *Chapter 2*. While the bulk of *Chapter 3* should be seen as a reference source, it also discusses some of the more important differences between polymer composites and conventional engineering materials. The heart of the book is *Chapter 4*, which discusses all commercially significant manufacturing techniques as well as some less common, but likely emerging, techniques. The treatment of each technique starts with a rather detailed description of the most prevalent solutions to how to make the raw material conform to the mold, followed by discussions of common raw materials and molds, as well as crosslinking and consolidation requirements. The treatment of each technique is rounded off by qualitative comparisons of technique and component characteristics, and finally application examples with reference to *Chapter 1*. *Chapter 5* treats the important issues of machining, joining, surface treatment, and repair. Of these issues, it is often only joining that is of significant relevance to the designer. A course of limited scope could end after *Chapter 4*, plus *Section 5.2* on joining, and still cover the most important post-design issues. The other parts of *Chapter 5* and the entire *Chapter 6*, which deals with quality control and characterization, often are not of primary interest to the designer, but certainly most relevant for production and quality control departments; most issues in these chapters follow as logical consequences of manufacturing. *Chapter 7* treats the not yet critical issue of composite recycling and provides an introduction to design for recycling, which already is or soon will be commonplace—at least in the automobile industry. *Chapter 8* finally deals with the important, but sometimes disheartening, aspects of the manufacturing environment, but points out that with proper knowledge, routines, and equipment, health and safety risks are minor.

While this book is primarily written as a textbook for engineering students, it will prove a good guide to post-design issues also for the practicing engineer, since a detailed subject index is included and most sections of the book can be read in isolation. For the reader desiring more detailed information than given herein, recommended sources of further reading are listed at the end of *Chapters 2–8*.

It is inevitable that this first edition contains errors and omissions and readers are therefore encouraged to point out such shortcomings to the author.

B. Tomas Åström
Sollentuna
March 1997

ACKNOWLEDGEMENTS

In the course of writing this book I have indebted myself to numerous individuals. First of all I am grateful to Jan Bäcklund for giving me the opportunity to really make some headway on the writing by agreeing to my taking a sabbatical. During this most enjoyable time at the Department of Mechanical Engineering, University of Auckland, New Zealand, I had the pleasure of being hosted by Peter Jackson and Debes Bhattacharyya and managed to write the bulk of the book—not to mention avoid a Swedish winter. Parts of a couple of chapters have also been written and edited during visits to the Center for Composite Materials, University of Delaware, USA, where I have been hosted by Karl Steiner. The conception of the book as well as the hectic final work nevertheless took place in Stockholm.

I am also indebted to a range of individuals and organizations throughout the world for being most helpful in providing photographs, as well as to Jakob Kuttenkeuler and Magnus Burman for creating several of the best-looking drawings. While not directly obvious from the book, a number of individuals have also helped me proofread the text. By far the most diligent proofreader is Krishnan Jayaraman who miraculously survived an entire draft version of the manuscript, followed by Jan Bäcklund and Dan Zenkert who each read several chapters, and Ulf Gedde who provided particularly useful and detailed suggestions for *Chapter 2*. Additional helpful proofreaders include Leif Carlsson, Michael Carlsson, Ingvar Eriksson, Anna Hedlund-Åström, Göran Isaksson, Clas-Åke Johansson, Per Jonsson, Staffan Lundström, Tönu Malm, Jan Nordfeldt, Kurt Olofsson, Erik Persson, Joachim Pettersson, Mikael Skrifvars, Karl Steiner, Staffan Toll, Bernt Åström, three anonymous reviewers, and a number of students in Stockholm, Gothenburg, and Auckland, who have all provided corrections and suggestions. While this book reflects their expertise, I have for reasons of my own not always heeded their advice and the remaining errors are my own.

The computer enthusiast may be interested to know that the bulk of the book was created on a Macintosh PowerBook 540c using Microsoft Word 5.1, Claris MacDraw Pro 1.5, Aldus Photoshop 3.0, CambridgeSoft ChemDraw Pro 3.5.1, and CambridgeSoft Chem3D Pro 3.5.

NOMENCLATURE

Notation	Property	Unit
C_p	Specific heat	kJ/kg °C
d	Cell size (for honeycomb core)	m
D	Flexural rigidity of sandwich	Nm²
E	Modulus (tensile, compressive, flexural)	Pa
G	Shear modulus	Pa
k	Coefficient of thermal conductivity (CTC)	W/m °C
ℓ	Length	m
r	Mandrel radius in filament winding	m
t	Thickness	m
T_g	Glass-transition temperature	°C
T_m	Melt temperature	°C
T_{max}	Maximum continuous-use temperature	°C
T_{proc}	Processing temperature	°C
v	Volume	m³
V	Volume fraction	—
w	Weight, width	kg, m
W	Weight fraction	—
α	Coefficient of thermal expansion (CTE), angle in filament winding	°C⁻¹, °
β	Constituent efficiency factor	—
γ	Working normal rake (in composite machining)	°
ε	Strain (tensile, compressive, flexural)	%
η_0	Zero shear-rate viscosity	Pa·s
ν	Poisson's ratio	—
ξ	End tab taper	°
ρ	Density	kg/m³
σ	Stress (tensile, compressive, flexural)	Pa
σ^*_{lt}	Interlaminar shear strength	Pa
τ	Shear stress	Pa

Subscripts

b	Boss of mandrel in filament winding	
c	Compressive, composite	
$core$	Core	
f	Fiber, flexural	
$face$	Face	

g	Gage
l	Longitudinal
length	Length (for honeycomb core)
m	Matrix
t	Transverse, tensile, tab
v	Void
width	Width direction (for honeycomb core)
$\perp$	Out-of-plane

Superscript

*	Ultimate (stress or strain)

In notations with multiple subscripts, the first one refers to lamina reinforcement orientation and the second to type of testing. σ_{lt}^* thus is the longitudinal tensile strength.

ABBREVIATIONS AND ACRONYMS

3D	Three dimensions, three-dimensional
4HS	Four-harness satin (weave)
5HS	Five-harness satin (weave)
8HS	Eight-harness satin (weave)
AD	Anno Domini
AFRP	Aramid-fiber reinforced plastics
AFV	Alternative fuel vehicle
Al_2O_3	Alumina
ASC	American Society of Composites
ASM	American Society for Materials
ASTM	American Society for Testing and Materials
ATR	Avions Transport Regionale
AWJ	Abrasive waterjet
BC	Before Christ
BMC	Bulk molding compound
BMI	Bismaleimide
BPO	Benzoyl peroxide
CAD	Computer-aided design
CAT	Computer-aided tomography
CEN	Comité Européen de Normalisation, European Committee for Standardization
CEU	Commission of the European Union
CFC	Chlorofluorocarbon
CFM	Continuous filament mat
CFRP	Carbon-fiber reinforced plastics
CMC	Ceramic-matrix composite
CNG	Compressed natural gas
CNS	Central nervous system
CO_2	Carbon dioxide
CSM	Chopped strand mat
CT	Computed tomography
CTC	Coefficient of thermal conductivity
CTE	Coefficient of thermal expansion
CVD	Chemical vapor deposition
DBP	Double-belt press
DETA	Diethylenetriamine

DFA	Design for assembly
DFD	Design for disassembly
DFR	Design for recycling
DGEBA	Diglycidylether of bisphenol A
DGEBPA	Diglycidylether of bisphenol A
DIN	Deutsches Institut für Normung, German Institute for Standardization
DMC	Dough molding compound
DSC	Differential scanning calorimetry
ECM	Electro-chemical machining
EDM	Electro-discharge machining
EN	European norm (standard)
EP	Epoxy
EPA	United States Environmental Protection Agency
ESPI	Electronic speckle pattern interferometry
EV	Electric vehicle
FE	Finite element
FRP	Fiber reinforced polyester (or plastic)
GFRP	Glass-fiber reinforced plastics
GMT	Glass-mat reinforced thermoplastic
GPRMC	Groupement Européen des Plastiques Renforcés/Materiaux Composites, Organisation of Reinforced Plastics/Composite Materials
HAZ	Heat-affected zone
HDT	Heat deflection temperature
HM	High modulus
HS	High strength
HS/S	High strength/strain
HSRTM	High-speed resin transfer molding
HSS	High-speed steel
IARC	International Agency for Research on Cancer
ILSS	Interlaminar shear strength
IM	Intermediate modulus
IMC	In-mold coating
IR	Infrared
ISO	International Organization for Standardization
LCA	Life-cycle assessment
LCP	Liquid crystalline polymer
LSE	Low styrene emission
MCMV	Mine counter-measure vessel
MDI	Methylene diphenyl diisocyanate
MEK	Methyl ethyl ketone
MEKP	Methyl ethyl ketone peroxide
MIBK	Methyl isobutyl ketone

MMC	Metal-matrix composite
MSDS	Material safety data sheet
MSW	Municipal solid waste
NDE	Nondestructive evaluation
NDI	Nondestructive inspection
NDT	Nondestructive testing
OCF	Owens-Corning Fiberglas
OEL	Occupational exposure level
PA 12	Polyamide 12
PA 6	Polyamide 6
PA 6,6	Polyamide 6,6
PAI	Poly(amide imide)
PAN	Polyacrylonitrile
PAS	Poly(aryl sulfone)
PBT	Poly(butylene terephtalate)
PC	Polycarbonate
PCD	Polycrystalline diamond
PE	Polyethylene
PEEK	Poly(ether ether ketone)
PEI	Poly(ether imide)
PEK	Poly(ether ketone)
PEKK	Poly(ether ketone ketone)
PEL	Permissible exposure level
PEL-C	Ceiling PEL
PES	Poly(ether sulfone)
PET	Poly(ethylene terephtalate)
PI	Polyimide
PMI	Polymethacrylimide
PMMA	Poly(methyl methacrylate)
PP	Polypropylene
PPE	Personal protective equipment
PPS	Poly(phenylene sulfide)
PS	Polystyrene
PSU	Polysulfone
PUR	Polyurethane
PVC	Polyvinylchloride
QA	Quality assurance
QC	Quality control
RIM	Reaction injection molding
RRIM	Reinforced reaction injection molding
RT	Room temperature
RTM	Resin transfer molding
SACMA	Suppliers of Advanced Composite Materials Association
SAMPE	Society for the Advancement of Material and Process

	Engineering
SCRIMP	Seeman composites resin infusion molding process
SES	Surface-effect ship
SiC	Silicon carbide
SiO_2	Silica
SMC	Sheet molding compound
SMC-C	Sheet molding compound with continuous and aligned reinforcement
SMC-C/R	Sheet molding compound with a blend of random and continuous, aligned reinforcement
SMC-R	Sheet molding compound with random reinforcement
SPE	Society of Plastics Engineers
SPI	Society of the Plastics Institute
SRIM	Structural reaction injection molding
SRM	SACMA recommended method
STEL	Short-term exposure limit
TDI	Toluene diisocyanate
T_g	Glass-transition temperature
T_m	Crystalline melting temperature
T_{max}	Maximum continuous-use temperature
T_{proc}	Processing temperature
TWA	Time-weighted average
UD	Unidirectional
UHM	Ultra-high modulus
UP	Unsaturated polyester
UV	Ultraviolet
VARI	Vacuum assisted resin injection/infusion
VE	Vinylester
ZEV	Zero-emission vehicle

CHAPTER 1

INTRODUCTION

Composite materials are the most advanced and adaptable engineering materials known to man. While composite materials in the sense used in this book—a conscious combination of two or more distinct material phases into one engineering material where the phases are still discernible—are only about five decades old, it is instructive to consider the evolution of materials and composites in a historical perspective. In school we were probably all taught that (past) civilizations are characterized by the most advanced engineering materials used at the time, e.g. stone, bronze, and iron ages, to indicate the fundamental importance of the state of engineering materials to mankind. It has always been—and to a significant degree still is—the properties of engineering materials that set the limits to man's engineering achievements; the more capable the materials, the greater the scope for groundbreaking engineering achievements.

1.1 The Evolution of Materials

During the stone age flint, which is a ceramic, was the best cutting material known to man. Metals were of little importance, whereas natural polymers, composites, and ceramics dominated in engineering applications (houses, weapons, boats, etc.), see *Figure 1-1*. From then on, development of metal-working skills strongly dominated most engineering achievements well into the 20th century. In particular the position of steel has been remarkable since the middle of the 19th century. However, in the 1960s things started to change and the development in engineering materials is no longer dominated by steel and other metals, which are in relative decline worldwide. Instead polymers, composites, and ceramics are regaining their relative importance to man. The difference is that while natural polymers, composites, and ceramics dominated in the past, the current developments are due to man-made, or synthetic, materials. While these new materials, which are being developed rapidly, open up exciting and sometimes mind-boggling new opportunities, they also require new methods and capabilities in design and manufacturing [1].

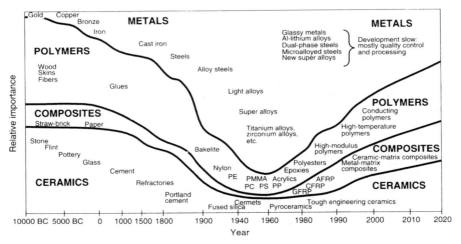

Figure 1-1 Schematic of the evolution of materials used in mechanical and civil engineering applications and their relative importance to mankind. PE, PMMA, PC, PS, and PP are synthetic polymers; most of them are introduced in *Chapter 2*. GFRP, CFRP, and AFRP are abbreviations for glass-, carbon-, and aramid-fiber reinforced plastics, or composites, which are the subject of this book. Note that the relative importance is not in terms of tonnage or value and that the time scale is nonlinear. Redrawn from reference [1]

The preceding discussion and figure allude to naturally occurring polymers and ceramics, but there are also numerous examples of natural composites, including wood, bone, insect exoskeleton, mollusk shell, and countless others (although *Figure 1-1* places wood, and probably would have placed all other natural composites, in the polymer category; it is a matter of definition). Natural composites tend to be weaker and less stiff than synthetic composites, but are often superior in terms of design and manufacture. Natural composites are also "smart materials" in that they adapt to the environment. As an example, a tree grows in such a way that it becomes stronger in the direction where strength is required and any incurred damage is gradually repaired. The composite concept is obviously not an invention.

As far as conscious human use of the composite concept goes, the earliest known example is the early Egyptians, who used straw to reinforce clay from the Nile to make brick. (There is even a biblical reference to this practice: Exodus 5:7.) In the third millennium BC, the Egyptians also made papyrus "paper" from the papyrus reed by placing strands of the reed parallel to each other to form layers, which then were stacked perpendicular to each other in alternate directions; the stack was then allowed to dry under pressure to form the paper-like sheet. Also the Chinese used straw-reinforced brick and in 108 AD invented paper (the word paper is, incidentally, derived from the word papyrus), which is a random, planar arrangement of individual cellulose fibers held together by a binder. Although predominantly known as Chinese and Japanese art forms, the invention of metal-wire reinforcing of ceramics has also been attributed to the Egyptians [2].

In the preceding paragraphs the noun composite (which is derived from the Latin verb *componere*, meaning to put together) has been used in a very generous sense. Moreover, in contemporary language, composite is used in several other contexts than that of this book. The meaning used herein is well established in engineering terminology and refers to a combination of two or more distinct materials into one with the intent of suppressing undesirable properties of the constituent materials in favor of the desirable properties. This definition obviously does not include metal alloys or polymer blends, which are material combinations on the atomic level. The key feature of a composite is that it offers a combination of properties that are not available in any isotropic material; composites are an entirely new and unique material family.

1.2 The Composite Concept

In load-bearing, or structural, applications, composites in most cases comprise a bulk phase enclosing a fibrous reinforcing phase; in conventional terminology one talks of matrix and reinforcement. The objective of the matrix is to integrally bind the reinforcement together so as to effectively introduce external loads to the reinforcement and to protect it from adverse environmental effects. While the matrix gives a composite its shape, surface appearance, environmental tolerance (to high temperature, water, ultra-violet light, etc.), and overall durability, it is the fibrous reinforcement that carries most of the structural loads and thus largely dictates macroscopic stiffness and strength.

Although numerous examples of the composite concept may be distinguished in everyday life, including paper, Kraft paper, particle (chip) board, and reinforced concrete to mention just a few, the most specialized incarnations are those that are the main subject of this book. Practically all successful structural composites have man-made constituents. The matrix may be metallic, ceramic, or polymeric in origin. While metal and to some degree ceramic matrices are used in structural composite applications, polymer matrices are currently by far the most significant matrix category and this book therefore concentrates on polymer composites. There are two main polymer families; thermosets and thermoplastics. The vast majority of current composite applications utilize thermosets, which solidify through a chemical reaction called crosslinking or cure. Thermoplastics differ in that they can be melted and therefore solidify through cooling of the melt without any chemical reaction taking place. In polymer-matrix composites, the common reinforcement types are (in order of decreasing importance) glass, carbon, and polymer. The composite reinforcement may be discontinuous ("short fibers") or continuous ("endless fibers") and randomly oriented or aligned, see *Figure 1-2*.

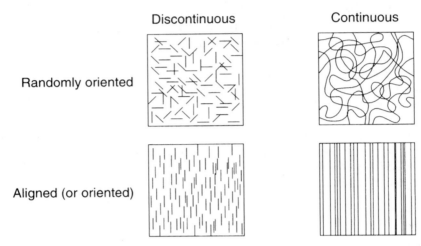

Figure 1-2 Schematic of different reinforcement configurations

Since the reinforcement is the primary load-bearing constituent, its configuration (form and degree of orientation in the matrix) is critical to the macroscopic properties of the composite. It should therefore come as no surprise that the most impressive mechanical properties are found in composites with continuous and aligned reinforcement, see *Figure 1-3*. While clearly structurally inferior, composites with randomly oriented reinforcement are common in both structural and semi-structural applications; whether the reinforcement in such composites is discontinuous or continuous is of secondary importance from a structural point of view. In many applications, different reinforcement configurations are mixed within one and the same component.

Probably the most common reason why polymer composites are used in structural applications is that they offer a given property, often stiffness, at a lower weight than the competition, which most of the time is a metal, but sometimes also plastic, wood, or concrete. One therefore talks of excellent *specific* stiffness and strength, i.e. property normalized with (divided by) density. There are numerous potential advantages of composites other than their excellent structural capabilities, including corrosion resistance, electrical insulation, reductions in tooling and assembly costs, and many more.

Figure 1-3 Fracture surface of continuous carbon-fiber reinforced epoxy composite. The fiber diameter is 7 μm. Photograph courtesy of Jakob Kuttenkeuler, Department of Aeronautics, Royal Institute of Technology, Sweden

These advantages, illustrated with a number of application examples, are further discussed in *Section 1.5*. It is, however, important to realize that these potential advantages assume that appropriate constituents are chosen, since if *in*appropriate constituents are chosen, the result instead may be that the composite alternative turns out to be vastly inferior.

While it is easy to become impressed with the potential advantages of composites, the disadvantages must not be ignored. It is probably essentially true that if one disregards cost, a polymer composite can outperform any other engineering material in all respects except temperature tolerance. It is nevertheless rare that cost is not important, thus one also has to take into account less desirable traits of composites, such as high raw material cost, lack of knowledge and experience, and difficult manufacturing. Composites are therefore certainly not appropriate in all applications; a critical assessment of the material alternatives must always be made in terms of performance-to-cost ratio. It is highly unlikely that decreases in raw material cost due to increasing production volumes alone will make composites directly cost-competitive with other materials in any wide range of applications, but in terms of knowledge, experience, and manufacturing matters can be significantly improved to further enhance the competitiveness of composites.

1.3 Design and Manufacturing

Almost all designers have been educated on and still only work with materials that are essentially isotropic, e.g. metals, and only the odd one has any real knowledge and experience of composites, which quickly make the design task vastly more difficult due to the inherent anisotropy of the material. There are centuries' worth of compiled experience and material data on conventional construction materials to be found in the literature. In contrast, there is at best decades' worth of experience to draw upon for composites and openly available material data bases are fragmented and incomplete. So why abandon known materials that are comparatively easy to design with? The answer is that the potential rewards in product performance, and thus ultimately cost, are vast. However, while composites potentially offer numerous advantages, they require a lot from the designer. Since few designers and other individuals who specify materials feel comfortable with composites, it is unlikely that these will see widespread acceptance until the gospel is spread to a wider audience. In short the answer is education, both at university level and as continuing education of practicing engineers. While countless universities worldwide educate engineers to be capable of designing with conventional materials, only a handful provide any composite-related education worth the name. Although composite design has been a field of active research for a couple of decades and some good textbooks and design guides are available, there has been remarkably little impact in terms of the number of applications. The exceptions are the aerospace and boat-building industries, where composites are well established since they offer clear advantages over conventional materials.

With composites manufacturing is more critical than with conventional construction materials, since material and component are normally manufactured simultaneously. The simultaneous manufacture of material and component adds to the burden of the designer, since he* must also have an understanding of how the component he is designing should be manufactured. While composites offer design opportunities unbeknown to metal design (e.g. far-ranging parts integration and geometrical complexity), there are also numerous restrictions to keep in mind (e.g. sharp corners and significant thickness differences). There is not only the issue of what is technically manufacturable, but also what can be economically manufactured; all aspects of manufacturability are strongly dependent on manufacturing technique and material. As if this were not enough, the properties of the component are strongly dependent on how it was manufactured, both through which technique and under what conditions. There is consequently a strong need for the composite designer to have an

---

* While male pronouns (instead of both male and female) are used throughout this book in the name of readability, this is not intended as an affront to female readers, who are encouraged to make the field of composites less male-dominated.

understanding of both materials and manufacturing. While a university student may find the odd course on composite design, chances are that he will learn little or nothing on manufacturing techniques and processing requirements of raw materials. In fact, he may look into the curricula of universities worldwide and still find extremely few opportunities to learn about composite manufacturing.

Design difficulties apart, several of the other disadvantages of composites one way or another also have their origin in manufacturing. Most proven manufacturing techniques are labor-intensive and thus costly. Partly due to the extent of manual labor, partly due to variations in raw material quality, and partly due to the aforementioned property dependency on processing conditions, repeatability may be poor, leading to a high scrap ratio or expensive post-manufacturing repair to further add to final component cost. Although some manufacturing techniques are certainly well established and reasonably well understood, it is probably fair to say that composite processing science is far from completely investigated. Over the past decade, significant research and development efforts have resulted in many enlightening investigations that have created a more fundamental understanding of important mechanisms in composite manufacturing, but countless unsolved issues remain.

It is the aforementioned shortcomings in composite-related education, the apparent lack of comprehensive textbooks on manufacturing-related topics, and the many unresolved issues in composite processing science that are the ultimate reasons why this book has been written.

1.4 Market

Just as with most new materials (cf. *Figure 1-1*), the development of high-performance polymer composites, often called advanced composites, has been driven by military, and later aerospace, needs, where performance often is more important than cost. While the search for new and improved materials is likely to continue to be driven by military needs and desires, most types of polymer composites have now become commodity materials. This transition from specialized "high-tech" to commodity material concept does not mean that improvements in material performance will cease, but the emphasis of the development has shifted to improvements in design, manufacturability, and ultimately cost.

While the main developmental efforts in the field of advanced composites have been military in origin, a largely parallel development of "low-tech" composites has taken place in for example the electrical and boat-building industries. In electrical applications, the interest was spurred by a need for a nonconductive engineering material, whereas composite boats early on were found to have advantages in terms of manufacturing cost, durability, and reduced need for maintenance.

That composites have become commodity materials in many areas becomes obvious when studying statistics on composite shipments in terms of weight, see *Figures 1-4* and *1-5*. Both in Europe* and in the United States transportation is the major applications area, followed by construction. The figures further illustrate the importance of consumer products, including sporting goods, and applications where electrical insulation and corrosion-resistance are valued. The figures also illustrate the apparent lack of significance of aerospace and defense. However, the data of the figures refer to weight and if value instead were to be considered, aerospace and defense would certainly become significant, since these industries tend to use very expensive materials and manufacturing techniques. Studying *Figures 1-4* and *1-5* it is apparent that direct comparisons between Europe and the United States are difficult since categorizations differ significantly.

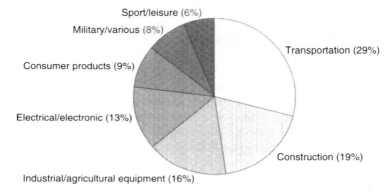

Figure 1-4 Market share of composite shipments in Europe in 1996 (by weight). Application examples of the respective categories include Transportation: all road, rail, water, and air applications, except industrial and agricultural equipment; Construction: general construction, civil engineering, kitchen, and bathroom interiors; Industrial/agricultural equipment: tanks, vats, silos, pipes; Electrical/electronic: circuit boards, equipment housings, insulating materials; Consumer products: furniture and electric household appliances; Sport/leisure: skis, boats, pools. Data from reference [3]

Having considered the relative distribution between application areas, it is instructive to consider the overall historical development of composite shipments (and some predictions). *Figures 1-6* and *1-7* illustrate that composite shipments have been increasing at a rather healthy pace in both Europe and the United States; the developments are in fact rather similar. In 1996 the overall composite shipments were approximately equal in Europe

---

* For the purpose of the statistics presented here, "Europe" is synonymous with the 15 nations of the European Union as well as Norway and Switzerland.

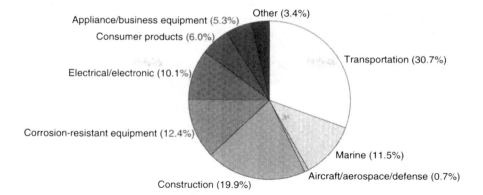

Figure 1-5 Market share of composite shipments in the United States in 1996 (by weight). Application examples of the respective categories include Transportation: road and rail applications; Consumer products: sporting goods and garden equipment; Appliance/business equipment: electric household appliances; Other: medical applications; other categories largely coincide with those of *Figure 1-4*. Data from reference [4]

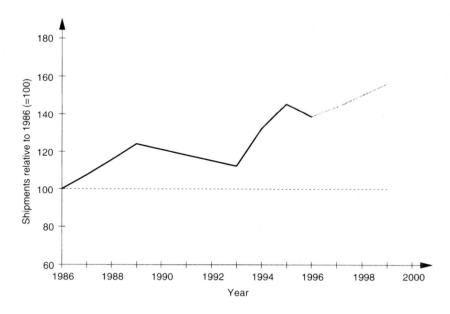

Figure 1-6 Market development for composite shipments in Europe 1986–1999 (by weight). Shaded line indicates predictions. Data, which have been normalized with respect to 1986, from reference [3]

and the United States at 1.5 million tons [3,4]. It is noteworthy that composite shipments appear to follow the general economic trends; the data of the figures clearly indicate downturns approximately coincident with the recession in the beginning of the 1990s. *Figure 1-7* illustrates that in the United States transportation and electrical applications are ever more dominating at the expense of aerospace, defense, and marine applications.

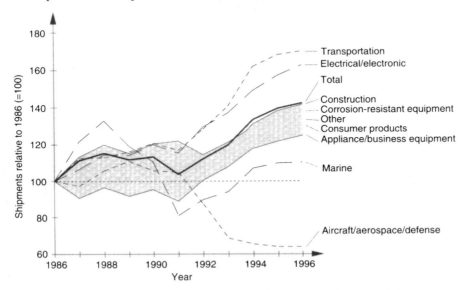

Figure 1-7 Market development of composite shipments in the United States 1986–1996 (by weight). For clarity, the shaded region denotes the maximum spread of construction, corrosion-resistant equipment, other, consumer products, and appliance/business equipment. Data, which have been normalized with respect to 1986, from reference [4]

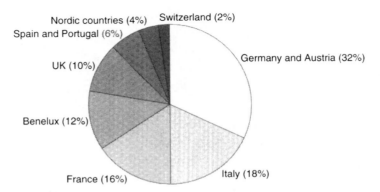

Figure 1-8 Breakdown of composite shipments in Europe in 1996 by country/region (by weight). Data from reference [3]

Taking a closer look at Europe, the dominance of the German-speaking parts is clear, followed by Italy, and France, see *Figure 1-8*. While composite shipments to parts of Europe other than those included in the figure are not insignificant, they are nevertheless not included in these statistics.

1.5 Application Examples

This section attempts to provide both representative and exceptional composite application examples and discusses the likely reasons why polymer composites were chosen instead of competitive materials. *Table 1-1* lists some of the many potential advantages of composites over more conventional engineering materials.

The intent of this section is to give the reader a feeling for the scope of possibilities of composites, but also to provide a common compilation of applications that subsequent chapters may refer to. This section therefore inevitably introduces terminology and refers to material systems and manufacturing techniques that have not yet been defined. For the purposes of the treatment in this chapter, these references can be ignored until they are defined in subsequent chapters.

1.5.1 Transportation Applications

From *Figures 1-4* and *1-5* it is apparent that transportation is the dominant application sector for composites and, according to *Figure 1-6*, the fastest growing one (at least in the United States, but probably in most areas of the world). In this section, "transportation" is taken to include all road and rail applications, whereas marine and aerospace applications have been awarded their own sections below.

The predominant reason for use of composites in transportation applications perhaps ought to be the high specific stiffness and strength obtainable, so as to manufacture a lighter vehicle that therefore can be equipped with weaker drivetrain, suspension, etc., to become even lighter; this is a so-called positive weight spiral. Such weight savings can be exploited to reduce fuel consumption or to increase payload (mainly for commercially used vehicles). The specific-stiffness advantage is indeed exploited where large weight reductions are valued, but mostly in somewhat unconventional applications. Such applications include Formula One and motorcycle racing, where no effort is spared to save weight and thus gain performance. In Formula One cars basically all parts that can be polymer composites are and the body is a self-supporting aircraft-like structure employing aerospace materials, design methodologies, and manufacturing techniques. Another transportation area where composites are slowly being applied to save weight is trains. High-speed trains in Europe and Japan particularly, but also freight and commuter trains contain increasing amounts of composites. There is even an example of entire railway cars being filament wound (see *Section 4.2.5*) as square tubes.

Table 1-1 Potential advantages of composites

Potential advantage	Typical applications or application areas
Specific strength and stiffness	All applications, but particularly in aerospace, defense, sporting goods, industrial machinery, bridges
Fatigue properties	Aircraft, fan blades, pressure vessels, drive shafts
Damping properties	Drive shafts, sporting goods, musical instruments
Corrosive resistance/ low maintenance requirements	Industry, marine, off-shore, bridges
Thermal insulation	Refrigerated containers, building panels, fire walls and doors
Electrical insulation	Electrical housings, transformer spacers, cable trays, ladders
Radar transparency	Aircraft radomes, patient supports for X-ray equipment
Nonmagnetic	Mine sweepers, patient supports for magnetic resonance scanners
Biological inertness	Replacement joints, tendons
Energy absorption	Helmets, bulletproof partitions, armor, crash barriers
Fire and smoke properties	Fire walls and doors, aircraft interiors
Geometrical complexity/ parts integration	All applications, but particularly in automobiles and aircraft
Anisotropy	Aircraft, springs, pressure vessels, piping
Shape flexibility	Used to achieve shapes that cannot (at similar cost) be made with metals, for example in automobile body panels
Thermal expansion/ dimensional stability	Used to eliminate thermal expansion, for example in satellite antennae
Tooling cost	All applications, but particularly in automobiles and aircraft
Manufacturing cost	All applications, but particularly in automobiles and aircraft
Low scrap value	Used to eliminate theft, for example of reusable beer kegs and road signs (where steel/aluminum designs are stolen)
"High-tech" image	Sporting goods, automobiles

However, in most parts of the world fuel still does not cost enough for anyone to be prepared to pay a higher up-front cost for their new car in order to reduce operating costs over a number of years, although the trend is certainly slowly changing towards lighter vehicles. While many car components indeed do tend to get lighter, there is a counteractive trend in that we expect cars with increased performance and more and more electronic gadgets, so the potential net weight loss is largely lost along the way.

One emerging motivation to produce lightweight vehicles is the alternative fuel vehicle (AFV), which runs on non-petroleum energy forms. Naturally, the ideal solution would be a zero-emission vehicle (ZEV), which with current technology essentially is synonymous with an electric vehicle (EV). Due to the unimpressive capacity of available battery types, EVs end up slow, have poor range, and are terribly expensive when compared to conventional automobiles. EVs also may take hours to recharge. Although battery technology is constantly improving it appears that no commercially viable breakthrough in battery technology to solve this technical inferiority is imminent. For the time being, greater improvements in performance may be obtained through reduction in vehicle weight and most reasonably successful EVs consequently rely heavily on composites.

Much work is underway worldwide to produce EVs with acceptable performance at a reasonable cost. In the United States the work is perhaps the most urgent, since in 1990 the California Air Resources Board decided that by 1998, 2 percent of the vehicles sold in the state by the seven companies with the largest market share would have to be ZEVs; by 2003, the figure should be 10 percent. At the time of writing, late 1996, the board has withdrawn its 1998 requirement after recognizing that it would have been impossible to meet, but still stands by the 2003 figure.

While specific stiffness and strength are rarely valued by the automobile industry (nor its customers), there is still the most convincing argument of them all for use of composites—reduction in production cost. A set of molds for sheet metal stamping is extremely expensive and such an investment only pays off if production series are long. For vehicles manufactured in short series, such as sports cars and new models for which the sales potential is not known, such investments may be prohibitively large. If composites are used instead of sheet metal, mold costs are much lower since fewer molds are required (sheet metal stamping is incremental, thus requiring several sets of molds, while compression molding of composites (see *Sections 4.2.4* and *4.3.3*) requires one set only). Molds for composites may also be much less rigid, since pressures in many manufacturing techniques are much lower than in sheet metal stamping. In fact, for short series it is quite feasible to use molds that themselves are polymer composites.

Another major cost saver, which is valid for any length series, is to employ the possibility of extensive geometrical complexity to integrate what in steel would be several smaller parts into one composite component;

far-ranging parts integration may lead to significantly reduced assembly costs. Cost reduction without doubt is the most persuasive argument, but the aforementioned possibility of geometrical complexity also may be used to manufacture, for example, body panels that cannot (at least at a reasonable cost) be made from sheet metal, thus allowing a greater design freedom and potentially more appealing vehicle aesthetics.

There are by now numerous examples of the use of composites in automotive applications, but the best known is probably still the venerable Chevrolet Corvette. The Corvette has had composite body panels since its introduction in 1953 and continues to be General Motors' testbed for composites, where novel composite applications are tried before being introduced in other models. While there are ample examples worldwide of the design freedom and parts integration that composites offer, examples of composite body panels fastened to a conventional steel space frame include several Lotus cars, the now no longer manufactured BMW Z1 and Pontiac Fiero (see *Figure 1-9*), the Dodge Viper (see *Figure 1-10*), and the Renault Espace (see *Figure 1-11*).

Figure 1-9 The Pontiac Fiero GT with composite body panels, some manufactured using reinforced reaction injection molding (see *Section 4.2.3.5*)

Whether a composite solution is chosen or not strongly depends on series length, partly due to the previously mentioned differences in mold costs and partly due to the fact that composite manufacturing in most cases is more time-consuming than sheet metal stamping. On the other hand, the marginal cost of manufacturing a composite component is often higher than the metal counterpart, so if a newly introduced vehicle becomes a commercial success and sales increase, sheet metal stamping may gradually become a financially sounder solution than composites. As series grow longer a short

Figure 1-10 The majority of the glass-reinforced body panels of the Dodge Viper RT/10, with an 8-liter V10 engine providing 400 hp, are manufactured through resin transfer molding (see *Section 4.2.3.2*). Photograph courtesy of Harry Karlsson Bil, Partille, Sweden

Figure 1-11 All body panels of the new Renault Espace (except for the hood, which is sheet steel) are compression molded from glass-reinforced unsaturated polyester sheet molding compound (see *Section 4.2.4*). Photograph courtesy of Renault Presse

cycle time also becomes more important. These may very well be the reasons why the body panels of the General Motors minivans (conceptually akin to the Renault Espace in *Figure 1-11*), which originally were composites, now are manufactured from steel.

Series length similarly determines which composite manufacturing technique is the most economical. The Renault Espace originally had composite body panels that were resin transfer molded (see *Section 4.2.3.2*), but as sales have picked up compression molding (see *Section 4.2.4*), which is less labor-intensive and more rapid, has become a more sensible manufacturing route. For really short production series, a significant degree of manual labor may make sense; the body panels of the Dodge Viper are resin transfer molded, which requires a moderate degree of manual involvement, while the body panels of the Lotus Elise are hand laidup (see *Section 4.2.1.1*), which requires even more manual labor.

Composites are also used in some truck and bus applications, usually in the form of body panels (see *Figure 1-12*). Other applications include a dashboard (see *Figure 1-13*), and even complete buses (see *Figure 1-14*).

Figure 1-12 Body panels on heavy vehicles that see considerable wind loads are often composites. For example the spoiler at the top, part of the grille, and the forward corner panels of the driver's cab of this Scania truck are compression molded from a glass-reinforced unsaturated polyester sheet molding compound (see *Section 4.2.4*). Most other body panels, as well as dashboard and several other interior structures, are unreinforced thermoplastics. Photograph courtesy of Scania Sverige, Södertälje, Sweden

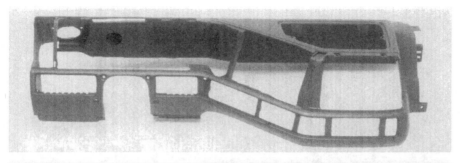

Figure 1-13 Dashboard of a heavy Volvo truck compression molded from glass-reinforced polypropylene (see *Section 4.3.3.1*); the molded and painted component is shown at the top and the complete dashboard below it. Photographs courtesy of Volvo Truck Corporation, Göteborg, Sweden

Although not yet very common, composites are also used in truly structural vehicle applications. The previously discussed Formula One body is self-supporting and some rare sports cars have also eliminated the traditional metal frame. Most major automobile manufacturers have also one way or another evaluated the concept of using a self-supporting composite base plate in conventional automobiles, but so far this concept has not been made commercial. Self-supporting composite bodies are, however, used in some buses (see *Figure 1-14*), truck trailers, and specialty vehicles (see *Figure 1-15*). Structural components in vehicle applications include wheels, battery trays, drive shafts, pressure vessels for compressed natural gas (CNG), leaf springs, flywheels (for energy storage on AFVs), and front ends (see *Figure 1-16*), to mention but a few. In many of these cases it is the reduction in the number of parts and thus the assembly that is the reasons for use of composites. For example, the one-piece composite front end shown in *Figure 1-16* would have required eleven steel parts, while a one-piece composite drive shaft for a light truck may replace a two-piece steel construction with a joint. Composites also turn up in an increasing number of non-structural or semi-structural under-the-hood applications, where temperatures are on

Figure 1-14 Neoplan bus with self-supporting sandwich* body of glass- and carbon-reinforced unsaturated polyester on expanded polymer foam core. This bus weighs a third less than any comparable bus and offers a load capacity slightly larger than its empty weight. Photograph courtesy of Gottlob Auwärter, Stuttgart, Germany

Figure 1-15 All-terrain vehicle with self-supporting sandwich body. The sandwich panels, which have polyvinyl chloride and polyurethane foam cores and glass-reinforced unsaturated polyester faces, are manufactured through resin transfer molding (see *Section 4.2.3.2*). The six panels that make up each unit are adhesively joined using an epoxy adhesive. Photograph courtesy of Hägglunds Vehicle, Örnsköldsvik, Sweden

---

* In composite terminology, a sandwich is a structure where a relatively weak and light material, such as a polymer foam or honeycomb material, is interleaved between and integrally bonded to two stiff sheets of material, most often composite laminates. One talks of a core being sandwiched between two faces. The purpose of using a sandwich structure is generally to obtain high flexural rigidity; the sandwich concept is further discussed in *Section 2.4*.

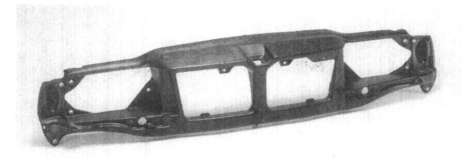

Figure 1-16 One-piece front end of the Volvo 850 compression molded from glass-reinforced polypropylene (see *Section 4.3.3.1*). In this case, the alternative steel solution would have comprised eleven stamped components spot-welded into one structure. Substantial reductions in molding and assembly costs were the deciding factors in favor of the composite alternative, although the weight was also lowered by 40 percent. Photograph courtesy of Volvo Car Corporation, Olofström, Sweden

the order of 150°C and sometimes higher. New materials and manufacturing improvements enable applications such as noise shields, heat shields, engine top covers, oil sumps, and air intake manifolds, although experimental engines have been made almost entirely from composites.

1.5.2 Marine Applications

The fiberglass, or glassfiber, boat was the first large-scale application of composites and the marine industry is still one of the major consumers of composites (cf. *Figure 1-5*). Leisure boats are probably also the application where the layman is most likely to know that composites are extensively used. Almost all leisure craft—from the smallest dingy to competition powerboats and sailing yachts—have been manufactured from composites for many years. The reasons for this relative dominance of composites in leisure craft are several, but the dominating impetus is once again cost; composites boats are relatively inexpensive. One of the most significant reasons for the lower cost is that manufacturing of composite boats is amenable to mass-production (which with boats still translates into small numbers if one compares with the automobile industry) and that fewer craftsman's skills are required than to work with wood or metal. Production costs are also positively influenced by the fact that a mold to manufacture a boat is inexpensive and relatively simple to construct. It therefore makes economical sense to manufacture very short series or even one-of-a-kind boats. Compared to other material alternatives, the initial costs involved in starting to manufacture composite boats are very low, since few specialized tools or facilities are needed. The basic knowledge needed to get started with composites is quite quickly obtained, although it may take a lifetime to become a professional. Yet another advantage is that composites place few restrictions on the all-important shape of the hull. From the owner's point of view it is finally extremely attractive that composite boats, relatively speaking, require very little maintenance since they do not rot or rust. Indeed, a great many of the very first composite boats made in the 1950s are still in use, largely unaffected by the elements.

Figure 1-17 shows a powerboat typical of an entire class of boats sprayed up from discontinuous glass-reinforced unsaturated polyester (see *Section 4.2.1.2*). While the hull of this type of rather small boat almost always is a single-skin laminate, deck and floorboards are often sandwich structures in order to obtain sufficient stiffness. Larger powerboats, such as that shown in *Figure 1-18*, are generally hand laidup; sandwich structures tend to be used throughout hull, deck, and floorboards of such boats (see *Section 4.2.1.3*).

At the other end of the spectrum are extreme competition powerboats and sailing yachts, such as that shown in *Figure 1-19*. In this category, no effort (or money) is spared in order to build the winning boat and the aim is almost always weight reduction. Both hull and deck of such boats are normally hand laid-up sandwich structures with high-performance continuous-fiber reinforced laminates and cores (see *Section 4.2.1.3*). While

Figure 1-17 Powerboat sprayed up from discontinuous glass-reinforced unsaturated polyester (see *Section 4.2.1.2*). The hull is a single-skin laminate, while deck and floorboards are sprayed up sandwich structures. The sandwich structures have an expanded polyvinyl chloride foam core, except at loading points where marine plywood is used. Some hatches are manufactured through resin transfer molding (see *Section 4.2.3.2*). Photograph courtesy of Ryds Båtindustri, Ryd, Sweden

Figure 1-18 Powerboat hand laid-up from glass-reinforced unsaturated polyester (see *Section 4.2.1.3*). Hull, deck, and most floorboards are sandwich structures with polyvinyl chloride foam core and laminates are reinforced with a combination of mats and fabrics. Photograph courtesy of Aplicator System, Mölnlycke, Sweden

Figure 1-19 Black Magic, New Zealand's challenge in the 1995 America's Cup in San Diego, is an example of advanced composites usage. Black Magic trashed the competition and won all races to ensure that the 1999 America's Cup will be held in Auckland, New Zealand. Photograph by Christian Février

manufacture of leisure boats is a relatively uncomplicated procedure, manufacture of competition boats more and more resembles techniques used in the aerospace industry. The same kind of trends largely apply also to the materials used; glass-reinforced unsaturated polyester is gradually replaced by high-performance materials such as carbon- and aramid-reinforced vinylester or epoxy, while low- to medium-performance cores are replaced by high-performance expanded polymer foams and sometimes even honeycomb cores. More recently and particularly in competition yachts, composites are becoming the material of choice also in masts, booms, and rigging, where true aerospace-grade materials (pre-impregnated carbon-reinforced epoxies) and manufacturing techniques (hand layup followed by autoclave crosslinking, see *Section 4.2.2*) are becoming the norm.

Composites have also made inroads into some specialized shipbuilding applications, where the sandwich concept with glass-reinforced unsaturated polyester laminates on expanded polymer foam or balsa cores is utilized. One example is surface-effect ships (SES), which are catamarans with curtains between the keels in bow and stern allowing an air cushion to be maintained between the keels; the air cushion lifts the hull out of the water to significantly reduce drag and thus allow high speeds to be reached (see *Figure 1-20*). Some SESs are capable of transporting up to four hundred passengers at speeds in excess of 50 knots. The largest composite sandwich SES built measures 40 by 15 meters. In this application composites help achieve a low enough weight to suspend an entire ship on an air cushion and hand layup is highly competitive since large structures are manufacturable and few look-alike SESs are built.

Figure 1-20 SES capable of transporting 341 passengers at speeds up to 45 knots. This SES is a sandwich construction with hand laid-up glass-reinforced laminates on a polymer foam core (see *Section 4.2.1.3*). Photograph courtesy of Brødrene Aa, Hyen, Norway

Composite sandwich ships are also used in military applications. For example mine-sweepers, or mine counter-measure vessels (MCMV), longer than 50 meters, wider than 10 meters and with displacements up to 400 tons have been built entirely from composites. In this application composite materials and wet hand layup offer substantial advantages. The non-magnetic material is desirable so as not to risk detonating magnetically sensitive mines, the foam core sandwich is damage tolerant to underwater detonations, and hand layup is once again highly competitive for size and series-size reasons.

Probably the most spectacular composite ship so far is the 72 m long single-hull stealth surface attack ship built for the Swedish Navy, see *Figure 1-21*. Apart from probably being the largest composite structure ever built, it is unusual as a marine structure in almost entirely utilizing carbon reinforcement in the sandwich laminates. Early analyses showed that for a given stiffness carbon-reinforced laminates are actually cheaper than glass-reinforced laminates, since the laminate thickness can be reduced to a third and

Figure 1-21 Artist's rendition of 72 m long surface attack ship built for the Swedish Navy. The ship is a sandwich structure with polyvinyl chloride foam core and carbon-reinforced vinylester laminates. Laminates and sandwich structure are manufactured simultaneously through vacuum-injection molding (see *Section 4.2.3.3*) for most of the structure, but hand layup (see *Section 4.2.1.3*) is also used to some degree. Drawing courtesy of Defense Materiel Administration, Stockholm, Sweden

labor costs are thus drastically lowered. Thinner laminates naturally also result in a lighter structure, allowing further weight savings in engines and drivetrain. The complete weight of the composite structure will be 150 tons of which 40 tons are carbon fibers, while the final displacement is expected to be around 600 tons. Four of these ships are currently on order and the first will be launched around the turn of the century.

1.5.3 Aerospace and Military Applications

Judging from *Figures 1-4* and *1-5*, aerospace and military applications would appear insignificant by weight, but both areas have been and certainly still are vital to the development of composites. One reason why aerospace and military applications are more relevant than they appear from statistics is that it is in these application areas that the most expensive raw materials are used. However, the main reason is that due to the extreme desire to obtain the ultimate in performance, most of our knowledge of composite design and all high-performance materials one way or another have their origins in these fields of application.

While the specific properties of composites are seldom seriously taken advantage of in road, rail, and water-bound vehicles, the opposite holds true in aerospace applications. Saved weight in a flying craft directly translates into increased load-carrying capability or performance enhancements and in fighter aircraft such performance enhancements may be literally the difference between life and death. Perhaps less obvious is the fact that composites can lower manufacturing cost in many applications due to the possibility of parts integration (as previously discussed for transportation applications). Yet another advantage is that composites have significantly better resistance to fatigue than light metals. Moreover, the radar transparency of glass-reinforced composites has long been used in aircraft radomes (nose cones), while a quite recently exploited aspect of composites is their inherent stealth properties (although much of the stealth technique lies in use of an angular geometry to reflect a minimum of radiation).

That the development of composites in aerospace applications is driven by military developments is particularly obvious in the United States (see *Figure 1-22*), although all modern military aircraft use significant amounts of composites (see *Figure 1-23*). The material choices (and often locations) in many military aircraft tend to be shrouded in secrecy, but as a rule composites are based on preimpregnated carbon-fiber reinforced epoxy that is crosslinked in an autoclave (see *Section 4.2.2*), although glass and aramid reinforcement as well as more exotic matrices than epoxies are also used to some degree. While many composites are single-skin laminates, sandwich components with honeycomb are in extensive use. Common polymer composite components on military aircraft include control surfaces (rudders, flaps, etc.), wing skins and substructures, leading edges, complete vertical and horizontal stabilizers, radomes, landing gear doors, access doors, weaponry, and external fuel tanks.

Figure 1-22 The Lockheed YF-22A Advanced Tactical Fighter consists of significant amounts of both thermoplastic (13 weight percent) and thermoset (10 weight percent) composites [5]. Composite components include wing skins and spars, as well as skins of vertical and horizontal tails. While the production F-22 was expected to have 35 weight percent composites [5], the first deliveries contain less than 20 weight percent composites, partly due to manufacturing difficulties [6]. Photograph courtesy of Boeing Commercial Airplane Group, Seattle, WA, USA

Figure 1-23 Up to 30 percent of the structural weight of the multi-role combat Saab Gripen (Griffin) is polymer composite. The Gripen is likely unique in that the entire wings, including all of the wing substructure, are composite. Other composite components include vertical stabilizer, canard wings, control surfaces, access doors, main landing gear doors, and radome. Components are hand laid up from carbon-reinforced epoxy prepregs and crosslinked in an autoclave (see *Section 4.2.2*). Extensive use is made of the sandwich concept using aluminum honeycomb core. Photograph courtesy of Saab Military Aircraft, Linköping, Sweden

Composites are also much used in civilian aircraft applications, but introductions of new composite components are for safety reasons considerably more cautious. Also in commercial airliners the potential weight saving is highly valued; it has been calculated that an excess kilogram in an airliner approximately requires another 130 liters of fuel per year [7], which instead could be used to increase range or payload. However, manufacture of airliners is such a cost-conscious business that weight savings are often not worth an increase in production cost, so unless weight savings are substantial, reduction in manufacturing cost is a more important motive. Consequently, the potential for parts integration translated into cost is one of the strongest arguments in favor of composites in civilian aircraft as well. As an example, the metal rudder of Airbus Industrie's A300 and A310 widebody jets was already in 1983, and at similar cost, replaced by a carbon-reinforced epoxy sandwich with honeycomb core. This change resulted in a 22 percent weight decrease and reduction in the number of parts from over 17,000 to 4,800 (since bonding eliminates rivets) [7].

Not surprisingly composite applications in airliners are not entirely unlike those in military aircraft; they include control surfaces, parts of wing skins, leading edges, entire vertical and horizontal stabilizers, fairings, engine nacelles, thrust reversers, propellers, radomes, landing gear doors, access doors, and large parts of cabin interior and cargo compartments. While American aircraft manufacturers tend to lead the way in terms of composites use in military applications, European aircraft manufacturers, particularly Airbus and Avions Transport Regionale (ATR), appear to be ahead in civilian applications (see *Figure 1-24*), although Boeing is increasing its use of composites for each new model (see *Figure 1-25*). The most advanced composite structures in civilian aircraft are the all-composite empennages, i.e. entire horizontal and vertical stabilizers, including control surfaces, found on all Airbus models since 1987. There is now talk of Airbus planning to make wings and eventually the entire fuselage from composites. Currently, Airbus's largest models, the A330 and A340, are said to contain in excess of 15 tons of composites each [7].

Virtually all composite components in civilian aircraft are hand laid up from epoxy prepregs with carbon, aramid, and glass reinforcement, although automated tape layup is used for some Boeing components (see *Section 4.2.2*). Many composites are sandwich components, almost always with Nomex honeycomb core. Hand layup strongly dominates manufacturing due to the relatively short series, but automated layup techniques are used to a limited albeit increasing degree.

Although their share of composite consumption is small compared to commercial airliners, light sport aircraft, such as gliders, small propeller planes, and various ultralight aircraft deserve to be mentioned. While mass-produced small propeller aircraft tend to employ conventional materials, a range of light sport aircraft are built almost entirely using composites. It has

Figure 1-24 Airbus has largely been at the forefront in terms of composites in commercial aircraft for a number of years. In 1983, Airbus introduced a single-piece composite rudder, two years later the entire vertical stabilizer was converted into composites, and since 1987 also the horizontal stabilizer has been composite. Now all Airbus models, including the A321 in the figure, have all-composite empennages. Composites are also used in control surfaces, fairings, engine nacelles, thrust reversers, radomes, floor support struts, and most of cabin interior and cargo compartments, to mention but a few applications. Photograph courtesy of Airbus Industrie, Blagnac, France

Figure 1-25 The Boeing 777 is the first American commercial aircraft with composite empennage. Skin panels, stringers, angles, brackets, etc. are manufactured through automated tape layup (see *Section 4.2.2*) [8]. Other composite components in Boeing aircraft include the types mentioned for Airbuses above. Photograph courtesy of Boeing Commercial Airplane Group, Seattle, WA, USA

been estimated that approximately 80 percent of all such composite aircraft, which number in excess of 13,000, have been built in Germany. It is noteworthy that most of these have been built using manufacturing techniques more similar to those used in boat building than those normally associated with aircraft manufacture [7].

Modern helicopters make even greater use of composites than aircraft. While for example helicopter main and tail rotor blades have been glass-reinforced composites for many years, an increasing number of metal components are replaced by composites to reduce manufacturing cost and save weight. In new helicopter models on both sides of the Atlantic, such as the Eurocopter Tiger attack helicopter and the NH 90 freighter (see *Figure 1-26*) as well as the Boeing V-22 tilt-rotor and Sikorsky RAH-66 Comanche helicopters, the concept of all-composite fuselages is being implemented, although the future of all these helicopters appears a little uncertain at the time of writing. That composites are being used extensively in helicopters, which jokingly are referred to as flying fatigue machines, is testament to the excellent fatigue characteristics of composites.

An application area that consumes very small amounts of polymer composites, but where gains from weight savings are even greater than in aircraft, is spacecraft. Considering the immense cost of launching a kilogram of payload into space, the efforts going into weight-optimizing the structures of both payload and launch vehicle (see *Figure 1-27*) are far-ranging and they

Figure 1-26 The Eurocopter NH 90 is a near-complete composite structure, which relies on sandwich structures with carbon-reinforced epoxy faces on Nomex honeycomb core for the primary structure, although glass- and aramid-reinforced faces are also used to some degree. One of the main reasons for using composites, which are hand laid up and autoclave consolidated (see *Section 4.2.2*), is that overall manufacturing costs are sharply reduced due to far-ranging parts integration. Photograph courtesy of Eurocopter Deutschland, München, Germany

thus contain a lot of composites. An extreme example of this is Orbital Sciences Corporation's launch vehicle Pegasus XL, which is a near-complete composite structure. The Pegasus, which is a three-stage solid propellant rocket, is air-launched from an aircraft and roughly halves the cost of boosting payloads into orbit compared to earth-launched competitors.

Figure 1-27 The European Space Agency's Ariane 5 launch vehicle makes extensive use of composites. Photograph courtesy of Aerospatiale, Les Mureaux, France

Composites are particularly well suited to applications where the primary load case is (internal) pressure. The manufacturing technique of filament winding (see *Section 4.2.5*) has been most successfully applied in manufacturing solid rocket motors, rocket nozzles, and fuel tanks for secondary propulsion systems (see *Figure 1-28*). This is the same concept as used for CNG tanks for AFVs and in auxiliary fuel tanks mounted under the wings of military aircraft, although in these two applications tanks are normally cylindrical instead of spherical. Nevertheless, it deserves to be pointed out that many components of a launch vehicle, not to mention a space shuttle which is to be reused, are exposed to such high temperatures that polymer matrices may no longer suffice. For such applications, ceramic and carbon matrices have been developed, but these unique classes of composite materials are only very briefly discussed in this book.

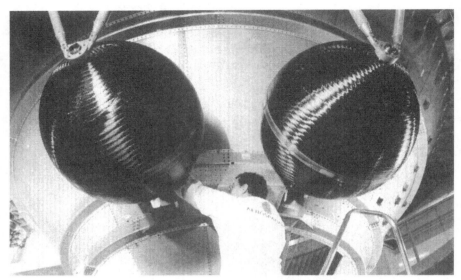

Figure 1-28 Tanks for high-pressure fuel on the central cryogenic stage of the Ariane 5 launch vehicle. The carbon-reinforced spherical tanks are filament wound onto a gas-impermeable metal liner (see *Section 4.2.5*). Photograph courtesy of Aerospatiale, Les Mureaux, France

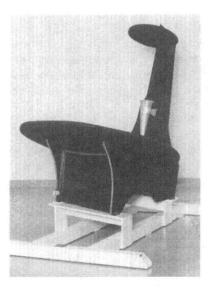

Figure 1-29 Satellite antenna designed to eliminate dimensional changes due to temperature variations. The structure is a hand laid up and autoclave consolidated (see *Section 4.2.2*) sandwich with honeycomb core and carbon-reinforced epoxy laminates. Photograph courtesy of Saab Ericsson Space, Göteborg, Sweden

Depending on whether they are exposed to direct sunlight or in the earth's shadow, satellites are exposed to very large temperature differences that could cause dimensional changes to antennae which could completely distort telecommunication signals for example. Satellite antennae are excellent examples of how composites may be designed to have virtually no thermal expansion over a very wide temperature range (see *Figure 1-29*). Solar arrays, general panels, tubular truss members, etc. on satellites are also often polymer composites.

Composites are also being introduced into a range of non-aerospace military applications. Arguments in favor of composites in such applications include lower weight, energy absorption, stealth (cf. *Figure 1-21*), and many others. The lower weight is mainly employed to improve portability of weapons (see *Figure 1-30*), bridges, radio antennae, etc., while the good energy absorption is used in helmets and in the armor of for example personnel carriers and even tanks.

Figure 1-30 One of the world's best-selling composite launch tubes for delivering antitank projectiles. The tube is filament wound from glass-reinforced epoxy without use of a liner (see *Section 4.2.5*). Although this is a disposable weapon, the tube would be capable of delivering 10–20 projectiles. Photograph courtesy of Bofors Weapon Systems, Karlskoga, Sweden

1.5.4 Construction Applications

While a precise definition of this category is difficult, it is herein taken to include the categories of "Construction", "Industrial/agricultural equipment", and "Corrosion-resistant equipment" of *Figures 1-4* and *1-5*. With this categorization, this is by far the most important area of composite applications (by weight). While civil engineers by and large appear to be quite uninterested in looking beyond conventional civil engineering materials, the situation is ever so slowly changing and some individuals in the profession are starting to see the advantages offered by composites.

In civil engineering applications light weight is seldom sought (after all, the standard is concrete and steel, so the industry is used to hauling a lot of weight), although there are some examples to the contrary. Light weight may prove advantageous when transportation is a problem and when heavy machinery is not available to aid in assembly. There are several examples of bridges being erected in national and state parks in the United States where trucks or helicopters could not or were not allowed to operate. Bridge components therefore had to be carried by man or mule to the site of erection, which hardly would have been possible with a steel or concrete design. Once the construction material is on site, the actual erection is significantly simplified if all components can be moved by workers instead of heavy machinery. Indeed, the most spectacular bridge application to date, the Aberfeldy footbridge over the river Tay in Scotland (see *Figure 1-31*), was

Figure 1-31 The Aberfeldy footbridge over the river Tay in Scotland is 113 m long, has a main span of 63 m, and a deck width of 2.2 m. The entire deck structure, hand rails, and A-frame towers are pultruded glass-reinforced unsaturated polyester (see *Section 4.2.6*) and the cable stays polyethylene-coated Kevlar ropes. The deck structure is assembled from a modular system of 0.6 m wide and 6 m long hollow sections weighing 66 kg each. With the exception of the small concrete foundations, the entire bridge is composite [9]. Photograph courtesy of Maunsell Structural Plastics, Beckenham, Kent, UK

erected in 1992 by eight students in ten weeks without the use of heavy machinery (except for the small concrete foundations) [9]. Light weight was also one of the main reasons for choosing a composite solution for the world's first drawbridge for vehicular traffic (see *Figure 1-32*), since this meant that both lifting tower and counterweight could be eliminated.

Figure 1-32 The Bonds Mill Bridge on the Stroudwater Navigation canal in England is the world's first drawbridge for vehicular traffic. The bridge is assembled from the same modular system as used for the Aberfeldy Bridge (cf. *Figure 1-31*). Photograph courtesy of Maunsell Structural Plastics, Beckenham, Kent, UK

Composites are being predicted to have a promising future in bridge applications in the United States based on the observation that its bridges, many of which were built in a rather short time frame, are decaying at a rapid rate and will need to be replaced at an equally rapid pace. Rather than closing off a highway overpass for months to build a new concrete deck, the argument goes that composite decks could be assembled beside the existing overpass (while it is still in use) and when ready, the old deck would rapidly be demolished and the complete composite deck could then be lifted in place by a single crane in a matter of hours. While this potentially enormous application area is indeed alluring and a few full-scale composite bridges are being built, it so far largely remains elusive.

A bridge-related application area that is seeing quite a lot of full-scale trials is so-called retrofitting of earthquake-damaged bridge columns, mainly in California. Following earthquakes, many steel-reinforced concrete columns of bridges and overpasses end up cracked. Unless seriously damaged, such columns can still perform satisfactorily if they are just laterally supported. Previously such retrofitting entailed enclosure of the column in two custom-made half-cylinder steel shells which then were welded. More novel approaches include filament winding (see *Section 4.2.5*) of a composite jacket onto the column using a machine traveling around the column. Similar concepts include wrapping of dry reinforcement onto the concrete followed by subsequent reinforcement impregnation using vacuum-injection molding (see *Section 4.2.3.3*). This latter concept, as well as that of bonding of already crosslinked composites onto decaying concrete structures, is also used outside of North America.

While light weight may often be a positive side effect for reasons already discussed, a more common argument in favor of composites in construction is that dramatic improvements in corrosion resistance, and indirectly reduction or elimination of the need for maintenance, are obtainable. These may prove very convincing arguments for structures in chemical plants (see *Figure 1-33*), waste water treatment plants (see *Figure 1-34*), harbors, and other corrosive environments. Many such construction applications, which consume large amounts of composites, employ general and often standardized construction members (such as those shown in *Figures 1-33* and *1-34*) in full analogy with the standard procedure for steel beams.

Figure 1-33 Standardized I-beams pultruded from glass-reinforced unsaturated polyester and vinylester (see *Section 4.2.6*) used as platform and walkway support in chlorine manufacturing operation.
Photograph courtesy of MMFG, Bristol, VA, USA

Figure 1-34 Grating and handrails pultruded from glass-reinforced vinylester (see *Section 4.2.6*) used in municipal waste water treatment plant. Photograph courtesy of MMFG, Bristol, VA, USA

Figure 1-35 Subsea structure to protect the wellhead of an oil well, for example from fishing trawls. This sandwich structure, which consists of glass-reinforced laminates on a polyvinyl chloride foam core, was hand laid up (see *Section 4.2.1.3*). Photograph courtesy of Brødrene Aa, Hyen, Norway

Other applications where large amounts of composites are used include building panels, piping, gasoline and water tanks, wine-maturing vats, silos, etc. In particular off-shore applications, i.e. oil-exploration platforms where the environment is extremely corrosive, are expected to become major application areas for composites in the near future. However, partially for fire safety reasons this is a very conservative industry and composites are only slowly being accepted. Current off-shore composite applications include seawater piping systems for fire fighting, fire-protection panels (sandwich structures with ceramic core), grating, ladders, handrails, and subsea wellhead protection structures (see *Figure 1-35*).

1.5.5 Electrical and Sanitary Applications

From a structural point of view some of the most mundane, albeit by consumption quite important (cf. *Figures 1-4* and *1-5*), composite applications are to be found in electrical and sanitary environments.

In electrical applications, the dielectric strength of glass-reinforced composites is of great interest and generally much more important than the structural capabilities (which are nevertheless valued in some cases). Applications in electrical and electronic areas include circuit boards, equipment housings, transformer spacers, and cable trays to mention a few. One particularly large order of cable trays was the 450 km required for the tunnel (or, rather, three tunnels) under the English Channel. For electricians, tool handles, booms for "cherry picker" trucks, and last but not least ladders are common applications; ladder rails are one of the biggest composite applications in North America (see *Figure 1-36*). The attraction of eliminating direct electrical contact between electrician and potentially hot wires as well as between electrician and ground is obvious.

Sanitary applications are dominated by shower enclosures (see *Figure 1-37*), bath tubs (see *Figure 1-38*) and spas, sinks, and cabinets, although there are many other applications. Related applications include bird baths, fish ponds, small pools, and water slides. In such applications it is mainly the design freedom and the possibility of economically manufacturing short series that are the main advantages.

1.5.6 Sports Applications

Second perhaps only to "fiberglass" boats in terms of public awareness, sporting goods, such as "graphite" golf clubs, "Kevlar" tennis racquets, and "ceramic" skis, are well-recognized and accepted composite applications. Although a method to pultrude (see *Section 4.2.6*) composite fishing rods was already patented in 1951, the real breakthrough of composites in sporting goods has taken place in a very limited time frame and has all but eliminated traditional materials. It is now virtually impossible to buy a (new) tennis racquet, slalom ski, or surfboard made from wood.

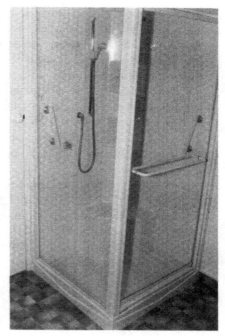

Figure 1-36 Ladder with (vertical) rails pultruded from glass-reinforced unsaturated polyester (see *Section 4.2.6*). Photograph courtesy of MMFG, Bristol, VA, USA

Figure 1-37 Two-piece shower stall with thermoformed acrylic surface film backed up by a sprayed-up glass-reinforced unsaturated polyester structure (see *Section 4.2.1.2*)

Figure 1-38 Corner bath tub with acrylic surface film backed up by a sprayed-up glass-reinforced unsaturated polyester structure (see *Section 4.2.1.2*)

While there are many reasons for this massive dominance of polymer composites, the main ones are fashion, vastly improved performance, lower weight, in some cases lower cost, and durability. In many cases it is difficult to ascertain whether the exclusive "high-tech" image of composites or real performance enhancements is the deciding factor since the vanity of man, particularly in sports, means that image often overshadows rational reasoning. Case in point, put a saw to your old slalom skis that proudly proclaim that they contain Kevlar or carbon reinforcement to enhance performance, and you may find that there are only a dozen such fibers in the ski, whose effect on performance thus is highly questionable. Nevertheless, improvements in performance and weight are in most sports applications real (the author has still not forgotten his first game of squash with a composite racquet; it was a completely different game).

While composite alternatives in most cases are more expensive than traditional solutions (where there are any still left), the reverse holds true in some cases. For example, split-cane flyfishing rods are significantly more expensive than carbon-reinforced epoxy alternatives, since so many craftsman hours go into painstakingly bonding precision-machined bamboo pieces into a split-cane rod. Cost is nevertheless often secondary; who is not prepared to pay an arm and a leg to beat his tennis partner or to improve his golf game?

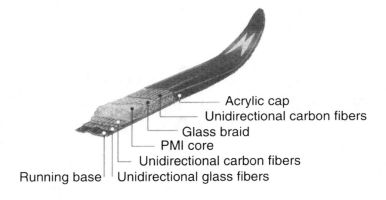

Figure 1-39 Virtually all modern skis are elaborate composite sandwich structures. The figure shows the construction of a cross-country ski, but in terms of materials and general construction principles slalom skis are similar. Apart from the running base and cap which are generally unreinforced thermoplastics, the most common material is glass-reinforced epoxy, although high-performance reinforcements are used in top-of-the-range models. Cores are expanded polymer foams, wood, or in some cases a mere cavity. The objective is generally to achieve highest possible flexural and torsional stiffness at lowest possible weight and cost. Drawing courtesy of Madsus, Biri, Norway

The example of flyfishing rods illustrates that the fashion factor can be reversed since the traditional solution is the most exclusive and, if you ask a flyfishing aficionado, also the better. It also illustrates how a move from natural to synthetic materials reduces the degree of craftsman skills that go into manufacture; composites have moved sporting goods manufacture from the craftsman's shop into an industrial environment (although considerable craftsman skills are certainly still required). The same goes for the raw materials used in the respective cases; synthetic materials are clearly more amenable to mass production. The durability advantage is also relevant, since for example wooden squash racquets and bamboo flyfishing rods are not only expensive, they are also very fragile.

There are clearly countless composite application examples in sporting goods. A non-exhaustive list includes golf clubs, snow skis (see *Figure 1-39*), ski poles, water skis, surf boards, windsurfing boards, kayaks and canoes, bicycles (see *Figure 1-40*), helmets, fishing rods, crossbows and arrows, all kinds of racquets, various hockey clubs, and shoes.

Figure 1-40 Prototype bicycle manufactured from carbon-reinforced polyamide. The tubes of the frame are filament wound (see *Section 4.3.4*), while the connecting joints are hand laid-up and autoclave consolidated (see *Section 4.3.1*). Photograph courtesy of Institut für Verbundwerkstoffe, Kaiserslautern, Germany

1.5.7 Other Applications

There is little doubt that one could devote an entire book to composite applications, but only two more areas will be mentioned. The first one is in medicine, where composites for quite some time have been used in patient support tables in X-ray machines. These tables offer great stiffness while still being transparent to the radiation in order not to obscure the radiograph. Composites are also being investigated for artificial joints to replace the more conventional metal components used in, for example, artificial hip joints. A problem with inserting a titanium hip implant into the femur is that the stiffness of the implant is so much greater than that of the bone, which gradually leads to bone deterioration. With composites the stiffness can be tailored to that of the bone to eliminate this problem. Some composite material systems are also biocompatible, i.e. they are not repelled by the body.

Another application area where composites are being used to a greater degree is in external joint supports tailored to the individual; such supports have become popular among skiers with bad knees. A similar application area with great potential is in prostheses where light weight and stiffness are essential. In both these areas the possibility of manufacturing one-of-a-kind custom-fit components is critical to the success of the composite alternative.

A somewhat unusual composite application is in musical instruments, where string instruments so far have received the most attention. With careful design and manufacture an acoustic guitar, or for that matter any string instrument, can be made to satisfy the pickiest musician. Apart from the possibility of tailoring the instrument to whatever desire the customer may have, there are advantages in that the instrument is not sensitive to climate changes or transportation damage and that no rare woods on the brink of extinction are used.

1.6 Book Contents

The background to this book is the author's belief that "designers should wear a manufacturer's hat," to quote Burt Rutan, who is one of the world's most imaginative aircraft designers [10]. While there are several textbooks and engineering guides dealing with various aspects of polymer composite design, few texts deal with the aspects following the design process. This book provides an introduction to the most significant issues of the post-design process and most notably manufacturing-related aspects. The treatment is limited to composites with polymer matrices. While the emphasis throughout the book is on load-bearing, essentially meaning continuous-fiber reinforced, composites, some techniques and applications relevant to non-structural composites are included where they are closely related to those relevant to structural composites.

To be able to comprehend dominant mechanisms in composites manufacturing, a basic understanding of the behavior of the constituents is required; such an understanding may be obtained from *Chapter 2, Constituent Materials,* which emphasizes processability and property issues. Although the chapter inevitably involves some polymer physics and chemistry, only a very rudimentary prior knowledge of chemistry is required to follow the treatment. *Chapter 3, Properties,* dresses the qualitative material characteristics discussed in *Chapter 2* in numbers. *Chapter 3* is mainly intended to be a rudimentary data base for preliminary design studies or homework problems, but also briefly touches on the anisotropy of composites.

The main chapter is *Chapter 4, Manufacturing Techniques,* where the commercially most relevant and some emerging manufacturing techniques are treated in some detail. The emphasis of this treatment is on providing the reader with sufficient insight into the different techniques to enable him to select technically and economically feasible techniques for a composite design, but also to create an understanding for the limitations a given material system and manufacturing technique place on component design. It is not the intention of this chapter to provide such detailed knowledge of the treated techniques that the reader can start manufacturing composites without delving deeper into the literature; the intention is rather to provide a comprehensive overview of possibilities and limitations of different materials and techniques.

Chapter 5, Secondary Processing, deals with the normally inevitable issues of machining, joining, and surface finishing as well as repair. An integral part of any manufacturing operation is *Quality Control and Characterization,* which is the topic of *Chapter 6.* This chapter concentrates on issues likely to arise in a manufacturing environment and largely ignores the equivalent issues related to composite constituents.

The two final chapters deal with topics that are becoming ever more relevant. While recycling may not yet be a major concern for designer and manufacturer, it soon will be in many applications due to legal requirements. *Chapter 7, Recycling,* gives an overview of existing recycling techniques for composites. *Chapter 8, Health and Safety,* finally, provides an overview of the dangers that may be present in a composite manufacturing environment and discusses appropriate protective measures.

This is not the first textbook or engineering guide to composite-manufacturing related activities, but this one is believed unique in its comprehensiveness. The main intention of the author is to raise the awareness of present and future designers of composite-manufacturing related activities so that they can become more knowledgeable in these areas and thus better designers. It is also the author's hope that this book may serve to raise the interest in composite manufacturing and to provide a baseline for further work in the field.

1.7 Outlook

Looking into the crystal ball, there are several interesting developments on the horizon. These developments may essentially be divided into increasing use of existing materials and technologies, and development or invention of new materials and technologies.

Looking first at the former, there is great scope for increased use of current technology in existing applications as well as introduction into completely new areas. Considering the applications discussed in *Section 1.5*, the technology obviously is there to make the all-composite automobile, ship, aircraft, bridge, etc. Then why are such all-composite designs so rare? While part of the answer lies in lack of experience, general conservatism, legal requirements, and liability issues, the main reason is cost. Only in more or less cost-insensitive areas such as military applications and sporting goods is the buyer prepared to pay extra for small gains in performance. Moreover, if a composite solution may provide long-term savings for the owner but the up-front cost is notably higher, it probably will not sell well anyway; private owners and shareholders alike tend to expect a rather rapid return on invested capital.

There is consequently a great need for increased cost-effectiveness in all stages of design and manufacturing of both raw materials and composite components. Particularly in the automobile industry the drive to lower cost and increase production rate is strong. One common way to reduce cost is through economies of scale, i.e. mass production. A couple of application areas that are being seen as aiding in this development are general construction and off-shore applications. If for example the highway authorities in the United States were to adopt the previously discussed concept of composite bridge decks, or if significant portions of the mammoth oil platforms used off-shore were to be made from composites, material consumption would increase drastically with likely positive side-effects of lower raw material prices and a more widespread acceptance of composites in construction.

While lifetime fuel consumption of vehicles, and in particular privately owned automobiles, rarely is given significant consideration at the time of purchase (unless there are extreme differences between candidate vehicles), rising fuel prices and legal incentives will gradually mandate more fuel-efficient vehicles. This is a promising development for composites, since lighter vehicles are synonymous with improved fuel efficiency. There is also a movement worldwide to include the overall resource utilization of a component or structure in a cradle-to-grave analysis. Such a life cycle assessment (LCA), which is briefly discussed in *Chapter 7*, considers all energy and resources that go into, for example, a vehicle, from raw material extraction and refining, to vehicle manufacture, use, and post-life recycling or disposal. LCAs are usually in favor of composite solutions, both due to their lower weight and because relatively little energy goes into raw material extraction and refining of composite constituents when compared to, for example, aluminum (which offers another efficient way of lowering the structural weight of a vehicle).

A related development is the increased interest in recycling. While recycling of paper, aluminum, steel, glass, etc. has become common, recycling of composites is a very recent occurrence. The development has nevertheless been rapid, mainly prompted by German recycling laws, and it is clear that in the near future recycling will be required in a range of applications, particularly in the automobile industry. While there are already technically feasible recycling techniques for all types of composites, the main stumbling block is the economical feasibility. Much work can thus be expected in improving existing recycling techniques and invention of completely new ones, both from technical and economical perspectives.

The development of new constituent materials appears to have slowed down in recent years due to the end of the cold war, recessions, and general belt-tightening. There is nevertheless still quite some work going on and improvements are being seen in properties of both fibers and matrices. In terms of reinforcements these are most notable in carbon fibers which are starting to push the envelope of elastic moduli to in excess of an impressive 1,000 GPa (cf. steel at 200 GPa and aluminum at 70 GPa), while at the other end of the spectrum the price of standard carbon fiber grades is coming down to gradually make carbon a commodity reinforcement.

Some of the biggest problems with polymer matrices are their poor temperature tolerance (when compared to metals), that many are brittle, and that most are demanding in terms of processing requirements. Developments are therefore constantly underway to raise the continuous-use temperatures of matrices, particularly for use in aerospace applications. At extreme temperatures where polymer matrices no longer suffice ceramic, metal, and carbon matrices are seeing increased use. Most conventional polymer matrices are thermoset and therefore inherently brittle. Although much successful work has been and still is going into modifying thermosets to alleviate this problem, another increasingly interesting possibility is offered by thermoplastic matrices which are inherently ductile. Almost all polymers are difficult or time-consuming to process and there is consequently ample room for improvements in processability. Considerable efforts are going into reducing times and temperatures required for crosslinking of thermosets and into reducing melt viscosities of thermoplastics.

Another aspect of thermoplastics is that due to the absence of chemical reactions during manufacturing, processing can potentially be significantly more rapid than with thermosets, which must be given time to undergo chemical conversion. This opportunity has not been lost on the automobile industry in its quest for more rapid manufacturing, and compression molding of thermoplastic composites (see *Section 4.3.3.1*) is rapidly gaining ground. Although a major breakthrough for thermoplastic composites was predicted in the 1980s and early 1990s, this has not—with the exception of compression molding—been realized. It is nevertheless likely that thermoplastic composites will gradually see wider acceptance due to both manufacturing and property advantages.

While most techniques for manufacturing of thermosets are relatively mature technologies, there is still scope for improvements in existing techniques and invention of new or hybrid techniques. One such example is the emerging technique of embedding sensors within a crosslinking component to enable tailoring of processing conditions to each component. This technique only makes economical sense for very expensive components, say a wing skin for a military aircraft, but may lead to improved component properties, reduced scrap ratio, and more rapid processing. With thermoplastics the situation is the opposite; most technically feasible techniques developed to date require further research and development work before they can expect widespread commercial acceptance. Developments in manufacturing technology are normally gradual rather than incremental. Developments may be in terms of increased automation, reduction in cycle time, and lowering of mold cost, but are more commonly in terms of improvements in component properties, such as geometrical complexity, surface finish, fiber content, repeatability, etc.

Health and safety is becoming more relevant and the increasing awareness of hazards posed by active chemistry influences the developments in both materials and manufacturing techniques. In this respect thermoplastics offer significant improvements over thermosets, since no active chemistry is involved in processing and health hazards are thus essentially eliminated. With thermosets closed-mold techniques (where the matrix is contained within closed molds to sharply reduce operator exposure) are increasingly being favored over the more conventional open-mold techniques.

Particularly in the aerospace industry, work is underway to develop so-called smart materials, in this case smart composites, although they are probably still a fair way from commercial application. Smart composite concepts include components with embedded sensors and adaptive structures. In the former case sensors, for example optical (glass) fibers, are embedded within the reinforcement to enable monitoring of loads and structural integrity during use. These sensors may very well be the same ones used to optimize manufacturing, as discussed above. While a composite through clever reinforcement orientation can be designed to increase its stiffness as load increases, an adaptive component allows independent changes in some property, such as stiffness or eigenfrequency, based on input from a control system, in order to optimize structural performance.

Composites are indeed very exciting materials in the extensive opportunities they offer in terms of tailoring of component or structure properties. However, at the same time, these possibilities require that both designer and manufacturer possess extensive skills to be able fully to utilize the inherent properties and possibilities. There is little doubt that polymer composites will continue to gain ground at the expense of traditional engineering materials, but the virtual explosion in the use of composites predicted not too long ago will likely not happen. A more realistic scenario is that composites will gradually ease their way into an increasing number of applications as well as increase their share in existing application areas. To what extent this will happen depends on you, the reader.

1.8 Summary

A composite consists of a reinforcing phase embedded in a bulk phase called a matrix. The combination is carried out with the intention of suppressing undesirable properties of the constituents in favor of the desirable ones; the result is a new and unique material family. While nature excels in employing the composite concept and man has been using fiber-reinforced clay bricks for several millennia, synthetic composites are at most half a century old. There are several types of composites, but the commercially most important type is fiber-reinforced composites with polymer matrices and the structurally most capable form employs continuous fiber reinforcement.

The main advantages of polymer composites lie in excellent specific mechanical properties, corrosion resistance, parts integration possibilities, and potentially lower cost. The predominant disadvantages are designers' lack of experience, knowledge, and material data bases, as well as in low temperature tolerance, arduous manufacturing, and quite often higher cost. While the potential advantages of composites in engineering applications are alluring, the complexity they introduce in design and manufacturing demands a lot of both designer and manufacturer. It is therefore imperative that a composite designer has a substantial degree of knowledge of materials and manufacturing issues to be able to do a good job.

Polymer composites are found in all engineering fields where they compete with more traditional construction materials, such as wood, steel, aluminum, concrete, and plastic. Automobile, boat, aircraft, construction, and electrical are among the most important application areas. In most areas composites are in their infancy in terms of extent of application and the future will prove that polymer composites are worthy of becoming a commodity material.

References

1. M. F. Ashby, "Technology of the 1990s: Advanced Materials and Predictive Design," *Philosophical Transactions of The Royal Society of London*, **A322**, 393–407, 1987.

2. J. Delmonte, "Historical Perspectives of Composites," in *International Encyclopedia of Composites*, Ed. S. M. Lee, VCH Publishers, New York, NY, USA, **2**, 335–341, 1990.

3. E. Mangenot, "The European Composites Market: Results and Outlook," Vetrotex International, Chambéry, France, 1996.

4. *Semi-Annual Statistical Report*, SPI Composites Institute, New York, NY, USA, 1996.

5. R. Abrams and J. Miller, "F-22," Aerofax Extra 5, Aerofax, Arlington, TX, USA, 1992.

6. S. W. Kandebo, "F-22 Assembly on Schedule After Initial Teething Pains," *Aviation Week and Space Technology*, 46–47, January 1997.

7. G. F. Turner, "Advanced Composite Materials in European Aircraft Present and Future," International SAMPE Symposium, **40**, 366–380, 1995.

8. V. P. McConnell, "Aerospace Applications: Affordability First," *High-Performance Composites*, 27–33, March/April 1995.

9. A. E. Churchman, P. R. Head, and J. M. Philips, *Composites* (French journal), **3**, 103–109, 1993.

10. B. Rutan, C. Hiel, and B. Goldsworthy, "Design with Composite Materials; From Complete Chaos to Clear Concepts—Part I," *SAMPE Journal*, **32**(5), 18–23, 1996.

CHAPTER 2

CONSTITUENT MATERIALS

The materials that make up, or constitute, a composite are normally considered to include at least the matrix and the reinforcement. Oftentimes the matrix is not homogeneous but rather mixed with inert fillers, performance-enhancing additives, etc. Likewise, the reinforcement is normally surface treated or coated with some substance to improve properties. These fillers, additives, and substances may or may not be considered constituents on their own right. When dealing with sandwich components the core and the face–core adhesive, if used, are additional constituents. This chapter introduces the commercially most relevant constituents and describes available forms and types, applications, and characteristics.

2.1 Matrices

The matrix of a composite has several functions: it is a binder that holds the reinforcement in place, it transfers external loads to the reinforcement, and it protects the reinforcement from adverse environmental effects. Moreover, the matrix redistributes the load to surrounding fibers when an individual fiber fractures and laterally supports the fibers to prevent buckling in compression. Whilst polymer matrices clearly dominate, other matrices are used to a limited degree in specialized applications; a brief overview of inorganic-matrix composites is given in *Section 2.6*.

To fully appreciate differences between candidate polymer matrices in terms of properties and processing requirements, it is essential to at least have a conceptual understanding of polymer physics and chemistry. However, this treatment is primarily intended for readers with little chemistry background, for which reason the following sections refrain from delving too far into the vast field of polymer science; the interested reader is referred to the literature specializing in this topic. The following treatment thus merely aims to provide a basic understanding of the polymer-science aspects most relevant to composite properties and the matrices' manufacturing requirements. It is the intention that the following sections on polymer physics and chemistry should provide the reader not only with a knowledge such as "this type of polymer is stiffer than that," but—much more importantly—an understanding of why this is the case.

2.1.1 Polymer Morphology

A polymer is a high molecular-weight compound that is composed of a multitude (poly) of repeated small segments (mers). Polymers are organic compounds primarily based on carbon and hydrogen atoms bound to each other by primary, or covalent, bonds.

Carbon is capable of forming four covalent bonds located tetrahedrally around the atom. Carbon atoms may form covalent bonds to each other both through single and double bonds, the former being referred to as saturated and the latter unsaturated. (Carbon atoms can also bond to each other through triple bonds, but such bonds are seldom encountered in polymer science.) In contrast, hydrogen can form one covalent bond only. *Figure 2-1* illustrates the tetrahedral bond arrangement in methane, which consists of carbon and hydrogen only.

Figure 2-1 Two ways of illustrating the structure of methane (CH_4). The simplified albeit very common version to the right is used for convenience and assumes that the reader knows that the bonds of carbon are located tetrahedrally around the atom

The least complex polymers are the hydrocarbon polymers which contain carbon and hydrogen only. The simplest hydrocarbon polymer is polyethylene (PE), the repeating unit of which is illustrated in *Figure 2-2*, where "n" indicates a large number (10^3–10^6) of repeating units to form a practically useful polymer and "—" indicates a covalent bond.

Another important and common hydrocarbon polymer is polypropylene (PP), which has a slightly more complex structure, see *Figure 2-3*.

Several hydrocarbon polymers other than PE and PP exist, although they are rarely of interest in composite applications. However, if one includes elements other than carbon and hydrogen, virtually endless other combination possibilities arise. Apart from carbon and hydrogen, the most common ele-

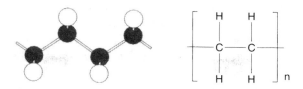

Figure 2-2 Two ways of illustrating the structure of PE

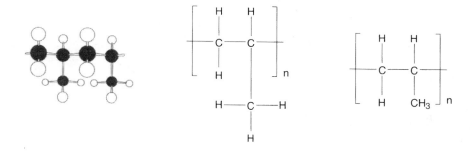

Figure 2-3 Three ways of illustrating the structure of PP

ments are oxygen which can form two covalent bonds; nitrogen with three; sulfur with two, four, or six; fluorine with one; chlorine with one; and silicone with four. As long as only carbon makes up the polymer backbone, one talks of carbon-chain polymers, whereas polymers having some non-carbon backbone atoms are referred to as heterochain polymers.

2.1.2 Structural Aspects Influencing Properties

The structure of a polymer molecule (such as those shown in *Figures 2-2* and *2-3*) is called configuration and the way in which the molecule is arranged is referred to as conformation. There are strong relationships between configuration and conformation of a polymer and its macroscopic properties in both solid and liquid form. In most cases these dependencies can be attributed to the molecular mobility, i.e. the degree to which the molecules can move relative to one another. These relationships may be hierarchically divided into intramer, intramolecular, and intermolecular structures. The next three sections discuss some molecular aspects of relevance and their influence on the polymer's macroscopic properties.

2.1.2.1 Intramer Structure

The properties of the polymer are greatly influenced by the structure of the repeating unit, i.e. what elements are present and how they are bound to one another. Single bonds allow rotation around the bond axis (torsion), while double bonds do not. Double bonds and cyclic structures, such as the common aromatic ring (see *Figure 2-4*), in the backbone and large side groups leave the molecule inflexible and bulky, thus impeding molecular motion which significantly influences macroscopic properties.

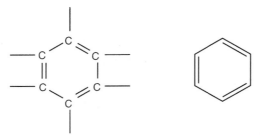

Figure 2-4 Three ways of illustrating the structure of an aromatic group

Polyisoprene is a good example to illustrate the strong influence of apparently subtle differences in configuration, see *Figure 2-5*; cis-polyisoprene (natural rubber) is soft whereas trans-polyisoprene (gutta percha) is a hard plastic. Note that these two configurations are different since the double bond does not allow rotation. The cis- prefix denotes that the backbone continues on the same side of the double bond and the trans- prefix that the backbone extends on different sides. Cis-polyisoprene and trans-polyisoprene are said to be geometric isomers.

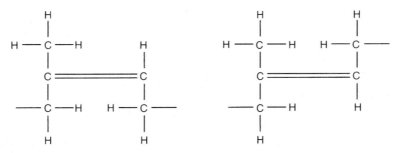

Figure 2-5 Cis-polyisoprene (left) and trans-polyisoprene (right)

2.1.2.2 Intramolecular Structure

While the repeating unit may be fully defined, there are still many combinations in which it may form a polymer. Taking PP as an example, the methyl (CH$_3$) group may be located in different positions, see *Figure 2-6*. Isotactic PP has all methyl groups on the same side, syndiotactic on alternating sides, and atactic in a random manner resulting in an equal number of methyl groups on both sides. Even though single bonds permit rotation, these three forms of PP are different. The regular structures of isotactic and syndiotactic PP, which are referred to as stereoisomers, translate into quite different macroscopic properties compared to atactic PP.

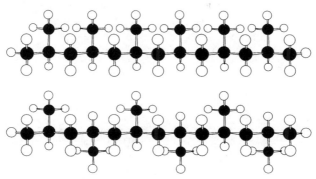

Figure 2-6 Isotactic (top) and syndiotactic (bottom) PP

The polymer examples so far shown have been linear with small side groups only; however, branching may occur to form star-shaped molecules. Further, no attention has yet been paid to the ends of the molecules, which obviously must deviate from the idealized polymer structure, but since the number of repeating units is so large, possible effects of the end groups are usually ignored. The molecules may also contain impurities, i.e. some flaw inconsistent with the idealized polymer structure. To improve certain properties of a polymer it is possible to copolymerize it from more than one type of repeating unit, see *Figure 2-7*.

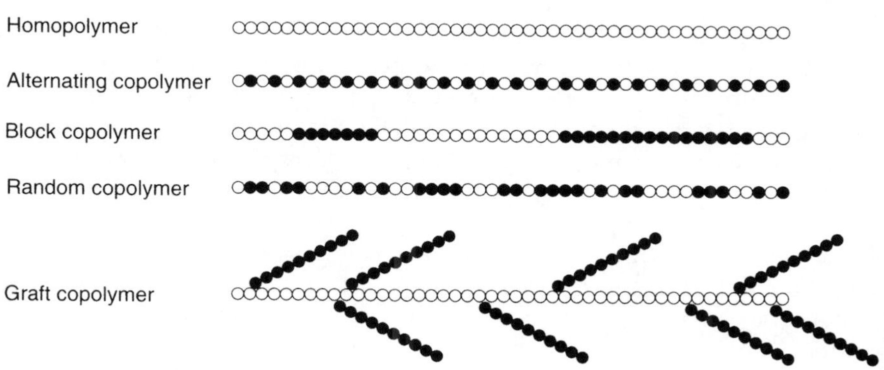

Homopolymer

Alternating copolymer

Block copolymer

Random copolymer

Graft copolymer

Figure 2-7 Different classes of copolymers

So far the molecules have been assumed straight, but since single bonds allow rotation there are nearly endless conformation possibilities. Thus, a 120° rotation around a bond in a PE molecule changes the conformational state without changing the configuration, see *Figure 2-8*. While single bonds allow rotation, certain torsion angles are preferred since they represent lower energy states; these are called stable conformational states. Given that each single carbon–carbon bond of a carbon-chain polymer permits rotation, a polymer with 10^3–10^6 repeating units may attain a nearly endless number of conformational states. In the general case a polymer is thus randomly arranged into a so-called Gaussian chain, but may under certain conditions attain a near-straight conformation.

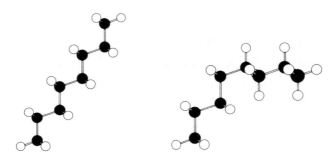

Figure 2-8 Two different conformational states of PE

Of great significance for the properties of the polymer is the number of repeating units or, alternatively, the molecular weight. However, there is no way of producing polymers of a single molecular weight and in reality it may vary by three to four orders of magnitude within the same polymer sample.

2.1.2.3 Intermolecular Structure

The interaction between molecules is a function of the intramer and intramolecular structures. It was earlier pointed out that the intramolecular bonds are covalent. There are also secondary (van der Waals) bonds, e.g. hydrogen, dispersion, and dipole–dipole bonds, which act between molecules. The reason for referring to them as secondary bonds is that they are of an order of magnitude weaker than the covalent bonds and therefore are of secondary, albeit significant, importance.

At a sufficiently high temperature a polymer sample melts; the individual molecules of the melt are randomly arranged, interlaced, and undergoing constant rearrangement. The covalent bonds ensure the integrity of the molecules and the molecular movement caused by the heating dwarfs the forces of the secondary bonds. (This state may be likened to a pot of boiling spaghetti, where the spaghetti strands are the molecules.)

If the polymer sample is then very quickly cooled, or quenched, the random molecular arrangement essentially is frozen in place by secondary bonds. (This is the not-so-appetizing bowl of solidified leftover spaghetti in the refrigerator.)

However, if instead the polymer sample is cooled slowly, the individual molecules may under certain conditions attempt to align themselves into a regular crystal formation. The crystal conformation is preferred since it represents the lowest possible energy state for the molecules. (This is akin to the original state of the spaghetti in the bag, i.e. aligned, although in addition to the orientational order of the spaghetti, the polymer molecules also possess positional order; cf. stacked spoons to stay with the kitchen analogy.)

Due to the precise close-packing of the molecules in the crystal state, the crystal structure has higher density than the random molecule arrangement, which usually is referred to as amorphous or glassy. Further, since the crystal conformation is preferred due to the lower energy state, energy is released upon crystallization, i.e. it is an exothermal process.

In reality it is not possible to achieve complete crystallinity and amorphous regions therefore surround the crystals; hence, polymers possessing the ability to crystallize are referred to as semicrystalline. The degree to which a polymer may crystallize is to a large degree dependent on the molecule being regular in structure and flexible. This is the reason why isotactic and syndiotactic PP may form crystals whereas atactic PP may not, and why PE, due to its ultimately simple, regular, and flexible structure, may achieve the highest degree of crystallinity of any polymer. Likewise, molecules that do not contain double bonds and cyclic structures in the backbone or bulky side groups are more likely to achieve a greater degree of crystallinity than if they did. Although many polymers of relevance as composite matrices are semicrystalline, most polymers are unable to achieve any appreciable degree of crystallinity under any circumstances.

It is believed that individual molecules create a folded lamellar crystal structure when they crystallize, as shown to the left in *Figure 2-9*. Note that molecules are likely to be part of more than one crystal; these interlamellar tie chains fix crystals to each other in lamellar stacks, as illustrated to the right in *Figure 2-9*. The lamellar stacks then become part of some supermolecular structure, e.g. a spherulite containing numerous lamellar crystals, as shown in *Figures 2-10* and *2-11*; note the difference in scales between the micrographs.

The resulting degree of crystallinity in a semicrystalline polymer is enhanced by higher pressure during crystallization as well as by lower molecular weight. However, most important of all is the temperature during crystallization, see *Figure 2-12*; there is an optimum temperature range for crystal growth. The temperature dependency of the resulting degree of crystallinity is often assessed in terms of cooling rate, see *Figure 2-13*. As the figure shows, it is possible either to obtain a more or less entirely amorphous

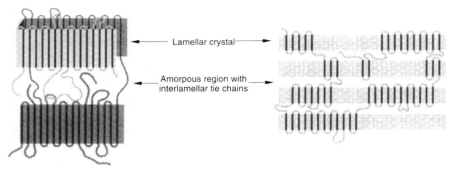

Figure 2-9 Idealized schematic of crystalline morphology

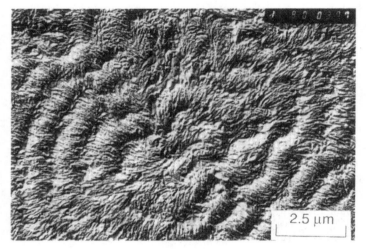

Figure 2-10 Micrograph of banded spherulite in PE. The sample has been etched to remove amorphous regions and reveal the lamellar stacks. Micrograph courtesy of M. Hedenqvist, Department of Polymer Technology, Royal Institute of Technology, Stockholm, Sweden

Figure 2-11 Polarized micrograph of banded spherulites in PE. Micrograph reprinted from reference [1] with kind permission from Elsevier Science Ltd, The Boulevard, Langford Lane, Kidlington OX5 1GB, UK

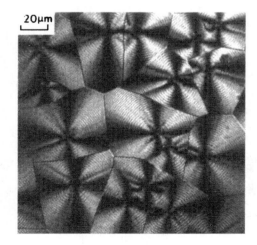

structure through quenching of the melt or a highly crystalline structure through very slow cooling.

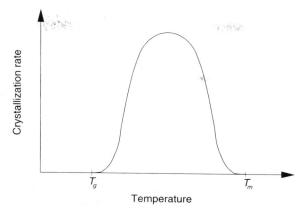

Figure 2-12 Schematic of the dependency of crystal growth on temperature. The temperatures T_g and T_m are defined in *Section 2.1.3.1*

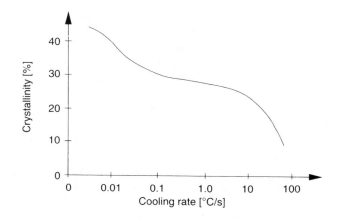

Figure 2-13 Crystallinity of poly(ether ether ketone) (PEEK) matrix in carbon-reinforced PEEK composite as function of cooling rate. Note the logarithmic cooling rate scale. Redrawn from reference [2]

Upon gradual heating of a solid semicrystalline polymer, the amorphous regions melt before the crystalline regions and a melt with the random molecular arrangement is eventually regained. If a crystalline phase is present, the melting of the crystals consumes energy in order to dislodge the molecules from their preferred low-energy state; the process is consequently endothermal. The reason why it is possible to go from solid to melt to solid etc. is that only secondary bonds act between molecules. A polymer with such a reversible behavior is called thermoplastic. In theory the melting and solidification cycles can be repeated an infinite number of times without affecting the molecules. In reality some molecules will react chemically, i.e. covalent bonds will be broken or formed, and the polymer and thus its properties will gradually degrade.

Under certain conditions covalent bonds may also form between molecules. One of the more notable locations that readily become sites of such crosslinks is the unsaturated carbon–carbon double bond, which through a chemical reaction may open up leaving a (saturated) single covalent bond within the molecular backbone and creating two new covalent bonds to other molecules. Starting with a polymer liquid consisting of molecules capable of forming crosslinks, any of a number of means may be employed to set off the chemical reaction which creates covalent crosslinks between neighboring molecules, thus creating a gigantic three-dimensional (3D) molecule. As crosslinks are formed, the polymer liquid gradually loses its ability to flow since the molecules can no longer slip past one another. The 3D network translates into a lower energy state than the random molecular orientation of the liquid, so the crosslinking process is (just like crystallization) exothermal. Since the 3D network created is bound together by covalent bonds it may not be melted through reheating. The kind of polymer having the ability to crosslink is called thermoset. Although the covalent crosslinking bonds dominate over the secondary bonds, they are not without importance in terms of macroscopic behavior. (While conventional wisdom has it that the covalent crosslinks of thermosets cannot be broken, recent work brought on by an increased interest in recycling of polymers has shown that thermosets indeed may be broken down into low molecular-weight building blocks for use in polymerization of new materials, although this is not yet in widespread commercial practice; see *Chapter 7* for further discussion.)

In conclusion there are two very different polymer families—thermoplastics and thermosets. Thermoplastics consist of long molecules with only secondary bonds between them, which allow thermoplastics to be melted. If irregular in structure and stiff, the molecules of a thermoplastic are randomly arranged both in the melt and in the solid; the polymer is said to be amorphous. On the other hand, if the molecules of the thermoplastic are regular and flexible, the molecules (while still randomly arranged in the melt) may form crystals as the thermoplastic solidifies; the polymer is semicrystalline. However, the degree of crystallinity of a semicrystalline polymer is dependent on cooling rate; with a high rate the solid polymer ends up more or less amorphous, whereas with a low rate it will be partly crystalline and partly amorphous. Initially thermosets also consist of long molecules with only secondary bonds between them. However, under certain conditions, such as the presence of carbon–carbon double bonds in the molecular backbone, covalent bonds may form between molecules resulting in solidification of the polymer resin. Since the intermolecular covalent bonds cannot be broken without simultaneously breaking the intramolecular covalent bonds, thermosets cannot be melted. As the molecular arrangement in solid thermosets (as well as in the liquid state) is random, thermosets are amorphous. While thermosets may be regarded as a special case in polymer science, it is a very important special case as far as composites go since thermoset matrices clearly dominate over thermoplastic matrices; the reasons for this dominance is further discussed in *Section 2.1.4*.

2.1.3 Polymer Properties

The following sections outline some of the macroscopic properties of polymers as functions of the structural aspects discussed in previous sections.

2.1.3.1 Thermal Properties of Solid Polymers

At sufficiently low temperatures an amorphous polymer is a glassy solid and any macroscopic deformation likely is due to stretching of secondary bonds and angle deformation of covalent bonds and only involves segments consisting of a few atoms. As temperature increases a region of rapid loosening of the secondary bonds is encountered and significantly larger segments of the molecules become free to move through rotation of covalent bonds. The temperature at which this change takes place is referred to as the glass-transition temperature, T_g, see *Figure 2-14*, and is accompanied by a decrease in stiffness of several orders of magnitude. As the temperature increases further, amorphous thermoplastics often have a so-called rubber plateau where the stiffness does not decrease significantly and deformations are due to molecules sliding past one another. Further heating leads to complete melting.

Thermosets (which are always amorphous), on the other hand, tend to have an extended rubber plateau, but since covalent bonds hold the polymer network together the polymer never melts. Nevertheless, if the temperature is increased enough, the polymer will start to degrade, i.e. covalent bonds will be broken or formed, and the molecular structure decomposes and may even ignite, to leave a charred carbon structure.

Semicrystalline thermoplastics also show a drop in stiffness at T_g (see *Figure 2-14*), since the amorphous regions lose so much stiffness. However, the crystalline regions remain unaffected and act as physical crosslinks meaning that the polymer retains much of its macroscopic stiffness. Not until the crystalline melting temperature, T_m, is reached does the polymer completely melt from a macroscopic perspective.

Since the transition temperatures (T_g and T_m) are functions of the molecular mobility in the bulk polymer it is easy to appreciate that a rigid and bulky molecular configuration requires a higher temperature to permit the

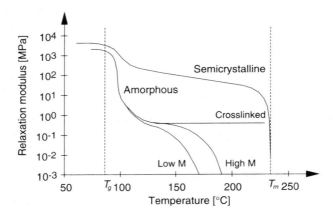

Figure 2-14 Relaxation modulus of polystyrene (PS) as function of temperature. Note the logarithmic stiffness scale. Approximate locations of T_g and T_m added. "M" denotes molecular weight. Redrawn from reference [3]

same degree of mobility. Consequently, rigid thermoplastics with bulky side groups and strong (intermolecular) secondary bonds have high transition temperatures. Likewise, the very strong (intermolecular) covalent bonds of thermosets, which clearly restrict mobility, cause them to have higher T_gs than many thermoplastics. By the same token, a higher degree of crosslinking results in a higher T_g. Moreover, the aforementioned changes in T_g that result from differences in chain flexibility and side group bulkiness also apply to thermosets.

Although transition temperatures are given as exact numbers, it is important to realize that in reality softening and melting take place over a rather wide temperature range; 50–80°C is not uncommon. The reason is that a polymer sample has a wide statistical spread in molecular weight, thickness of crystals, and configuration, e.g. degree of tacticity and branching. The transition temperatures therefore represent statistical averages.

Heat transfer in polymers is due to thermal agitation across intramolecular and intermolecular bonds; the stronger the bond, the higher the conductivity. The coefficient of thermal conductivity (CTC) therefore increases with molecular weight, molecular alignment, crystallinity, and crosslinking. The CTC may increase or decrease with temperature depending on the polymer. The specific heat, or heat capacity, arises from the freedom of movement of the molecules and thus decreases with crystallinity and crosslinking and increases with temperature (i.e. molecular mobility) and accordingly increases rapidly as T_g is passed.

It was previously mentioned that since the crystalline morphology represents close-packing of the molecules the density of the crystal is higher than in the amorphous phase, see *Figure 2-15*. The figure further shows that the density dependency on temperature changes considerably as T_g is passed. This behavior may also be assessed in terms of the coefficient of thermal expansion (CTE). The CTE of polymers tends to be a linearly increasing function of temperature both below and above T_g, but with a stronger temperature dependency above T_g. Closely related to the CTE is the total volumetric shrinkage from liquid (processing temperature) to solid (service temperature), which if not taken into account may lead to poor dimensional tolerances and sink marks. Partly due to the difference in volumetric shrinkage for the amorphous and crystalline phases, semicrystalline thermoplastics often have poor surface finish.

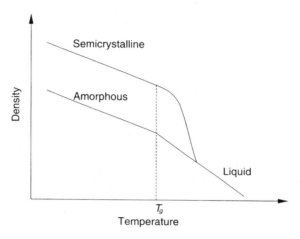

Figure 2-15 Polymer density as function of temperature

2.1.3.2 Mechanical Properties of Solid Polymers

In attempting to correlate the molecular structure of a polymer to its macroscopic mechanical properties, it is instructive to first contemplate how a molecule may deform. The molecular movement that requires the least degree of energy is torsion of a single covalent bond. Another order of magnitude of energy is required to change the angle of a covalent bond, whereas yet another order of magnitude of energy is required to stretch a covalent bond. The amount of energy required to stretch secondary bonds is of the same order as that required for torsion of covalent bonds. This hierarchy in deformation modes approximately translates into a corresponding hierarchy for the macroscopic elastic moduli and this is why the molecular conformation is so important.

As an example, one may consider PE processed in such a way that the molecules are essentially aligned in a certain direction. One such process is melt spinning where fibers are drawn from the melt and in doing so the molecules are aligned in the fiber direction, whereupon the fibers are quenched to freeze the desired axial molecular conformation in place. With such a conformation, the covalent bonds within molecules govern the properties in the fiber direction and the secondary bonds between molecules govern the properties perpendicular to the fiber direction. The spinning process also stretches the molecules, meaning that tensioning of the fiber essentially means stretching of covalent carbon–carbon bonds, which is the reason for the very high modulus of PE fibers. In terms of macroscopic properties, the tensile modulus is two orders of magnitude greater in the fiber direction than perpendicular to it.

Nevertheless, few manufacturing processes for thermoplastics result in a degree of molecular orientation significant enough that it can be seen as an advantage, i.e. the material mostly becomes amorphous or semicrystalline, but with a local degree of molecular orientation only. Since crystals are stiffer than amorphous regions, the macroscopic stiffness increases with degree of crystallinity, see *Figure 2-14*. However, higher stiffness may also be obtained using polymers featuring increased molecular rigidity, bulky side groups, and to a lesser degree increased molecular weight, although it should be remembered that such configurational traits impede crystal growth. Thus, if both high stiffness and high temperature tolerance are desired there will be a trade-off between high transition temperatures and a high degree of crystallinity, i.e. stiffness. Significantly, the thermoplastics with the highest mechanical properties and temperature tolerance owe these traits to stiff molecular structures, which also deprives them the ability to form crystals to any significant degree, i.e. they are amorphous.

Thermosets typically are stiffer than thermoplastics due to the 3D molecular structure bound together by covalent bonds, meaning that any macroscopic deformation is probably due to stretching of covalent bonds. Consequently, the stiffness of thermosets increases with crosslink density.

The strength of a material may be measured in so many different ways that sweeping statements about strength are difficult to make. However, the strength of polymers is highly dependent on and increases with the strength of intramolecular and intermolecular bonds, degree of crystallinity, and to a lesser degree molecular weight. Consequently, courtesy of the 3D molecular structure thermosets tend to have higher strength than thermoplastics.

In general the relatively weak intermolecular forces of thermoplastics translate into ductile materials with high strain to failure, toughness, and damage tolerance, since the molecules to a certain degree can slip relative to each other without breaking covalent bonds. Since crystalline regions act as physical crosslinks, ductility decreases with increasing degree of crystallinity. Correspondingly, thermosets tend to be brittle and have low strain to failure, toughness, and damage tolerance, since the covalent bonds cannot yield much.

For both thermoplastics and thermosets there is a trade-off between high stiffness, which dictates high degree of crystallinity or crosslink density, respectively, and high toughness, which requires the opposite. A common way of obtaining a useful compromise both with thermoplastics and thermosets is to copolymerize a stiff but brittle material with a ductile material to improve toughness.

As discussed in the previous section, amorphous polymers—both thermosets and thermoplastics—lose most of their stiffness and thus usefulness at T_g, whereas semicrystalline thermoplastics may be used up to temperatures close to T_m unless the requirements on stiffness and dimensional stability are very strict.

2.1.3.3 Other Properties of Solid Polymers

Several properties other than thermal and mechanical may be of importance depending on the intended application, e.g. electrical properties, optical properties, and tolerance to environmental exposure. Degradation of a polymer may occur through unwanted crosslinking (in thermoplastics) or chain scission, where the latter produces two electron-deficient molecules which readily react chemically. Oxidative degradation involves reaction between atmospheric oxygen and the polymer. In radiative degradation and photooxidation (ultraviolet (UV) light), degradation occurs due to radiation or light causing covalent bonds to be broken if subjected to energies in excess of the bond energy. Chemical attack on a polymer may cause swelling or dissolving (the latter for thermoplastics only). In hydrolysis the polymer reacts with water resulting in chain scission. Increased molecular mobility results in a higher rate of diffusion and absorption of chemicals, meaning that polymers are more susceptible to attack above T_g. Predictably, crystal regions and crosslinks reduce this susceptibility. All kinds of degradation are likely to affect polymer properties in an undesirable fashion. Although polymers generally contain antidegradants to slow down degradation, they nevertheless gradually degrade—it is merely a question of whether the degree of degradation becomes significant during the intended life of a component.

2.1.3.4 Properties of Liquid Polymers

The fluid mechanics of polymer solutions and melts is treated in the field of rheology (the study of deformation and flow of matter) and is of utmost relevance to composite processing since all composite manufacturing techniques involve some kind of material flow. In composite processing it is the shear viscosity of the polymer liquid that is of primary interest; it may be defined as the proportionality constant between shear stress and shear rate. Since polymer liquids, in contrast to water, for example, can sustain tensile loads, there is also an equivalently defined elongational viscosity that normally is neglected in (the modeling of) processing of continuous-fiber reinforced composites. However, it is most relevant to consider the elongational viscosity in unreinforced plastic processing, where the polymer often is significantly stretched.

In the unstressed state the molecules of a high molecular-weight polymer liquid are heavily intertwined. For a liquid to flow requires that molecules move in relation to one another and the resistance to flow in a high molecular-weight polymer liquid therefore is significant. However, if the liquid is sheared, the molecules are aligned, fewer entanglements remain, and the resistance to flow (i.e. the viscosity) is reduced. On the other hand, if the molecular weight is moderate, i.e. the molecules are not excessively long, the resistance to flow is lower since there are not as many entanglements. In this case shear is unlikely to significantly lessen the degree of entanglement and thus will not affect the viscosity much. Besides temperature, the molecular weight of a polymer is therefore the single most important factor in determining its viscosity. The viscosity and the so-called shear-thinning, or pseudoplastic, tendency (i.e. that the viscosity is reduced as shear rate increases) increase rapidly with molecular weight. Since a higher temperature manifests itself in increased molecular mobility it strongly facilitates flow.

Here a return to the spaghetti analogy may be warranted. If one sticks the fork into a plate of cooked spaghetti lubricated with sauce and raises it to one's mouth, one is liable to end up with the entire contents of the plate on the fork since the spaghetti strands are so heavily intertwined. However, if one twirls the fork, one is more likely to end up with an appropriately sized mouthful, since the twirling motion through shear unentangles the intertwined strands. Correspondingly, if the spaghettis have been cut it is possible to pick up a mouthful without twirling and one is liable to find that twirling makes little difference. Finally, when the spaghetti and the sauce is still piping hot it is easy to unentangle the spaghetti strands through twirling, whereas if one is late for dinner one finds that it is much more difficult to separate mouthfuls from the cold spaghetti without using the knife, i.e. the viscosity of the sauce is temperature-dependent.

Increased pressure reduces the free volume within a material and therefore reduces the molecular mobility in much the same way as a reduction in temperature. The pressure dependency is nevertheless often neglected in processing of continuous-fiber reinforced composites, though it should be considered in unreinforced and short-fiber reinforced plastic processing where the pressures encountered may be significant. A rule of thumb states that for unreinforced thermoplastics, a pressure increase of 100 MPa is approximately equivalent to a temperature decrease of 50°C.

As will be further discussed below, not yet crosslinked thermosets normally are of moderate molecular weight and dissolved in a solvent. Courtesy of their low molecular weight, thermosets behave as Newtonian liquids for most practical purposes, meaning that the viscosity is independent of shear rate although it is strongly dependent on temperature. The significant temperature dependencies of two different types of epoxy (which is a common high-performance thermoset) are illustrated in *Figure 2-16*.

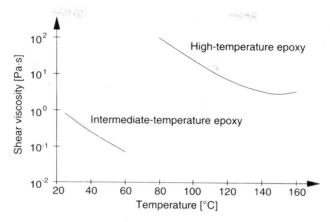

Figure 2-16 Shear viscosities of epoxies as function of temperature. The lower viscosity epoxy is intended for in-manufacturing reinforcement impregnation, while the higher viscosity epoxy is used in aerospace grade applications. Note the logarithmic viscosity scale. Data from references [4] and [5]

Since a thermoplastic melt consists of high molecular-weight molecules it is usually shear-thinning, see *Figure 2-17*, which shows the behavior of the high-performance thermoplastic PEEK. As further illustrated by the figure there is also a temperature dependency, though usually not as strong as for thermosets.

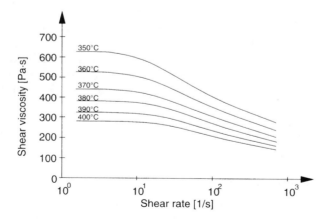

Figure 2-17 Shear viscosity of molten PEEK as functions of shear rate and temperature. Note the logarithmic shear-rate scale. Data from reference [6]

In most composite manufacturing operations a low viscosity is desirable, meaning that it may be tempting to increase the processing temperature based on the temperature dependencies displayed in *Figures 2-16* and *2-17*. Although the temperature dependency is indeed exploited for this purpose, it cannot be done indiscriminately. When increasing the temperature of a not yet crosslinked thermoset the crosslinking reaction is stimulated, which may lead to a premature increase in molecular weight and thus viscosity. *Figure 2-18* illustrates the so-called pot life (effective use time), in this case somewhat arbitrarily defined as when a viscosity of 1.5 Pa·s is reached, for the lower-viscosity epoxy of *Figure 2-16*. The figure also shows the gel time, which is defined as the time when the liquid has turned into a rubbery solid.

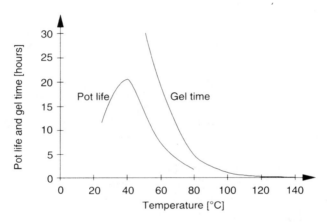

Figure 2-18 Pot life and gel time of epoxy intended for in-manufacturing impregnation as a function of temperature. Data from reference [4]

As previously mentioned, solid polymers eventually degrade through chain scission or crosslinking. This is even more so for molten thermoplastics since the molecular mobility is so much greater than in solid form, as well as the fact that the higher temperature stimulates chemical reactions. *Figure 2-19* illustrates the allowable exposure time before onset of degradation for molten PEEK as function of temperature. The tolerance to degradation of PEEK is greater in the absence of oxygen, which is a trait shared with most thermoplastics that readily oxidize at elevated temperature if given the opportunity.

2.1.3.5 Healing of Polymers

When the two mating crack surfaces of a thermoplastic sample that has been fractured are brought into intimate contact, the sample heals due to molecular interdiffusion across the interface, see *Figure 2-20*. The resulting bond strength is proportional to the molecular interpenetration depth across the

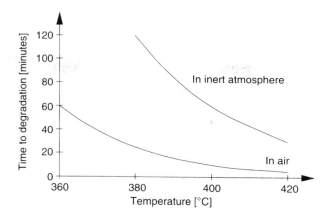

Figure 2-19 Time to surface degradation of PEEK as a function of temperature in presence and absence of oxygen. Data from reference [7]

interface. Given sufficient time, this diffusion process will lead to a strength equal to that of the virgin material and the interface is then no longer distinguishable, as shown in the last step in the figure. This healing process is referred to as autohesion, i.e. self-adhesion between two surfaces of the same thermoplastic.

The time required to achieve virgin material strength is strongly dependent upon the permissible degree of molecular diffusion which is proportional to temperature, see *Figure 2-21*. However, the temperature must be in excess of T_g for any molecular diffusion at all to be possible. The theory behind autohesion, called reptation theory due to the supposed snake- or reptile-like movement of the molecules, only strictly holds for amorphous thermoplastics, where the molecular movement is essentially unrestricted above T_g. In semicrystalline thermoplastics some molecules are at least partially locked into the crystal structure and their movement therefore is not unrestricted until above T_m; in this case reptation theory does not

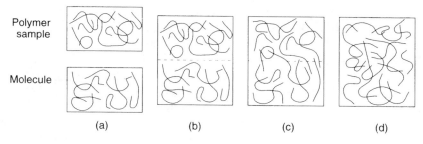

Figure 2-20 Autohesion of a thermoplastic polymer; (a) Surface approach (b) Intimate contact (c) Beginning of interdiffusion (d) Randomization completed and virgin strength achieved. Redrawn from reference [8]

apply below T_m. However, the same general trends in terms of temperature- and time-dependencies for bonding should be expected also for semicrystalline thermoplastics.

At normal processing temperatures, full bonding between two thermoplastic surfaces may be achieved in a fraction of a second provided the surfaces instantaneously come into intimate contact. In all realistic situations (except possibly healing of a fractured sample where the surfaces may match) surface waviness and roughness prevent instantaneous intimate contact. Pressure is thus required to achieve transverse material flow and smoothing out of the surfaces to realize intimate contact (cf. *Figure 2-20b*).

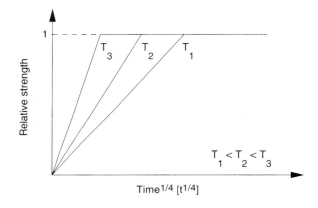

Figure 2-21 Time-temperature dependencies for autohesion of a thermoplastic polymer. Relative strength equal to unity corresponds to virgin strength. "T" denotes temperature. Note that the time scale shows the fourth root of time. Redrawn from reference [9]

2.1.3.6 Liquid Crystalline Polymers

There is a kind of thermoplastic polymer which may be viewed as an intermediate between amorphous and semicrystalline polymers. Liquid crystalline polymers (LCP) have long rigid sections in the molecular backbone, typically containing several aromatic groups, that under certain circumstances enable the molecules to maintain an ordered structure also in melts and high-concentration solutions. Under such circumstances the rigid sections of neighboring molecules possess directional and in some cases also positional order. In this state the molecules easily slip past one another due to few entanglements between molecules, and the polymer thus is easily processed. LCPs are used in fiber and unreinforced forms and are sometimes referred to as rigid-rod or self-reinforcing polymers due to their molecular structure. While not used as matrices in composite applications, LCP fibers are used as composite reinforcement as further discussed in *Section 2.2.3*.

2.1.3.7 Elastomers

Elastomers bridge the gap between thermoplastics and thermosets. Elastomer molecules are essentially linear, but are bound to each other by the occasional covalent crosslink. An elastomer can be elastically extended to several times its unstressed length and will upon removal of the load instantaneously return to its unstressed configuration. At intended use temperatures, which are above T_g, an elastomer is an elastic rubber, i.e. the use temperatures correspond to the rubber plateau of the material, cf. *Figure 2-14*. The rubber plateau of elastomers may extend over several hundred degrees Celsius. While synthetically derived elastomers dominate, cis-poly-isoprene (natural rubber; cf. *Figure 2-5*) is an example of a naturally occurring elastomer. Thermoplastic elastomers are copolymers containing hard segments which act as physical crosslinks instead of the covalent ones of regular elastomers. An advantage of thermoplastic elastomers is that they may be melt-processed in the same type of processes as used for conventional thermoplastics. Regardless of whether the crosslinks are chemical or physical the properties of elastomers can be widely varied by altering the crosslink density. Elastomers are not used as matrices in composite applications, but are often copolymerized with naturally brittle matrices, such as epoxies, to improve ductility.

2.1.4 Polymers as Composite Matrices

So far this chapter has dealt with generic aspects of unreinforced thermoplastics and thermosets and it should be kept in mind that usage of polymers is several orders of magnitude greater in unreinforced than in reinforced, i.e. composite, applications. Commodity plastics are used in non-structural applications, such as food packaging and toys, while other polymers are classified as engineering plastics, meaning that they posses properties that enable them to replace traditional construction materials, e.g. metals and wood, in load-bearing applications even without reinforcement. Thermoplastics are almost exclusively used when no reinforcement is included and dominate also when short fibers are incorporated. However, thermosets clearly dominate in structural composite applications, although thermoplastics have lately received increased attention also in continuous-fiber reinforced composite applications due to a number of attractive potential advantages.

While numerous engineering plastics are available, only a few are used as composite matrices, since an engineering plastic with excellent properties is not necessarily an appropriate composite matrix. Issues that are of importance when selecting a polymer for use as a composite matrix are reinforcement-matrix compatibility in terms of bonding, mechanical properties, thermal properties, cost, etc., though perhaps the most important aspect may be its processability, i.e. how easy it is to deal with it in manufacturing situations.

Table 2-1 Qualitative comparison between thermoset– and thermoplastic–matrix composites. "+" denotes an advantage

Property	Thermosets	Thermoplastics
Cost	+	
Temperature tolerance	+	
Thermal expansion	+	
Volumetric shrinkage	+	
Stiffness	+	
Strength	+	
Toughness		+
Fatigue life	+	
Creep	+	
Chemical resistance		+
Available material data	+	
Raw material storage time (shelf life)		+
Simplicity of chemistry		+
Viscosity	+	
Processing temperature	+	
Processing pressure	+	
Processing time		+
Processing environment		+
Mold requirements	+	
Reformability		+
Recyclability		+

Among the many issues that may be considered part of the processability are viscosity, processing temperature, processing time, and health concerns. Low viscosity is important in achieving reinforcement impregnation, where each reinforcing fiber ideally should be surrounded by the matrix without voids present. Not yet crosslinked thermosets have shear viscosities at processing temperature on the order of 10^0 Pa·s, while melt viscosities at processing temperature of thermoplastics are on the order of 10^2 Pa·s or higher (for comparison the shear viscosity of water at room temperature is 10^{-3} Pa·s), meaning that it is much easier to complete impregnation with thermosets than with thermoplastics. While some thermosets may be crosslinked at room temperature, other thermosets and all thermoplastics require an increased processing temperature that in general must be well controlled and may be as high as several hundred degrees Celsius for some polymers. While thermoplastics only need to be melted, shaped, and then cooled to achieve dimensional stability in a matter of seconds at the one extreme, thermosets may take several days to fully crosslink at the other extreme. The very nature of thermosets makes them unpleasant to work with from a health point of view, since chemical reactions involving volatile and potentially toxic substances are required to crosslink the polymer. In contrast, the molecular structure of thermoplastics makes them chemically inert if processed correctly, meaning that no hazardous substances need to be considered. On the other hand, the molten thermoplastic and the heated machinery may cause severe burns. (Health and safety issues are further treated in *Chapter 8*.)

While thermosets heavily dominate as matrices in structural composite applications for reasons of good mechanical and thermal properties, low cost, and low viscosity to mention a few, the interest in thermoplastics is driven by several potential advantages, as qualitatively outlined in *Table 2-1*. Among the prime reasons behind the increased interest and usage of thermoplastic matrices are advantages in areas such as toughness, processing time, recyclability, and work environment. When studying *Table 2-1* it is important to keep in mind that no such sweeping comparison can be absolutely fair. Naturally, a high-performance thermoplastic will outdo a standard-performance thermoset in most respects except cost and vice versa.

2.1.5 Thermoplastic Polymer Matrices

A thermoplastic is usually fully polymerized when delivered from the supplier, meaning that all chemical reactions are complete and the user can concentrate entirely on physical phenomena, such as heat transfer and flow. However, there are some rare exceptions to this rule. The user may choose to take care of part of the polymerization starting off with low molecular-weight prepolymer, thus avoiding the high-viscosity disadvantage during reinforcement impregnation. Courtesy of the low molecular weight, the polymer fluid may have a viscosity comparable to that of a thermoset resin. After the reinforcement is impregnated, the final polymerization process takes place and the molecular weight thus drastically increases. Depending on the type of polymer, the high molecular-weight polymer may or may not decompose into lower molecular-weight polymer molecules upon remelting. Whilst the technology behind this kind of thermoplastics is known, it is not yet of commercial significance although further research and development will likely change the situation in the near future.

One of the main features of amorphous thermoplastics is that they are dissolvable in common industrial solvents. This means that the reinforcement can be impregnated with a low-viscosity solution, thus avoiding the problem of high melt viscosity, but it also means that the solidified polymer (and composite) is not solvent resistant. For solvent-impregnated reinforcement, the residue solvent that was not completely driven off after impregnation is a serious concern since it impairs the quality of the composite. Amorphous thermoplastics have very good surface finish since they do not shrink much when they solidify and there is no differential shrinkage from the presence of crystalline regions.

Table 2-2 Repeating units of thermoplastic polymers used as composite matrices

Polymer	Repeating unit
Polyethylene, PE	
Polypropylene, PP	
Polyamide 6, PA 6	
Polyamide 12, PA 12	
Polyamide 6,6, PA 6,6	
Poly(ethylene terephtalate), PET	
Poly(phenylene sulfide), PPS	
Poly(ether ether ketone), PEEK	
Poly(ether imide), PEI	

Continued on next page

Table 2-2 continued

Polymer	Repeating unit
Poly(ether sulfone), PES	
Poly(amide imide), PAI	

Semicrystalline polymers usually have good solvent resistance due to the crystalline regions which prevent dissolution of the entire molecular structure. The crystallinity also improves high-temperature performance and long-term properties, such as creep. If the crystallinity is too low, these benefits are not seen and if it is too high the material loses toughness and becomes brittle, although it usually gains in stiffness; hence, there is an optimum degree of crystallinity. Useful semicrystalline polymers have 5 to 50 volume percent crystallinity, with an optimum of 20 to 35 percent for composite applications [2]. When processing semicrystalline polymers one thus must consider and preferably control the cooling rate (cf. *Figure 2-13*), which rarely is an issue with amorphous matrices. Semicrystalline polymers shrink more than amorphous ones upon solidification; the higher the final crystallinity, the higher the density change between melt and solid, see *Figure 2-15*. Due to the difference in shrinkage between amorphous and crystalline regions, the surface of semicrystalline thermoplastics is not as good as for amorphous ones. Since solvents normally cannot be used to dissolve semi-crystalline polymers (there are some rare exceptions), reinforcement impregnation is extremely difficult.

With the above section on polymer science fresh in memory, the following sections provide a brief overview of some of the more common thermoplastic polymers used as composite matrices; the repeating units of the polymers discussed are compiled in *Table 2-2*. When reading these sections, it is helpful to keep in mind the influence of the different configurations and conformation states discussed above. The properties discussed herein are merely family characteristics and large variations are the rule. While the following sections largely concentrate on qualitative properties, *Chapter 3* provides detailed quantitative information.

2.1.5.1 Polyethylenes

PE can be both commodity and engineering plastic depending on grade, but is rarely used as composite matrix due to low temperature tolerance and modest mechanical properties. However, PE fibers may be used as composite reinforcement, as is further addressed in *Section 2.2.4*. PE has the highest degree of crystallinity of any polymer due to its simple, regular, and flexible molecular structure, thus enabling PE to be used well above its T_g.

2.1.5.2 Polypropylenes

Just like PE, PP may be both commodity and engineering plastic depending on grade. PP is the chemically least complex and cheapest polymer commonly used as composite matrix. One generally talks of homopolymer and copolymer PP, where the homopolymer version consists of PP units only and the copolymer version is copolymerized with PE units to improve toughness (cf. *Figure 2-7*). Homopolymer PP has a T_g in the range –20 to –10°C, while the T_m is in the range 165–175°C. Obviously the service temperature of composites normally is above PP's T_g; just as for PE this is possible due to PP's high degree of crystallinity but naturally also due to the reinforcing fibers.

In structural composite applications PP is usually reinforced with glass fibers and such composites are often hidden from view since the surface finish tends to be poor. In recent years PP has become the most common thermoplastic matrix in mass-produced structural composite applications, including automobile components such as various under-the-hood applications and seatback frames (see *Figures 1-13* and *1-16*).

2.1.5.3 Polyamides

One of the best-known thermoplastic polymer families is the polyamides (PA), often called nylons. (The Nylon designation (capitalized) originally was Du Pont's trade name for PA fibers for use in ladies' stockings, but nylon (not capitalized) has become an accepted designation for PAs regardless of manufacturer.) In contrast to PE and PP, PA may be used at moderately increased temperatures, thus greatly improving its usefulness as matrix. PAs are characterized by the presence of amine groups (-CONH-). A number of different PA grades, e.g. PA 6, PA 6,6, PA 6,10, and PA 12, are available, where the numbers indicate the number of carbon atoms in the repeating unit, see *Table 2-2*; the properties naturally vary accordingly. The biggest drawback of some of the more common PA grades is that they are hygroscopic, i.e. absorb water. Some PA grades are among the aforementioned thermoplastics that may be obtained as low-viscosity prepolymers intended for final polymerization following reinforcement impregnation. Depending on grade T_gs are in the range 45–80°C, while T_ms are in the range 180–265°C. In composite applications PAs are normally reinforced with glass fibers and used in applications similar to glass-reinforced PP, but where higher temperature tolerance and improved mechanical properties are required. Where higher performance is required carbon reinforcement may be used, see *Figure 1-40*.

2.1.5.4 Thermoplastic Polyesters

Although perhaps chiefly recognized as thermosets, polyesters also are available in thermoplastic forms, e.g. poly(ethylene terephtalate) (PET) and poly(butylene terephtalate) (PBT). The properties of PET and PBT are similar to those of PAs, but lacking the hygroscopic disadvantage. PBT's T_g is 60–70°C while T_m is 225–235°C; the corresponding transition temperatures of PET are 80°C and 260–265°C. In composite applications thermoplastic polyesters are reinforced with glass fibers and used in applications similar to glass-reinforced PP and PA.

2.1.5.5 Poly(phenylene Sulfides)

The most common member of the poly(arylene sulfide) family is poly(phenylene sulfide) (PPS), which has good tolerance to most chemicals and fire. PPS exhibits moderate mechanical properties and temperature tolerance; T_g is 85°C and T_m 285°C. In composite applications PPS is reinforced with glass or carbon fibers and used in high-performance applications.

2.1.5.6 Polyketones

Whilst there are numerous aromatic polyketones, including poly(ether ketone) (PEK), poly(ether ketone ketone) (PEKK), the most common is poly(ether ether ketone) (PEEK). The polyketones possess high mechanical properties, high temperature tolerance, good solvent resistance, and a high price; T_g of PEEK is 145°C and T_m 345°C. In composite applications PEEK is reinforced with glass or carbon fibers and used in critical high-performance applications.

2.1.5.7 Polysulfones

Polysulfone (PSU), poly(ether sulfone) (PES), and poly(aryl sulfone) (PAS) are high-performance amorphous polymers with good tolerance to high temperatures and fire. These properties come at a high price and the melt viscosities are also very high. Since polysulfones are amorphous, they are not resistant to all solvents although their resistance to many chemicals nevertheless is very good. The T_gs are 190°C for PSU and 220–230°C for PES. In composite applications polysulfones are reinforced with glass or carbon fibers and used in the same type of applications as polyketones.

2.1.5.8 Thermoplastic Polyimides

The polyimide family includes poly(ether imide) (PEI), polyimide (PI), and poly(amide imide) (PAI), which are all amorphous. Polyimides have the highest temperature tolerance of the thermoplastics mentioned herein; the T_gs for PEI, PI, and PAI are 215, 255, and 250 – 290°C, respectively. Despite being amorphous they are very tolerant to solvents and environmental exposure and offer very good mechanical properties with the disadvantages of very high melt viscosities and high price. In composite applications the members of the polyimide family are reinforced with glass or carbon fibers and used in the same type of applications as polyketones and polysulfones.

2.1.6 Thermoset Polymer Matrices

The most common thermosets used as composite matrices are unsaturated polyesters, epoxies, and vinylesters. Unsaturated polyesters clearly dominate the overall market, whereas epoxies are preferred in high-performance applications. When reading the sections below describing the characteristics and application areas of these thermoset matrices, it is important to note that the properties of a given type of thermoset may be varied within a very wide range to fulfill the requirements of a particular application merely through variations in chemical formulation and through use of various additives. It is therefore quite difficult to provide a truly fair comparison of candidate matrices and the characteristics of the thermoset polymer families given below should be taken as indicative of the families' characteristics only.

Relatively detailed descriptions of polymerization and crosslinking of an unsaturated polyester are given below and general manufacturing-related issues are discussed in some detail in an attempt to provide a representative example of how a solid thermoset is constructed from low molecular-weight liquid building blocks. The reason for this detailed treatment of unsaturated polyesters is that it is imperative to have a good understanding of the crosslinking process to be able to fully understand the subsequent treatment of manufacturing. The detailed chemical reactions for other thermoset polymers are considered beyond the scope of this treatment, whereas the manufacturing-related aspects discussed for unsaturated polyesters tend to be shared by thermosets in general.

2.1.6.1 Unsaturated Polyesters

The workhorse of thermoset matrices is the unsaturated polyester (UP), which offers an attractive combination of low price, reasonably good properties, and uncomplicated processing. Whereas basic unsaturated polyester formulations have drawbacks in terms of, for example, poor temperature and UV-light tolerance, additives may significantly reduce these disadvantages to suit most applications.

Structure

The basic building blocks for constructing an unsaturated polyester molecule are a diacid, which has two carboxyl groups, and a dialcohol, which has two hydroxyl groups. In a process called condensation polymerization, since water is produced as a byproduct, a diacid molecule reacts with a dialcohol molecule to create an ester, see *Figure 2-22*. Since both molecules originally have two identical groups and only one from each molecule is involved in the first reaction step, this process in theory can be repeated any number of times provided that an unlimited supply of building blocks is present. Through this condensation reaction an unsaturated polyester is created as shown at the bottom of the figure. In reality the diffusional mobility of the growing molecule becomes limited as it gets larger; for unsaturated polyesters the number of repeating units typically is in the range 10–100.

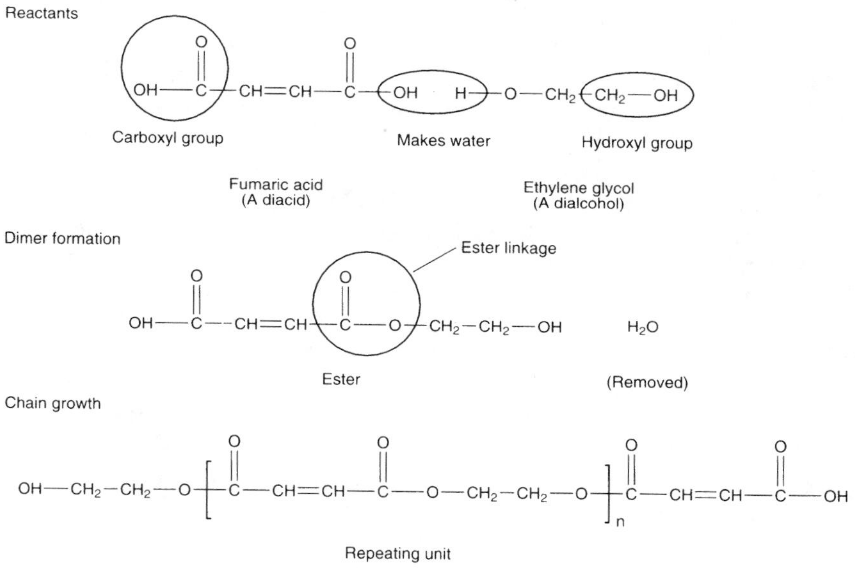

Figure 2-22 Schematic of condensation polymerization of an unsaturated polyester. Redrawn from reference [10]

Comparing the unsaturated polyester at the bottom of *Figure 2-22* with the structure of the thermoplastic polyester PET in *Table 2-2* one finds that they both contain the characteristic ester linkage (as highlighted in the middle of *Figure 2-22*). The fundamental difference between the two structures is that the polyester of *Figure 2-22* contains an unsaturated carbon–carbon double bond which is a potential site for crosslinks—thus *unsaturated* polyester.

Depending on unsaturated polyester type (there are numerous types of building blocks) and number of repeating units in the polymer, the polymerization product may be a highly viscous liquid at room temperature or a low melting-point solid. For further processing the polymer is dissolved in a low molecular-weight monomer, or reactive dilutent, and the result is a low-viscosity liquid usually referred to as resin. Unsaturated polyester resins contain between 35–50 weight percent monomer. By far the most common monomer is styrene, although vinyl toluene, chlorostyrene, and others may be used to modify processing characteristics or properties of the crosslinked unsaturated polyester.

With the addition of a small amount of initiator to the resin the crosslinking reaction, also called cure or curing reaction, is initiated. The initiator, which is an organic peroxide, e.g. methyl ethyl ketone peroxide (MEKP) or benzoyl peroxide (BPO), is added in small amounts, typically 1–2 percent by weight. The initiator is a molecule that produces free radicals, i.e. reactive chemical species with an unpaired electron. The free radical attacks the double bond of the unsaturated polyester molecule and bonds to one of the carbon atoms, thus producing a new free radical at the other carbon atom, see the initiation step of *Figure 2-23*. This newly created free radical is then free to react with another double bond. Since the small monomer molecules—in this case styrene—move much more freely within the resin than the high molecular-weight polymer molecules, this double bond very likely belongs to a styrene molecule, as illustrated in the bridging step of *Figure 2-23*. The bridging step creates a new free radical on the (unsaturated) styrene, which then is free to react with another double bond and so on. Obviously the styrene is not only used as solvent, but actively takes part in the chemical reaction to produce "bridges" between unsaturated polyester molecules. Monomers are consequently often called crosslinking or curing agents or, perhaps even more appropriately, reactive dilutents. Initiators are often called catalysts, but since they are consumed in the crosslinking reaction this terminology is obviously incorrect. (A catalyst is a substance that increases the rate of a chemical reaction without being consumed or permanently altered.)

As the molecular weight of the crosslinking polymer increases it gradually starts to impair the diffusional mobility of the growing molecules and the reaction rate slows down. When the movement of the free radicals is also impaired they are prevented from finding new double bonds to continue the reaction which then stops. The result of the crosslinking reaction is a gigantic, 3D molecule that from a macroscopic point of view leads to the transformation of the liquid resin into a rigid solid.

The crosslinking reaction is intimately linked to temperature. Since the crosslinked molecular morphology represents a lower energy state than the random molecular arrangement in the resin, the reaction is exothermal. Further, the free-radical production is stimulated by an increase in temperature and a higher temperature also promotes molecular mobility. Until diffusional limitations reduce the reaction rate, the crosslinking rate therefore increases; heat is released by the formation of new bonds, which promotes an increase in free-radical production and mobility, which promotes an increase in rate of bond formation, etc.

The reactivity of an unsaturated polyester may be modified using inhibitors that consume free radicals to postpone crosslinking and accelera-

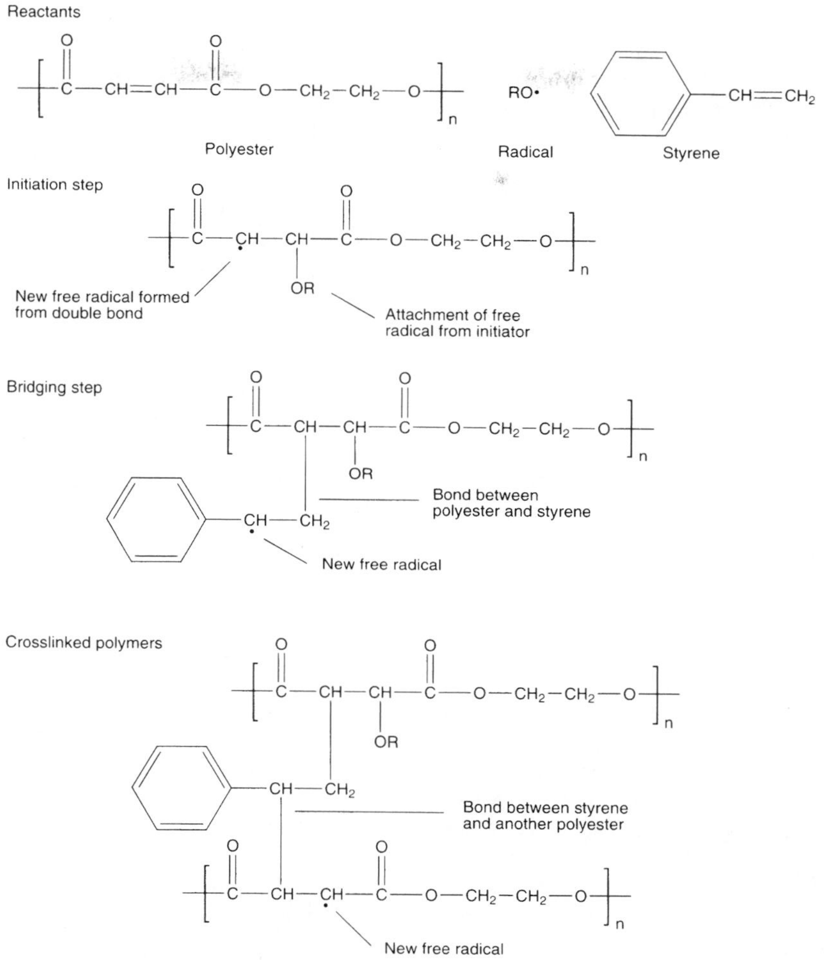

Reactants

Polyester   Radical   Styrene

Initiation step

New free radical formed
from double bond

OR

Attachment of free
radical from initiator

Bridging step

OR

Bond between
polyester and styrene

New free radical

Crosslinked polymers

OR

Bond between styrene
and another polyester

New free radical

Figure 2-23 Schematic of addition or free-radical crosslinking of an unsaturated polyester. Redrawn from reference [10]

tors that promote free-radical production. *Figure 2-24* illustrates typical temperature–time relationships for room-temperature crosslinking of an unsaturated polyester following addition of initiator. As illustrated by the figure, the temperature does not immediately increase after addition of an initiator despite free radicals being produced. The crosslinking reaction does not start (and the temperature does not increase) until all inhibitor molecules have reacted with free radicals, which corresponds to the inhibition time. As crosslinking commences, the resin viscosity soon starts to increase due to the increasing molecular weight and reinforcement impregnation rapidly becomes impossible; the pot life is over. Soon thereafter the resin becomes a rubbery solid, or gel, and the gel time is reached. The crosslink-ing activity now accelerates very rapidly until the increasing molecular weight of the crosslinking polymer starts restricting molecular movement, which occurs around the maximum temperature, and the crosslinking grad-ually tapers off.

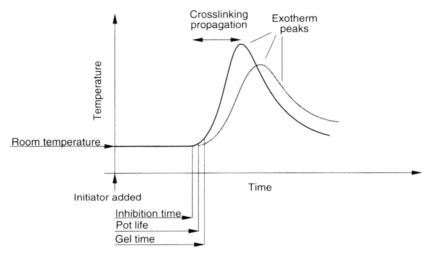

Figure 2-24 Temperature as a function of time for room-temperature crosslinking of an unsaturated polyester. The different temperature curves illustrate different proportions of initiator, inhibitor, and accelerator. A reduced amount of initiator and accelerator, as well as an increased amount of inhibitor, leads to later crosslink-ing at lower temperature, and vice versa. Redrawn from reference [11]

An unsaturated polyester with characteristics as illustrated in *Figure 2-24*, i.e. that starts to crosslink at room temperature after addition of the initiator, usually requires that an accelerator is present to promote the reaction. The use of accelerators may be an alternative to increasing processing temperature. However, many unsaturated polyesters will not crosslink satisfactorily unless subjected to elevated temperature or radiation to stimulate the crosslinking. For unsaturated polyesters that do not crosslink at room temperature it is important that the temperature of the crosslinking resin remains above the instantaneous T_g, lest the crosslinking be halted due to lack of molecular mobility. If the temperature falls below the instantaneous T_g without crosslink-ing having been completed this temperature becomes the final T_g. If the composite is subsequently postcured, i.e. subjected to increased temperature at a later point in time to allow for completion of crosslinking, this results in a higher T_g (provided that any reactive sites remain). Similarly, the highest temperature reached by a room-temperature crosslinking unsaturated poly-ester approximately becomes its T_g.

With time an unsaturated polyester resin spontaneously starts to crosslink even without the addition of an initiator, especially if stored at elevated temperature. Inhibitors are used also to prevent such premature crosslinking, thus increasing the permissible storage time, or shelf life. Since the production of free radicals is temperature-dependent, a lower storage temperature extends shelf life.

The addition of various amounts and kinds of initiators, inhibitors, and accelerators offers a significant flexibility in adapting the reactivity of a resin to a specific processing need, see *Figure 2-24*. However, it is important to note that the properties of the crosslinked resin are not independent of such variations. Although an unsaturated polyester may be crosslinked in a matter of minutes, the time frame in composite applications often extends into hours or even days. For a thin laminate the resin can be formulated and/or heated to crosslink rapidly without significant side effects. In contrast, thick laminates must be crosslinked slowly to avoid too high temperatures, which otherwise may degrade the polymer or create unnecessarily large residual stresses and even cracks due to differential shrinkage. The heat-generation issue in crosslinking of thick composites is aggravated by the poor heat conduction of polymers and active cooling therefore may be required in certain applications.

Types and Forms

The resin supplier delivers the polymerized but not yet crosslinked unsaturated polyester resin dissolved in monomer, almost always styrene, which apart from being a crosslinking agent also lowers resin viscosity to facilitate reinforcement impregnation; as-delivered shear viscosities are on the order of 1 Pa·s. It is common that the resin supplier adds inhibitor and accelerator to the resin before delivery and it is then up to the user to add appropriate amounts of initiator, accelerator, and inhibitor to suit the specific manufacturing situation. Other additives, such as fillers, pigments, and flame retardants, which are further discussed in *Section 2.1.8*, are normally also introduced by the user.

As will be further discussed in *Chapter 8*, one of the major drawbacks of unsaturated polyesters is that the volatile monomer readily evaporates during manufacturing, thus creating an unhealthy work environment. One way to limit this problem is to lower the molecular weight of the unsaturated polyester molecules to enable a reduction in monomer content of the resin without raising the viscosity. Another way to reduce styrene emissions are so-called LSE (low styrene emission) resins, which contain a substance that migrates to the surface of the resin to create a thin film impenetrable to styrene, but this approach is only effective when the resin is at rest since it takes some time for the film to form.

An unsaturated polyester is classified by the basic building blocks used to produce it. The cheapest form, the orthophthalic polyester, is considered a general-purpose resin, but is limited by modest mechanical properties and low resistance to temperature and environmental exposure. Isophthalic polyesters cost more, but offer significant improvements in mechanical properties, thermal stability, and environmental tolerance due to aromatic rings in the backbone. Several other, less common unsaturated polyester types are available for special applications.

Characteristics

As mentioned in the introduction to this section, unsaturated polyesters can be modified to suit a wide variety of applications, both through various additives and through variations in resin composition. Nevertheless, generic traits of unsaturated polyesters include good mechanical properties, low viscosity, uncomplicated crosslinking requirements, and low price. The main drawbacks are low temperature tolerance, significant shrinkage from crosslinking, and notable health problems. Orthophthalic polyesters typically have T_gs around 70°C while isophthalic polyesters have T_gs around 100°C.

Applications

The unsaturated polyester reinforcement of choice is glass fiber, since this combination is a good match in terms of both performance and price. In fact, the glass-unsaturated polyester combination has become so common that it often, but obviously incorrectly, is referred to as fiberglass (or glassfiber). Common acronyms used for glass-reinforced unsaturated polyester composites are GFRP, which stands for glassfiber-reinforced polyester (or plastic), or FRP. Typical applications include:

- Automobile, truck, and bus components (see *Figures 1-11, 1-12, 1-14,* and *1-15*)
- Leisure boats (see *Figures 1-17* and *1-18*)
- Mine counter-measure vessels and high-speed passenger ships (see *Figure 1-20*)
- Construction members, such as building panels, beams, gratings, pipes, walk ways, cable trays, etc. (see *Figures 1-31* through *1-33*)
- Storage tanks and silos
- Electrical equipment
- Bathroom interiors and pools (see *Figures 1-37* and *1-38*)

2.1.6.2 Epoxies

Where mechanical properties and temperature tolerance of unsaturated polyesters no longer suffice, epoxies (EP) are often used due to their significant superiority in these respects. Of course, these improved properties come at a higher price and epoxies are therefore most often seen in fields where the cost tolerance is the highest, e.g. aerospace, defense, and sports applications.

Structure

Whereas the building blocks and chemical reactions involved in producing and crosslinking of unsaturated polyesters are similar for different polyester types, the situation is rather more complex with epoxies. A wide range of building blocks are used in polymerization and numerous different compounds in crosslinking. In the following, only one epoxy configuration is considered and will have to serve as a representative for the entire epoxy family. Likewise, only one possible crosslinking reaction is outlined.

An epoxide, or oxirane, group consists of one oxygen and two carbon atoms arranged in a ring, see *Figure 2-25*. Often the epoxide group contains yet another carbon atom and is then referred to as a glycidyl group, see *Figure 2-25*. The most common epoxy is based on condensation polymerization of epichlorohydrin and bisphenol A creating diglycidylether of bisphenol A (DGEBPA or DGEBA), as illustrated in *Figure 2-26*. When compared to unsaturated polyesters, the number of repeating units in an epoxy molecule is much lower, typically on the order of 10; at an average n value of 2 DGEBPA is solid at room temperature. Although epoxies are normally referred to as polymers, they are therefore strictly oligomers ("few mers").

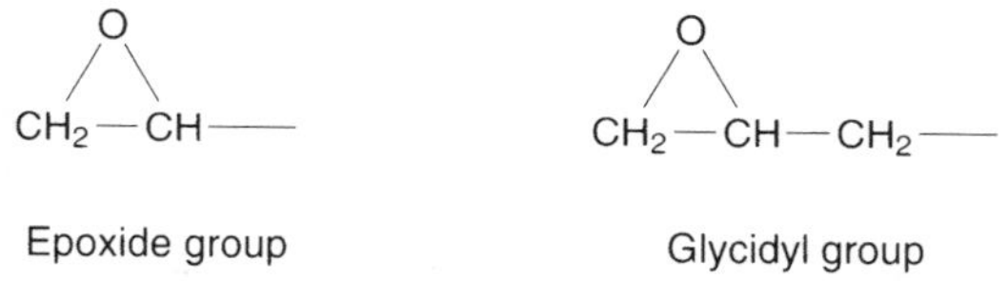

Figure 2-25 Structure of epoxide (oxirane) group (left) and glycidyl group (right)

The epoxide groups are the sites of potential crosslinks, i.e. the conceptual equivalent of the carbon–carbon double bonds in an unsaturated polyester. However, studying *Figure 2-26* one finds an important difference from unsaturated polyesters in that for DGEBPA there are only two possibilities for crosslinks per molecule regardless of the number of repeating units. It was previously mentioned that a wide range of building blocks may be used to create an epoxy. Whereas all epoxies have the epoxide group in common, the number of epoxide groups per molecule may vary; with three or more epoxide groups the molecular structure is branched, since the epoxide groups are always at the end of a branch. An epoxy with several epoxide groups is said to have high functionality.

Since polymerized epoxies very well may be solids at room temperature, reactive dilutents may be used to create a low-viscosity liquid amenable to processing. So as to be able to become an integral part of the crosslinked polymer, dilutents also contain epoxide groups, although their molecular weight, and often also the functionality, is lower than for the bulk resin. In addition to lowering the viscosity, dilutents also may be used to modify a

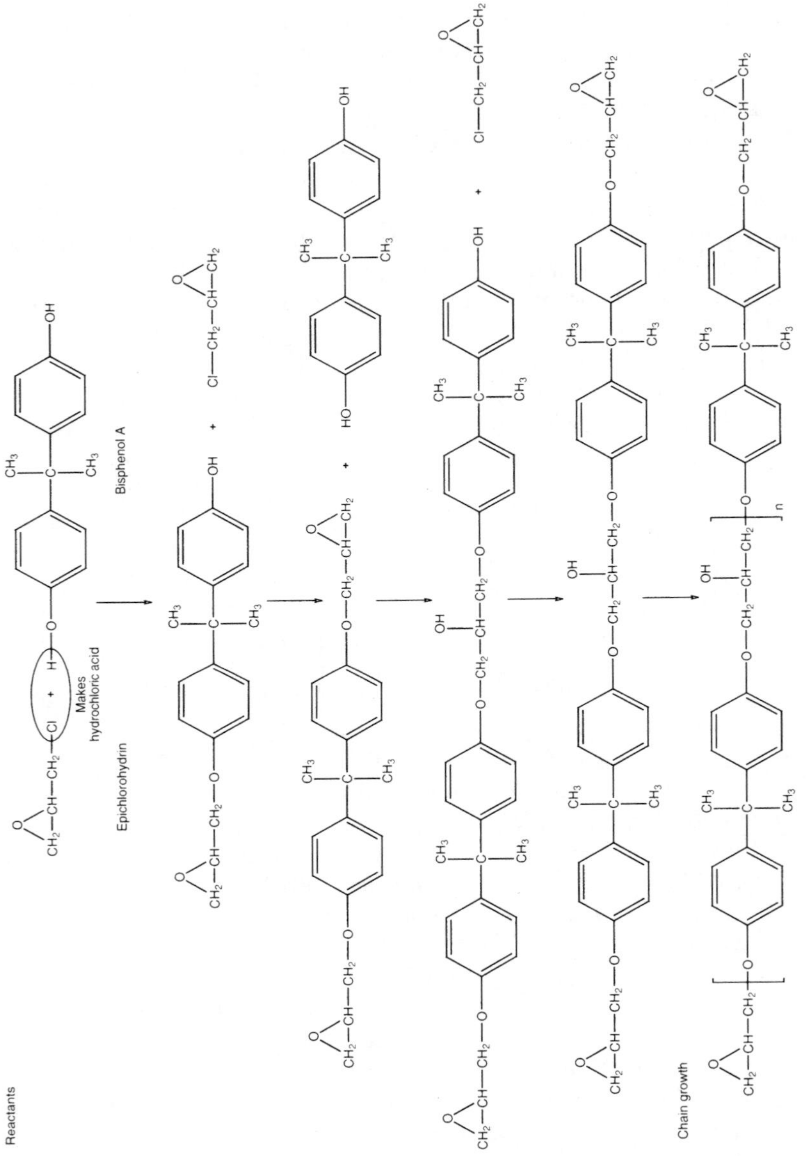

Figure 2-26 Schematic of condensation polymerization of DGEBPA. Redrawn from reference [12]

range of other properties, such as shelf and pot lives. From a manufacturing point of view, reactive dilutents may be regarded as the conceptual equivalent of the monomers used with unsaturated polyesters.

To complicate matters further, a large number of substances and reactions may be employed to create crosslinks between molecules. Two common types of crosslinking agents, often called hardeners, are amines and anhydrides; amines are most commonly used and are characterized by the NH_2 group at the end of a branch. *Figure 2-27* gives an example of the amine hardener diethylenetriamine (DETA).

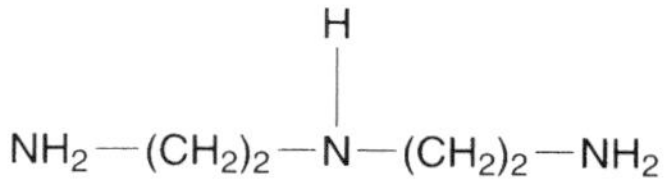

$$NH_2-(CH_2)_2-\overset{\overset{\displaystyle H}{|}}{N}-(CH_2)_2-NH_2$$

Figure 2-27 Structure of the amine hardener DETA

The following crosslinking example considers the reaction between a generic epoxy and a generic amine hardener. For the sake of brevity, *Figure 2-28* only shows the groups active in the crosslinking; "R" and "R′″" denote the remainder of the molecules. The epoxy at the top of *Figure 2-28* thus could represent the DGEBPA molecule at the bottom of *Figure 2-26* and similarly the amine hardener at the top of the figure could represent the DETA molecule of *Figure 2-27*. The figure illustrates how the reactive hydrogen of the hardener first reacts with one epoxide group and then another, thus creating a crosslink between two epoxy molecules. Since both R and R′ contain at least one more reactive group each, crosslinking may continue as long as more reactants are present, thus creating a gigantic 3D molecule.

With unsaturated polyesters the free-radical crosslinking reaction is a self-supporting chain reaction started by a small amount of initiator. For epoxies on the other hand, the hardener must be added in amounts such that the number of epoxide groups corresponds to the number of crosslinking sites available in the hardener. From a macroscopic point of view this means that the amounts of epoxy and hardener to be mixed are of the same magnitude. In the example of DGEBPA and DETA, the former has two sites for crosslinks (epoxide groups) per molecule, while DETA has two amine groups that may bond to two epoxide groups *each*; a proper blend of DGEBPA to DETA thus would ensure that there are twice as many DGEBPA molecules as there are DETA molecules. With anhydride and other hardener systems the crosslinking reactions are more complex, but from a manufacturing point of view a conceptual understanding of crosslinking with an amine hardener is sufficient. However, one notable difference is that with anhydride hardeners a small amount of accelerator may be required.

Although some epoxies are formulated to crosslink at room temperature, most epoxies used in composite applications require an increased temperature to initiate the crosslinking and a well-controlled temperature history throughout the process to crosslink as intended. *Figure 2-29* shows such a typical temperature history, often called cure cycle, for a high-performance epoxy. Many epoxies also require subsequent postcuring to completely crosslink and to achieve optimum properties. The crosslinking times required by epoxies normally significantly exceed those of unsaturated polyesters.

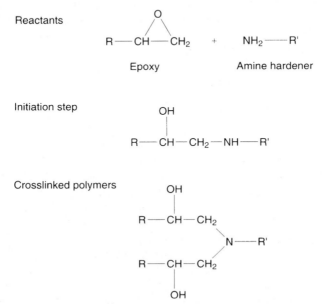

Figure 2-28 Schematic of crosslinking of epoxy using amine hardener. Redrawn from reference [13]

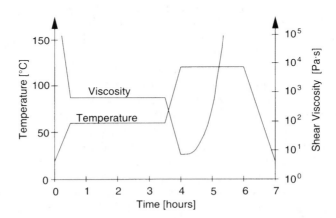

Figure 2-29 Typical cure cycle for high-performance epoxy, including a schematic illustration of the matrix viscosity development. Note the logarithmic viscosity scale. Data from reference [5]

Types and Forms

The parts of an epoxy system intended for use in reinforcement impregnation by the user is delivered in the same manner as unsaturated polyesters, i.e. as liquids. The epoxy resin, which consists of an epoxy dissolved in a dilutent, is mixed with carefully metered amounts of hardener and accelerator; typical epoxy:hardener:accelerator mix ratios could be 100:30:0 (with amine hardener) or 100:90:1 (with anhydride hardener) by weight. The room-temperature viscosity of a ready-mixed epoxy resin is on the order of 1 Pa·s. The user may choose to add more dilutent, fillers, pigments, etc.

There is such a vast array of epoxies and hardeners available that it is impractical to try to compile a comprehensive listing here. However, two common ways of characterizing epoxy systems are by hardener type, i.e. amine or anhydride, and temperature requirements during crosslinking. One thus talks of room-temperature epoxies, which may crosslink in a week at 25°C, as well as high-temperature epoxies which may crosslink in a matter of hours at 120°C, 177°C (350°F), or 190°C, etc. Many epoxy systems intended for in-manufacturing impregnation leave it up to the user to choose crosslinking temperature (higher temperature naturally leads to shorter crosslinking time), but with different component properties as a result. Formulations intended for room and moderately increased temperature crosslinking typically employ amine hardeners, whereas aromatic amines and anhydrides require higher temperatures to crosslink as intended.

Characteristics

The previous section on unsaturated polyesters pointed out that changes in resin formulation and additives may significantly change an unsaturated polyester's properties to suit a range of applications. The spectrum of properties that are achievable with changes in resin formulation and additives is even greater with epoxies, making sweeping statements about the epoxy family somewhat difficult. Mechanical and thermal properties of epoxies are nevertheless between good and excellent, which explains their extensive usage in a wide range of applications. Of significance are also low viscosity, low shrinkage, and the fact that epoxies adhere very well to reinforcement fibers. The major drawbacks of epoxies are high price, toxicity, and complex processing requirements, which often include elevated temperature and consolidation pressure, thus translating into costly manufacturing operations. Many epoxy hardeners are known to bring on health problems, most notably in the form of dermatitis and allergies, and rigorous health provisions are therefore necessary (see in addition *Chapter 8*).

Epoxy systems intended to crosslink at room temperature have T_gs on the order of 60–80°C, whereas high-performance epoxies requiring temperature cycles such as that shown in *Figure 2-29* may have T_gs around 120°C, although some epoxies have T_gs in excess of 200°C. A rule of thumb states that the maximum temperature capability of a composite is somewhat higher than the maximum temperature experienced by the matrix during crosslinking.

Applications

Epoxies are used with all kinds of reinforcement, but most commonly together with carbon fibers, since this combination offers an attractive blend of properties and cost. Common applications include:

- High-performance racing vehicles and yachts (see *Figure 1-19*)
- Aircraft control surfaces and skins (see *Figures 1-22* through *1-26*)
- Space craft (see *Figures 1-27* through *1-29*)
- Pressure vessels (see *Figure 1-28*)
- Launch tubes for portable grenade launchers (see *Figure 1-30*)
- Sporting goods (see *Figure 1-39*)

2.1.6.3 Vinylesters

Vinylesters (VE) are chemically closely related to both unsaturated polyesters and epoxies and in most respects represent a compromise between the two. Vinylesters were developed in an attempt to combine the fast and simple crosslinking of unsaturated polyesters with the mechanical and thermal properties of epoxies.

Structure

Vinylester resins are produced by reacting an epoxy with an unsaturated acid. The most common vinylester is based on the reaction between methacrylic acid and DGEBPA, see *Figure 2-30*. Note the similarity with the polymerization result at the bottom of *Figure 2-26* as well as the ester groups at each end of the bisphenol A vinylester (cf. *Figure 2-22*).

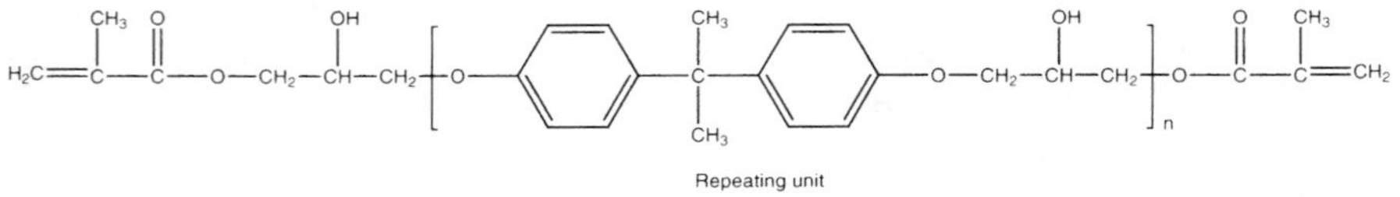

Repeating unit

Figure 2-30 Structure of bisphenol A vinylester

The major advantage of vinylesters is that the crosslinking reaction is identical to the free-radical crosslinking of unsaturated polyesters and the same crosslinking agents, initiators, accelerators, and inhibitors are used. However, the sites for crosslinks in unsaturated polyesters (the carbon–carbon double bond) occur in the repeating unit, meaning that n+1 crosslink sites are available. In vinylesters there are carbon–carbon double bonds at the end units only, meaning that vinylesters, just like epoxies, have few possible crosslink sites per molecule (two in this case). Vinylesters of high molecular weight will therefore have relatively low crosslink density and thus lower modulus than if the starting point is a lower molecular-weight polymer, once again a trait shared with epoxies. Vinylesters crosslink in time frames and under conditions similar to those of unsaturated polyesters, i.e. fairly quickly and often at room temperature.

Types and Forms

Everything that was said about delivery forms and additives for unsaturated polyesters also applies to vinylesters. Being based on epoxies, vinylesters are typically classified by the epoxy type used. Just like unsaturated polyesters and epoxies, vinylesters are available for room-temperature crosslinking or may require elevated temperature. In the latter case the chemical composition of the polymer is likely to be different from that of *Figure 2-30*.

Characteristics

The fact that vinylesters are processed the same way as unsaturated polyesters is their major advantage; a processor familiar with unsaturated polyesters is unlikely to encounter any difficulties in changing to vinylesters. In virtually all respects the characteristics of vinylesters fall between those of unsaturated polyesters and epoxies. T_gs of room-temperature crosslinking vinylesters are in the same range as room-temperature crosslinking unsaturated polyesters; the vinylesters requiring elevated crosslinking temperatures typically have T_gs 30–50°C higher.

Applications

Being an unsaturated polyester–epoxy compromise, vinylesters are more likely to be used in an application where an unsaturated polyester does not quite fulfill the requirements, rather than in an application where an epoxy represents an overkill. Applications and reinforcements are thus likely to be the same as for unsaturated polyesters, but where somewhat improved properties are required (see *Figure 1-21*). An application area in which vinylesters have been particularly successful is the corrosive industrial environment (see *Figures 1-33* and *1-34*).

2.1.6.4 Phenolics

Phenolics were invented in the beginning of the 20th century (witness the trade name Bakelite) and have since been used extensively in unreinforced and short-fiber reinforced applications. Whilst phenolics gradually have been replaced by other thermosets in many applications, they have lately experienced a revival in continuous-fiber reinforced applications mainly due to processability improvements.

Types and Forms

Phenolic resins are based on phenol and formaldehyde. Two broad types of phenolics are available; resole and novolac. Resoles crosslink when heated without the addition of a crosslinking agent, whereas the novolacs are brittle thermoplastics at room temperature and require a crosslinking agent to set. Novolacs further create problems with corrosion of metals, since ammonia is produced during crosslinking. Since they are inherently brittle, unreinforced phenolics are of little significance in engineering applications; the most commonly used material forms therefore are filled or short-fiber reinforced. Moreover, recent developments allowing in-process impregnation of continuous reinforcement using both resole and novolac phenolics have opened up the possibility of manufacturing structural composites.

Characteristics

The prime advantages of phenolics over other composite matrices are their excellent high-temperature and fire tolerances combined with low smoke emission and reasonable price. The major disadvantages include mediocre mechanical properties, especially brittleness, and the fact that until recently they have been impossible to pigment; colors usually range from yellow / red to black and darken with time. Depending on grade, processing temperatures are in the range 100–200°C. Continuous-use temperatures of standard grades are around 150°C, while certain grades can withstand up to 260°C.

Applications

Phenolic composites, which nearly exclusively are glassfiber reinforced, are most likely found in applications where the structural requirements are modest, but where high-temperature and fire tolerance is valued. Such properties may be desirable in:

• Automobile under-the-hood applications
• Aircraft interiors
• Corrosive industrial environments, especially off-shore oil platforms
• Electrical applications

2.1.6.5 Other Thermosets

Whilst the aforementioned thermosets are the most commonly used ones in composite applications, there are several other thermoset candidate matrices. Among these are polyurethanes (PUR), imide-based thermosets, and cyanate esters, where the latter are a fairly recent introduction at the high-performance end of the spectrum.

PUR is a much-used resin family which, though mainly used without reinforcement, also may be fiber-reinforced. Primary advantages are low viscosity, low crosslinking temperature, extremely rapid crosslinking if desired, and low price. Having viscosities on the order of 1 Pa·s, gel times down to less than a second if desired, and requiring crosslinking temperatures in the range 50–70°C, translates into very flexible and inexpensive processing where total manufacturing cycle times may be in the order of minutes. The drawbacks of PURs lie in that mechanical properties and temperature tolerance are less good than for all the aforementioned thermosets and that the resin toxicity is significant (see *Chapter 8*). Applications of PUR composites, which almost exclusively are glass-fiber reinforced, may for example be found in the automobile industry.

Probably the most prominent disadvantage of all the thermosets discussed above is their relatively low temperature tolerance, particularly when compared to metals. The quest for ever-higher service temperatures has led to much research into developing more temperature-tolerant matrices. Whilst many PIs are thermoplastics with excellent temperature tolerance (see *Section 2.1.5.8*), there are also thermoset versions with temperature tolerance in excess of 300°C that also offer outstanding mechanical properties. Thermoset PIs require high temperatures and long times for crosslinking and may be found in exotic high-performance applications, such as in military aircraft, and are almost certain to be carbon-fiber reinforced.

More common members of the PI family are the bismaleimides (BMI) that allow continuous-use temperatures of 230°C and short-term exposure to temperatures in excess of 300°C, while also offering outstanding mechanical properties. The major advantage of BMIs in relation to PIs is that they may be processed at temperatures common to high-performance epoxies (typically 177°C), meaning that the same manufacturing facilities may be used. However, processed this way BMIs require postcuring at temperatures up to 260°C to achieve optimum properties. Carbon-fiber reinforced BMIs are used in military aircraft.

2.1.7 Matrix Forms

To enable complete wetout during reinforcement impregnation the matrix must be liquid at some point in time, but this need not be the case at room temperature. Case in point, useful thermoplastics are solids at room temperature and are melted or dissolved for the impregnation phase. Delivered from the supplier, thermoplastics may be obtained in a wide variety of forms, e.g. as powder, pellet, fiber, and film, see *Figure 2-31*.

Most thermosets are liquids at room temperature (see *Figure 2-31* and cf. "intermediate-temperature epoxy" in *Figure 2-16*), but in most cases the low viscosity is obtained through the use of a dilutent, since the molecular weight often is so high that without it the resin is a solid. In contrast, other thermosets, such as some phenolics and PURs, may be room-temperature solids supplied in the form of powder or pellets. Some thermosets, most notably epoxies, may be partially crosslinked and the reaction interrupted, thus producing a very high-viscosity liquid (cf. "high-temperature epoxy"

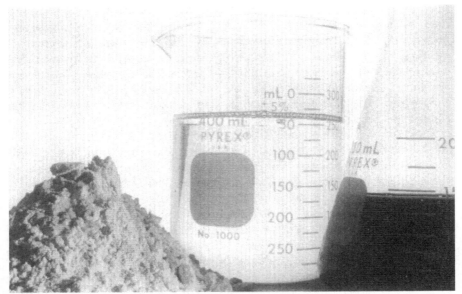

Figure 2-31 PPS powder (left), unsaturated polyester (center), and (pigmented) epoxy (right). Photograph courtesy of Center for Composite Materials, University of Delaware, Newark, DE, USA

Figure 2-16). These thermosets are delivered as solids, e.g. fiber-reinforced (see *Section 2.3*) or as film, and remain meltable solids as long as they are stored at −18°C to delay further crosslinking. Partially crosslinked epoxies are used to melt-impregnate reinforcement essentially in the same way as thermoplastics (see *Section 2.3.2*).

2.1.8 Additives

Given the fact that there appear to be countless different candidate matrices for composite applications, one may be tempted to draw the incorrect conclusion that there is always a resin for any given application. There are countless ways to alter a given property of a resin and thus the composite through the use of additives. Such alterations are most commonly taken care of by the resin supplier, but especially when using liquid thermosets the user may elect to take care of this himself. Indeed, for some manufacturing techniques it is the expertise in terms of resin formulation that distinguishes a successful processor from his hapless competitor. Among the properties that may be altered with additives are processability, mechanical properties, electrical properties, shrinkage, environmental resistance, crystallization, temperature tolerance, fire tolerance, color, cost, and volatile evaporation.

For thermosets the most obvious additives are those used in crosslinking, i.e. crosslinking agent (reactive dilutent), initiator, accelerator, inhibitor, etc.

Since these have already been discussed in the proper context, they will not be further discussed here.

Improved processability may mean different things in different processes and with different materials. Plasticizers, which are low molecular-weight compounds, may be added to thermoplastics to lower the viscosity, but also tends to have the undesired side effect of lowering temperature tolerance and decreasing mechanical properties. For lamination of vertical surfaces and upside down, it may be desirable to improve thixotropy of a thermoset matrix to prevent it from running before crosslinking. In many applications the addition of release agents into the resin, so-called internal mold releases, facilitate component removal from the mold.

Inert, i.e. chemically inactive, fillers may improve mechanical properties, such as stiffness, strength, impact, wear, and dimensional stability; electrical properties, such as conductivity, dielectricity, and static electricity; and decrease shrinkage from crosslinking (the matrix shrinks from crosslinking, while the inert filler does not). Another property commonly improved with additives is the resistance to UV light, i.e. sunlight, which otherwise tends to embrittle and discolor most polymers.

The fire-tolerance properties of a matrix often are not sufficient and flame retardants may improve the properties through several different mechanisms. They may produce water on the surface at ignition temperature, they may produce a surface char that is impenetrable to oxygen, or they may produce free radicals that react with the surface oxygen thus starving the fire.

To avoid costly painting it is often desirable to color the resin through the addition of very small amounts of colorants. Most of the colorants used in plastics are pigments, meaning that they are not soluble in the resin. The most common pigments are titanium dioxide (white) and carbon black (black); various heavy-metal oxides provide most colors in between.

In low-cost applications where structural strength is not of paramount importance the matrix may be mixed with an inert filler to reduce the use of resin, which costs more than the filler. It is important to note that addition of a compound to improve a particular property almost always inadvertently affects one or more other properties.

Two compounds that from a logical standpoint definitely belong in this section are blowing agents, which are used to produce expanded foams, and microspheres, which are used as fillers to produce syntactic foams. Expanded and syntactic foams are discussed in *Section 2.4.3*.

2.2 Reinforcements

The reinforcement is the constituent that primarily carries the structural loads to which a composite is subjected. The reinforcement therefore to a significant degree determines stiffness and strength of the composite as well as several other properties. Composite reinforcement may be in the form of fibers, particles, or whiskers.

A fiber, or filament, has a length-to-diameter ratio that approaches infinity and a diameter of order 10 μm. All common fibers are manufactured in a drawing process, where the liquid raw material is drawn from an orifice. The drawing process ensures that the molecules of fibers organic in origin are aligned and parallel to the drawing direction, translating into significantly higher strength and stiffness in the axial direction than transverse to it. The most common types of fibrous reinforcement used in composite applications are glass, carbon, and aramid. However, it is not only the fiber type that is of significance; equally important is its configuration or form, cf. *Figure 1-2*.

The linear density of fiber bundles, also known as yarns, tows, strands, and rovings, may be quantified in several different ways. The TEX number identifies the weight in grams of 1,000 meters of material, while the denier number denotes the weight in grams of 9,000 meters of material. The yield may be stated in yards per pound (yd/lb) or meters per kilogram (m/kg), while the size of some fiber types is merely denoted by the number of fibers per bundle. The designation used depends on fiber type, industry, and convention.

Particles have length-to-diameter ratios of order unity and dimensions that range from that of a fiber diameter to several millimeters. Particles have no preferred directions (such as that caused by fiber drawing) and are mainly a means to improve the properties or lower the cost of an isotropic material, whereas the use of fibers, particularly continuous fibers, creates an entirely new class of (anisotropic) material.

Whiskers have length-to-diameter ratios of order 10–1,000 and diameters of order 0.1–1 μm. Whiskers are pure single crystals manufactured through chemical vapor deposition (CVD) and thus have preferred directions. Whiskers are used to improve the properties of isotropic construction materials and since the whiskers are likely to be more or less randomly arranged in the matrix, whisker-reinforced materials are likely to be considered macroscopically isotropic as well. Due to the requirement of precise control over crystal growth in manufacturing, whiskers end up being too expensive for many potential applications.

2.2.1 Glass Fibers

Glass fabric was first produced in 1893, but it was not until 1931 that a commercial process for glass-fiber manufacture was introduced by the Owens Illinois Glass Company. In 1938, Owens Illinois merged with Corning Glass Works to form Owens–Corning Fiberglas (OCF), thus marking the beginning of today's glass-fiber industry [14]. Since then OCF has licensed its technology to manufacture continuous glass fibers to several companies around the world. Glass fibers clearly dominate as reinforcement in all but high-performance composite applications due to an appealing combination of good properties and low cost.

Manufacture

The major ingredient of glass fibers is silica (SiO_2), which is mixed with varying degrees of other oxides. The mixture is melted and extruded through minute holes in a platinum-alloy plate, or bushing, see *Figure 2-32*. The holes in the bushing, which have a diameter of 0.8–3.2 mm, usually number into the thousands.

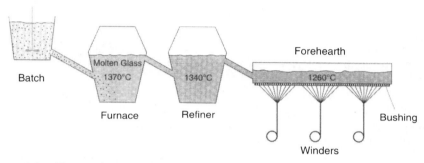

Figure 2-32 Glass melting. Redrawn from reference [14]

The glass fibers vertically emerging from the bushing are drawn at linear velocities of up to 60 m/s and are then quenched by air or water spray to achieve an amorphous structure, see *Figure 2-33*. The final fiber diameter is determined by hole size, temperature and viscosity of the melt, pressure drop, cooling rate, and drawing velocity. A protective coating, or size, is applied to the fibers before they are gathered together and wound onto forming packages or are chopped, see *Figure 2-33*. The wet glass is then dried for 10–15 hours at 120–130°C.

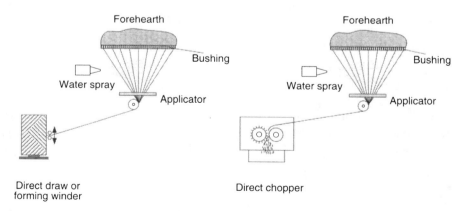

Figure 2-33 Fiberglass forming. Redrawn from reference [14]

Types and Forms

An assembly of collimated glass fibers is called a yarn, tow, or strand, and a group of collimated yarns is called a roving. When yarns are wound onto forming packages following drawing, as well as when rovings are spun from yarns, they are usually slowly rotated. This so-called twist promotes yarn and roving integrity and thus simplifies subsequent handling, but it also impedes impregnation since it becomes more difficult to separate the fibers.

The diameter of individual fibers varies between 3 and 25 μm and in composite applications is commonly in the range 10–20 μm. The linear density of glass yarns and rovings is given by TEX number or yield in yd/lb (in North America). Common TEX numbers are 600, 1,200, 2,400, etc., while common yields are 900, 450, 225, etc. (where 900 stands for 90,000 yd/lb; 1 yd/lb = 496,055/TEX).

Several different glass compositions are available, the most common being E and S glass, where "E" denotes electrical and "S" high strength. E glass offers excellent electrical properties and durability and is a general-purpose grade that heavily dominates consumption. S glass and R glass are similar and offer improved stiffness and strength as well as high-temperature tolerance. Not surprisingly, the latter glass types are considerably more expensive than E glass. ECR glass (corrosion resistant) and C glass are similar and, while having properties similar to E glass, offer improved corrosion resistance. In addition to these glass compositions, a few other specialized ones are also available.

Characteristics

Properties characteristic of glass fibers are advantages such as high strength, very good tolerance to high temperatures and corrosive environments, radar transparency, and low price. Disadvantages include relatively low stiffness, moisture sensitivity, and abrasiveness. A characteristic that sets glass fibers apart from other reinforcements used in composite applications is that they are amorphous (i.e. glassy) and therefore isotropic.

Applications

Apart from S and R glass which are used in some high-performance applications, glass reinforcement is mainly used where high stiffness is not required and where part cost is a critical factor. In most composite applications the matrix used is unsaturated polyester, or perhaps vinylester, so everything said in the unsaturated polyester *Applications* section also apply here. Likewise, with few exceptions the applications tend to be the same:

- Automobile, truck, and bus components (see *Figures 1-9* through *1-16*)
- Leisure boats (see *Figures 1-17* and *1-18*)
- Mine counter-measure vessels and high-speed passenger ships (see *Figure 1-20*)

- Aircraft radomes and interiors (see *Figures 1-22* through *1-26*)
- Construction members, such as building panels, beams, gratings, pipes, walk ways, cable trays, etc. (see *Figures 1-31* through *1-34*)
- Storage tanks and silos
- Electrical equipment
- Bathroom interiors and pools (see *Figures 1-37* and *1-38*)
- Sporting goods (see *Figure 1-39*)

2.2.2 Carbon Fibers

In his search for a suitable material for lamp filaments, Thomas Alva Edison in 1878 made carbon fibers from cotton fibers and later also from bamboo. In the late 1950s carbon fibers were produced for high-temperature missile applications and in the late 1960s were available in large quantities [15], first made commercial by Courtaulds in the United Kingdom. Carbon fiber has the highest strength and stiffness of any composite-reinforcement candidate and the development towards new carbon fiber types with improved stiffness, strength, and ultimate strain is rapid.

Manufacture

Carbon fibers are commercially manufactured from three different precursors: rayon, polyacrylonitrile (PAN), and petroleum pitch. In the first two cases, the starting point of the carbon-fiber manufacturing process is textile fibers, whereas fibers are spun directly from the melt when pitch is the starting point, see *Figure 2-34*. The method of fiber manufacture may be melt or solution spinning.

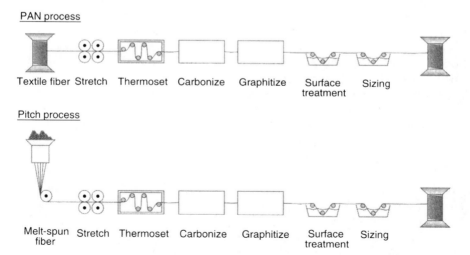

Figure 2-34 Schematic of carbon fiber production. Redrawn from reference [15]

The fibers are initially drawn and oxidized at temperatures below 400°C to crosslink them to ensure that they do not melt during subsequent processing steps; drawing and oxidizing may occur concurrently. The fibers are then carbonized above 800°C in a process called pyrolysis, i.e. heat treatment in the absence of oxygen, to remove non-carbon elements and create fibers virtually consisting of carbon only. Graphitization is then carried out at temperatures above 1,000°C to further eliminate impurities and enhance crystallinity. During both carbonization and graphitization further drawing may be employed to enhance orientation within the fibers. After graphitization fibers are surface treated and size is applied.

The carbon atoms are covalently bonded together in graphene layers which are held together by secondary dispersion bonds. The properties of carbon fibers are the result of the strong covalent carbon–carbon bonds within the graphene layers and it is therefore mainly the degree of orientation of these layers that determines the properties of the fiber. A higher temperature during graphitization promotes orientation of the graphene layers in the fiber direction, thus resulting in a higher tensile modulus. The highest tensile strength is reached at a maximum graphitization temperature of around 1,500°C for PAN- and some pitch-based fibers, while the strength continues to increase with increasing temperature for most pitch-based fibers.

Types and Forms

Carbon fibers are supplied in yarns denoted by 3k, 6k, 12k, etc., indicating the number of individual fibers they contain, where for example 3k stands for 3,000 fibers. Carbon fiber yarns receive little or no twist. Fiber diameters vary between 4 and 11 μm and are commonly around 7 μm. Carbon fibers are often quantitatively referred to as for example "ultra-high modulus," "high modulus," "intermediate-modulus," "high strength," etc., where the borders between categories constantly change due to the rapid development of new fiber types. Properties significantly depend on precursor type and heat treatments employed during manufacture.

Carbon fibers are often incorrectly referred to as graphite fibers. Strictly speaking, a graphite structure possesses three-dimensional orientation of the graphene layers, whereas the two-dimensional orientation of carbon fibers means that they should correctly be referred to as carbon fibers [15].

Characteristics

While carbon fibers have the highest strength and stiffness of any composite reinforcement candidate, only high strength or high modulus can be normally obtained in the same fiber and is also accompanied by fairly low strain to failure. Other advantages include tolerance to high temperatures and corrosive environments, as well as lack of moisture sensitivity. The major disadvantage of carbon fibers is their high price, while others include brittleness and conductivity. Whereas the conductivity of carbon may be advantageous in some rare instances it is generally a nuisance in that carbon reinforcement may cause galvanic corrosion of metal inserts and loose carbon particles suspended in the air may easily short out electrical and electronic equipment. In manufacturing operations dealing with unimpregnated carbon fibers which may be abraded, it is therefore necessary to provide costly shielding of electrical and electronic equipment.

While the transverse CTE of carbon fibers is positive, they are have a negative longitudinal CTE, which is due to bending of the predominantly longitudinally aligned graphene layers as temperature increases. The negative CTE therefore increases in numerical value as the modulus (i.e. graphene layer orientation) increases; the longitudinal CTE becomes positive at temperatures on the order of 700°C. Although never a concern with polymer matrices, it may be noteworthy that carbon starts to oxidize at 350–450°C (depending on precursor type) and the otherwise excellent environmental resistance is then impaired as temperature increases.

Applications

Carbon-fiber reinforcement dominates in high-performance applications due to its outstanding mechanical properties combined with low weight. The development of high-performance composites, largely made possible by carbon fibers, has been and to a significant degree still is driven by aerospace and defense needs. The negative longitudinal CTE is a most useful property in that it permits design of composites with virtually zero effective CTE over several hundred degrees. This property is exploited in space applications, where temperature variations are large and dimensional changes could have disastrous effects on the performance of for example satellite antennae (see *Figure 1-29*). Since carbon-fiber reinforcement is mostly combined with epoxies, applications tend to be more or less the same as for epoxies:

- High-performance racing vehicles, yachts, and ships (see *Figures 1-19* and *1-21*)
- Aircraft control surfaces and skins (see *Figures 1-22* through *1-26*)
- Space craft (see *Figures 1-27* through *1-29*)
- Sporting goods (see *Figures 1-39* and *1-40*)

2.2.3 Aramid Fibers

From a formal point of view one could argue that all commonly used composite reinforcements except glass are organic, but despite carbon fibers being made from organic fibers they are not themselves organic. Several different organic fiber types have nevertheless been used as composite reinforcement, but the category is dominated by aramid fibers. Kevlar is often assumed synonymous with aramid, but is in fact just the (Du Pont) trade name of the most common of a few commercially available aramid fiber types. When introduced in 1971, Kevlar was first targeted for used as reinforcement in automobile tires.

Manufacture

Aramids, short for aromatic polyamides, are members of the PA family; *Figure 2-35* illustrates the repeating unit of Kevlar (cf. *Table 2-2*). Due to the simple, regular structure and the aromatic rings in the backbone, this aromatic polyamide is an LCP. Aramid fibers are manufactured in a process called solution spinning. The polymer powder is dissolved in sulfuric acid and is extruded at about 80°C through small holes, called spinnerets, into a narrow air gap at a rate of 0.1–6 m/s. The fibers are quenched in a 1°C water bath to solidify the fibers and wash off most of the acid. The fibers are further washed, dried under tension, and then wound onto spools. Since aramid fibers are not brittle a protective size is not necessary.

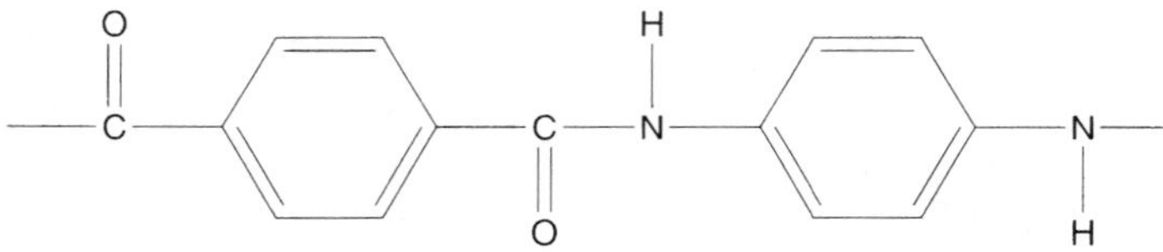

Figure 2-35 Repeating unit of Kevlar. Redrawn from reference [16]

When an LCP is extruded through a spinneret, the molecules align with the direction of shear and subsequent quenching ensures that the orientation remains in the final fiber. The degree of orientation may be further enhanced by heat treatment under tensioning, resulting in improvements in the longitudinal modulus. Due to high degree of crystallinity and rigid molecular structure the temperature tolerance of aromatic polyamide is very good for an organic material.

Types and Forms

The fiber diameter is typically 12 μm and yarns, which are not twisted, consist of anything from a couple of dozen to several thousand fibers per yarn. Yarns are normally designated either by the number of fibers or the denier count. Among several different aramid grades available for composite applications are Kevlar 29, 49, and 149 (the latter marketed as Kevlar H$_m$ in Europe), with increasing modulus as the number increases (although not in proportion). Kevlar 29 is referred to as having high toughness and is typically used in ballistic armor and in composites where high damage tolerance is required. Kevlar 49 and 149 are referred to as having high and ultra-high modulus, respectively, where Kevlar 49 is the most common in composite applications. Similar properties are available in "non-Kevlar" aramid fibers.

Characteristics

Advantages of aramid fibers are very good mechanical properties, especially toughness and damage tolerance, moderately high temperature tolerance and corrosion resistance, and good electrical properties. While possessing several attractive properties, aramid also brings on a few notable difficulties. From a design point of view it is important to realize that the strength in longitudinal compression is only a fraction of that in tension and that the fiber-matrix compatibility generally is poor. The outstanding toughness of aramid also creates a problem in that fibers are very difficult to cut and machining of aramid-reinforced composites therefore requires special tools and techniques (see *Section 5.1*). Other disadvantages of aramid fibers are moisture sensitivity and high price. Aramid fibers have positive transverse CTE, but negative longitudinal CTE.

Applications

The major advantage of aramid fibers lies in their outstanding toughness and damage tolerance which have given rise to energy-absorbing applications, such as in bullet-proof vests woven from (unimpregnated) aramid yarns. Since aramid fibers, just like carbon fibers, have negative longitudinal CTE they offer similar possibilities in designing structures with zero effective CTE. Whilst aramid reinforcement is significantly less common in composite applications than glass and carbon, partly due to its tendency to absorb water, there are several applications, including:

- Impact-prone areas of aircraft, e.g. leading edges of wings (see *Figures 1-24* and *1-26*)
- Ballistic armor
- Sporting goods

2.2.4 Polyethylene Fibers

High-modulus PE fibers may be manufactured through a process called gel spinning, which is a variant of solution spinning. In gel spinning the polymer is first dissolved in a solvent whereupon the temperature of the solution is lowered to increase the viscosity to a gel before it is extruded through spinnerets to align the molecules in the fibers. The process is conceptually identical to solution spinning of aramid fibers.

Since the tensile properties of both PE and aramid fibers are dictated by the properties of the covalent bonds of the molecular backbones, their mechanical properties are similar, but due to the lower density of PE fibers their specific strength and modulus are higher and comparable to carbon fiber properties. The main drawbacks of PE fibers are poor temperature tolerance and poor matrix compatibility. PE fibers have so far not been extensively used in composite applications, but current or potential applications include ballistic protection and sporting goods.

2.2.5 Other Fibers

Several specialty fibers are used in different applications, for example offering extra high temperature tolerance, radar transparency, etc. Fibrous reinforcements used with polymer matrices include boron and ceramic fibers, metal wires, as well as natural fibers, such as jute, copra, and wood.

Boron fibers and the most common ceramic fiber type, silicon carbide (SiC), may be manufactured through CVD onto a substrate fiber that is continuously drawn through a vertical reactor. In the manufacture of boron fibers, the substrate is a tungsten or carbon fiber and the resulting boron fiber has a diameter of 100 μm or more. In the manufacture of SiC fibers, the substrate is a carbon fiber and the resulting SiC fiber has a diameter of approximately 140 μm. SiC fibers may also be manufactured from organic-fiber precursors. Another reasonably common ceramic fiber is alumina (Al_2O_3).

Boron, SiC, and Al_2O_3 fibers are characterized by high stiffness and reasonably high strength. While being relatively uncommon in polymer-composite applications, boron and SiC fibers are used in combination with metal matrices. Silica (SiO_2), or quartz, fibers are on the other hand used with polymer matrices in radomes and stealth aircraft due to their excellent dielectric properties.

2.2.6 Reinforcement–Matrix Interaction

Conceptually a fiber-reinforced composite consists of (transversely) isotropic fibers in an isotropic matrix with a perfect bond in between. In reality, however, it is significantly more complicated. For a composite to be able to support external loads, fibers and matrix must cooperate. It is often assumed that the fiber–matrix bond should be perfect, i.e. have the same properties as the matrix (since it is the weakest constituent), and a strong bond is indeed often desirable to improve, for example, interlaminar shear strength, delamination resistance, fatigue properties, and corrosion resistance. However, in some cases a weak bond actually may be preferable; the damage tolerance of a composite with a brittle matrix usually is enhanced by a weak fiber–matrix bond [17]. Whatever the load case, the fiber–matrix interface is of crucial importance to the properties of a composite.

Due to the significance of the interface much effort has been and still is being spent trying to tailor the surface properties of the reinforcement to enhance compatibility with the matrix. In the previous sections, the references to surface treatments and coatings of reinforcing fibers was intentionally kept brief and low on detail since this topic is so complex. In the following the topic is treated in more detail.

Manufacturers of brittle fibers, such as glass and carbon, apply a size to the fibers to protect them from damage during subsequent handling, such as spinning, weaving, etc. Since a single reinforcement yarn or roving may contain tens of thousands of fibers and thus may be difficult to handle, the size also may be intended to promote integrity during such handling. However, since a size may interfere with the creation of a strong interface, and promotion of yarn integrity is detrimental to reinforcement impregnation, the size may be seen as a necessary evil. The size is therefore often burned off when subsequent handling has been completed. To enable application of the size to every fiber, it is in glass manufacture applied before the fibers have been gathered into a yarn (see *Figure 2-33*), whereas carbon fibers must be spread out as widely as possible before application (see *Figure 2-34*).

A good bond between fiber and matrix may be formed through several mechanisms that all require complete wetting of the fibers by the resin. The preferred bonding mechanism is chemical (covalent) bonding, whereas mechanical interlocking may work well if the fiber surface is irregular, which is the case with some carbon and organic fiber types. Assuming that the fiber is porous, interdiffusion of polymer molecules offers another possibility, whereas electrostatic attraction (secondary bonds) probably does not offer a strong enough bond to be alone responsible for satisfactory bonding [18]. Despite over thirty years of intensive research, the true mechanisms behind most successful fiber–matrix bonds are not well understood. Moreover, successful recipes are jealously guarded trade secrets. Nevertheless, to enhance bonding the following considerations should be addressed [17]:

- Removal of weak fiber surface layers or surface contaminants
- Improvement of wettability
- Creation or addition of chemically reactive groups
- Changes in surface topography

For glass fibers a coupling agent, also called a finish and adhesion promoter, is applied to the fiber surface to enhance fiber–matrix compatibility. When bonding to thermosets is required, this coupling agent usually is of the silane family, with one end group of the silane molecule being chemically compatible with the glass and the other end with the matrix. The silane coupling agent thus bonds covalently to the matrix and is thought to have an occasional covalent bond also to the glass [18,19]. Since each fiber preferably should be completely coated by the coupling agent and this is difficult to achieve with silanes, the coupling agent is usually applied together with some film-forming polymer that easily spreads on the fiber surface. However, film-forming polymers are polar and therefore attract water, which is detrimental to the interface strength. This is nevertheless the commercial practice and the film-forming polymer usually doubles as size. The coupling agent thus is applied to the fibers at the same time as the size and one therefore talks of a multi-functional size [19]. Well-functioning coupling agents are available for unsaturated polyesters and epoxies, whereas coupling agents for high-temperature thermosets and all thermoplastics are not as efficient or are even nonexistent.

For carbon fibers the situation is rather different in terms of achieving fiber–matrix compatibility. Instead of applying a coupling agent the fiber is surface treated so as to promote its chemical reactivity and compatibility with the matrix. Most commonly the fiber surface is oxidized to create surface oxygen groups that can form covalent bonds with the matrix. It appears as if mechanical interlocking also is important with carbon fibers, since bonding generally is less good for high modulus fibers which have a higher degree of orientation of the graphene layers and thus smoother surfaces. Alternative but much less common means of achieving good matrix compatibility are through plasma treatment and growing, or grafting, of whiskers on the fiber surface. In terms of size carbon fibers are treated in a fashion analogous to glass fibers, although the size typically is an epoxy if the intended composite matrix is an epoxy. The size is likely to bond to the fibers and, being an epoxy, to the matrix as well.

For aramid and PE fibers the situation further differs in that no size is needed to protect the tough fibers and no successful coupling agent has been developed. Coupling agents used with glass and carbon have proved ineffective. Although probably not yet commercial practice, plasma treatment to attach, for example, amine groups to the surface of aramid fibers to enhance compatibility with epoxies has proved successful.

A common misconception is that the properties of the matrix are uniform throughout the composite, but various studies have proved the existence of a matrix property gradient from the fiber surface into the bulk matrix, a so-called interphase (not to be confused with the interface; the interface is an area, while the interphase is a volume). Without an interphase there would be a huge discontinuity in modulus at the fiber–matrix interface. It has proved to be advantageous to some composite properties to have a ductile interphase with an intermediate modulus [18]. For example the epoxy size applied to carbon fibers contains no hardener (although some will probably diffuse in from the bulk matrix following reinforcement impregnation) and the interphase thus will have fewer crosslinks than the bulk and be more ductile. In contrast, with some semicrystalline matrices the fibers act as nuclei for crystallization, meaning that the interphase may end up stiffer than the bulk matrix.

One theory states that the silane coupling agent on glass fibers creates a porous polymer network around the fibers which the molecules of the matrix can penetrate, thus mechanically interlocking with the coupling agent and creating an interphase with properties different from the bulk matrix [20]. With carbon fibers which consist of convoluted but generally aligned graphene layers there is a possibility for diffusion of matrix molecules into the fiber, indicating that the interphase may extend into the fiber and not only into the matrix. Since the interphase is chemically nonhomogenous it is likely to be more susceptible to chemical degradation and water migration than the bulk matrix.

2.2.7 Fibrous Reinforcement Forms

Although essentially all fibrous reinforcement is produced as continuous fiber yarns, this is often not the form in which it is finally used. Equally as important as the type of reinforcement is the form. Further, it is not uncommon that for example carbon and aramid fibers are mixed into hybrid fabrics. The following sections describe the most common reinforcement forms and the term yarn is used throughout to refer to fiber bundles, strands, tows, and rovings.

2.2.7.1 Weaving

Weaving, see *Figure 2-36*, produces fabrics in a variety of shapes and weave configurations, see *Figure 2-37*. The longitudinal direction of the fabric is called warp and the transverse direction weft or fill. For composite purposes, woven fabrics may be characterized in terms of the fabric crimp, which is a measure of the yarn waviness. A yarn extracted from, say, a meter-wide fabric will be longer than one meter and a crimp of 2 percent thus indicates that the extracted yarn is 1.02 meters long. For the woven fabrics in *Figure 2-37* crimp decreases from left to right. A lower crimp improves drapeability, or formability, since it also means fewer warp–weft crossovers that may resist shearing. Lower crimp also means straighter fibers, which translates into better composite mechanical properties. The drawback of lower crimp is reduced fabric integrity, i.e. yarns may easily move during handling.

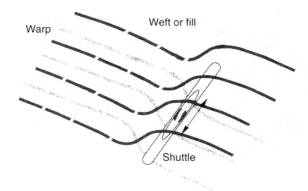

Figure 2-36 Schematic of weaving. Redrawn from reference [21]

The most common weave, from shirts to composite reinforcement, is the plain weave, which is the tightest possible, see *Figures 2-37a* and *2-38a*. The basket weave (*Figure 2-37b*) is a variation of the plain weave with two or more yarns per warp and weft and offers improved drapeability over the plain weave, while twill (*Figure 2-37c*) offers even greater drapeability. The satin weaves are a family of weaves categorized by the number of yarns in the repeating unit; *Figure 2-37d* shows an eight-harness satin weave (8HS) also known as eight-end satin weave. Through this minimum of interlacing and crimp, a very flexible weave is obtained. Other common satin weaves are 5HS, see *Figure 2-38b*, and 4HS, commonly called crowfoot. Unidirectional (UD) fabrics are fabrics that have heavy yarns in the warp direction and very light yarns in the fill direction, see *Figure 2-38a*. UD fabrics are used since they are much easier to handle and align than individual yarns.

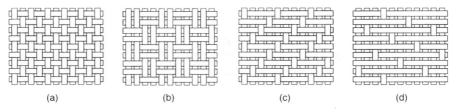

(a)   (b)   (c)   (d)

Figure 2-37 Examples of woven fabric forms (a) plain weave; (b) basket weave; (c) twill; (d) eight-harness satin weave

Figure 2-38 Woven fabrics (a) plain weave of a UD fabric; (b) five-harness satin weave. Photographs courtesy of Center for Composite Materials, University of Delaware, Newark, DE, USA

The important parameters to consider when dealing with woven fabrics is fiber type, yarn type, weave style, crimp, yarn count (yarns/length unit), and areal weight. Also of great importance for the weaving operation is the size, which ideally should eliminate any fiber damage. Following weaving the size is often burned off. Areal weights of woven fabrics typically range from 0.1 to 1 kg/m^2 and thicknesses from 25 to 500 μm, but much heavier and thicker are also available, particularly with glass fabrics.

Although woven fabrics are normally wide and straight, weaving can produce fabrics that are of virtually any width as well as curved. With specialized machinery it is further possible to obtain fabrics with ±45° yarn orientation (instead of the customary 0°/90°) and triaxial weaves with three yarns intersecting at 60° angles. Fully computer-controlled machinery has made 3D weaving a reality and 3D woven fabrics are being evaluated in some specialized applications. *Figure 2-39* shows an example of an interlinked multilayer fabric, where warp yarns hold several layers of weft yarns together. Other examples of 3D weaving are integrated T and I sections. 3D fabrics may provide interlaminar properties impossible to achieve with conventional ply-by-ply reinforcement placement (where no through-the-thickness reinforcement is present), but they are also quite difficult to impregnate since the weave is so tight.

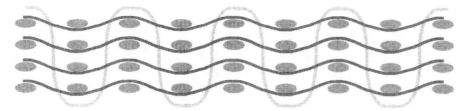

Figure 2-39 One type of interlinked multilayer fabric. Redrawn from reference [22]

2.2.7.2 Noncrimp Fabrics

In recent years noncrimp, or inlaid, fabrics have become popular. In noncrimp fabrics, yarns are placed parallel to each other in separate layers of different orientation which are then stitched together using polyester thread, see *Figure 2-40*. Apart from the obvious advantage of zero crimp, noncrimp fabrics allow rapid and precise layup of multilayered reinforcement.

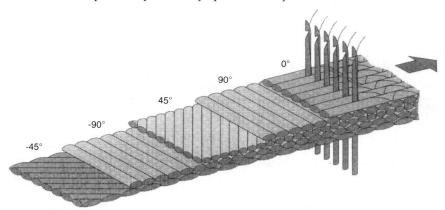

Figure 2-40 Schematic of noncrimp fabric. Redrawn from reference [23]

Another form of noncrimp reinforcement is where aligned fibers, usually all aligned in the same direction, are given handling integrity through the use of a fibrillar thermoplastic binder. This type of reinforcement provides an alternative to the aforementioned woven UD fabrics.

2.2.7.3 Knitting

An alternative to weaving is knitting, see *Figure 2-41*, which produces a looser and more flexible fabric. Although not shown in the figure, both warp and weft insertion of straight yarns is possible to enhance mechanical properties of the composite, while also resulting in impeded fabric drapeability. Though less common and often tailored to a specific application, knitted fabrics may be used in the same processes and applications as woven fabrics.

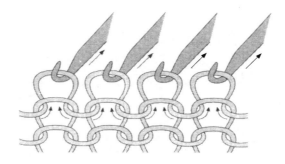

Figure 2-41 Schematic of knitting. Redrawn from reference [21]

2.2.7.4 Braiding

Braiding is one of the oldest textile manufacturing methods and is traditionally used to produce for example ropes, shoe laces, and reinforcement for garden hoses, see *Figure 2-42* for a schematic of flat braiding and *Figure 2-43* for a photograph of a circular braider. Braiding yields a tape or a tubular structure with the reinforcement oriented at $\pm\alpha°$ and, if desired, also in the longitudinal (0°) direction, see *Figure 2-44*. Among the features of braiding are high integrity and drapeability (unless longitudinal reinforcement is present), while the resulting composites have high torsional stiffness. Applications include (vehicle) drive shafts where tubular braids are used. 3D braiding, although not in widespread commercial use, is available.

2.2.7.5 Mats and Non-Wovens

A commonly used reinforcement form is the mat, which is produced either as continuous strand mat or chopped strand mat. While continuous strand mat sometimes is referred to as continuous filament mat (CFM) to distinguish its abbreviation from that of chopped strand mat (CSM), both mat types are more commonly and certainly confusingly abbreviated CSM. This

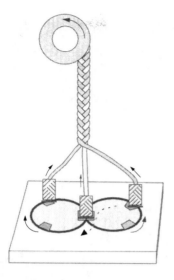

Figure 2-42 Schematic of flat braiding. Redrawn from reference [21]

Figure 2-43 72-carrier tubular braider. Photograph courtesy of Center for Composite Materials, University of Delaware, Newark, DE, USA

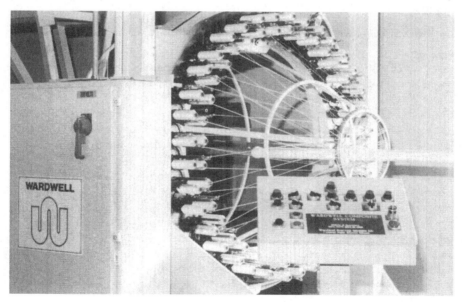

confusion is nevertheless not damaging since the two types are often used interchangeably and the resulting composite properties are not dramatically different. Continuous strand mat is produced by more or less randomly swirling continuous yarns onto a moving carrier film or belt and then applying a binder, which typically is a thermoplastic polymer, to loosely hold the mat together, see *Figure 2-45*. Chopped strand mats similarly are produced

Figure 2-44 Tubular ±45° glass braid with 0° carbon tracer yarn draped (from right to left) over a core in the braider of *Figure 2-43*. Photograph reprinted from reference [24] with kind permission from Technomic Publishing Co, Lancaster, PA, USA

by randomly depositing chopped fibers (of order 25 mm in length) onto a carrier film and then applying a binder. CSMs are widely used in lightly stressed applications since they allow significant draping and provide a certain degree of in-plane isotropy.

Non-wovens is a term applied to a two-dimensional arrangement of fibers that are given structural integrity through the use of some kind of binder. One form of non-woven is the veil, which is a thin material composed of very fine fibers, which may be conventional reinforcing fibers or some organic fiber. Veils are often used to create a polymer-rich and cosmetically appealing composite surface and may also be printed with a pattern or a logotype. Although non-wovens and veils are usually not referred to as CSMs and vice versa, they logically belong to the same category. The term CSM is usually reserved for glass, whereas the terms non-wovens and veils apply to any thin, planar arrangement of fibers of any other type.

2.2.7.6 Combination Fabrics and Preforms

It is common that different types of fabrics are stitched or needled together to facilitate both handling and reinforcement orientation. The most common combination is probably that of CSMs and woven fabrics, which are widely

Figure 2-45 Continuous strand mat (CSM). Photograph courtesy of Center for Composite Materials, University of Delaware, Newark, DE, USA

used in for example boat building to save time and money in reinforcement cutting and placement operations. More refined solutions to the handling and orientation problems are offered by the aforementioned interlinked and noncrimp multilayer fabrics. For example the use of $0°/±45°$ noncrimp fabrics substantially reduces cost in the manufacture of the surface attack ship shown in *Figure 1-21*.

While combination fabrics, at least for the purposes of this treatment, are supplied in the form of broadgoods (wide material rolled onto a bobbin) that require cutting to fit onto or into the mold, the term preform refers to a 3D net-shape reinforcement arrangement, meaning that no further cutting or trimming is needed. Preforms may be manufactured in several different ways, but the most common one is probably compression molding of CSMs; since CSMs are bound together by thermoplastic polymers, they may be heated, molded into net shape, and then cooled to set the shape, see *Figure 2-46*.

Another way to tailor reinforcement to a specific geometric shape is to use a preforming process, such as that illustrated in *Figure 2-47*. Chopped fibers and binder are sprayed onto a perforated screen mold and a custom

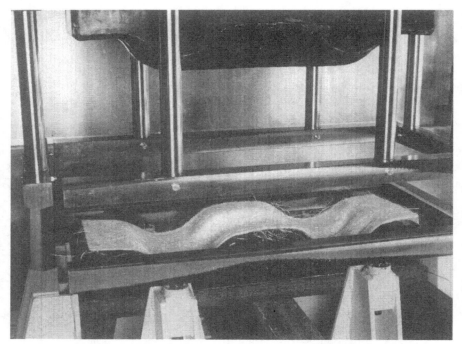

Figure 2-46 Preform compression molded from flat CSM still in lower mold half. Photograph courtesy of Swedish Institute of Composites, Piteå, Sweden

preform shape is thus gradually built up. More refined versions of such pre-forming processes utilize industrial robots to deposit chopped or continuous fibers and binder onto a stationary screen mold. Preforms often consist of several reinforcement layers, for example a couple of CSM layers sandwiching a UD fabric strip, which are stitched together to facilitate han-dling as well as to produce a certain degree of transverse reinforcement. Net-shape preforms may also be manufactured through weaving and knit-ting, for example, but this practice is likely to be limited to high-performance applications.

2.3 Preimpregnated Reinforcement

One of the most complex, difficult, and—not least—important aspects of com-posite manufacturing is the impregnation of the reinforcement, particularly if the matrix is highly viscous. Consider a carbon-fiber reinforced composite with a fiber volume fraction of 0.6, an average fiber diameter of 7 μm, and a volume of 1 ml (a cube with 10 mm sides). This volume contains 16 km of fiber, 0.4 ml of matrix, and 0.34 m² of interface. If the matrix is a high-viscosi-ty thermoset or a molten thermoplastic with even higher viscosity, it can

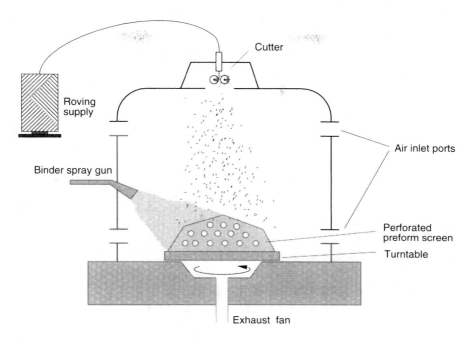

Figure 2-47 Schematic of preform manufacture. Redrawn from reference [10]

easily be appreciated that it is extremely difficult to evenly spread 0.4 ml of matrix onto a fiber surface area equivalent to almost ten pages of this book. Invariably, the impregnation will be imperfect and result in some dry fibers and entrapped gas as well as non-uniform fiber distribution (cf. *Figure 1-3*). This is the background to the widespread use of preimpregnated reinforcement (prepregs) in manufacturing of high-cost and high-performance composites. Commercial prepregs are manufactured with dedicated machinery under well-controlled conditions and the result is low void content, reasonably uniform fiber distribution, and the prepregs often contain matrices that are not for sale except in prepreg form. The significant convenience of prepregs naturally comes with a notable disadvantage in terms of high price.

Some thermosets, most notably epoxies, may under well-controlled conditions be lightly crosslinked and further crosslinking arrested through lowering of the temperature; a resin that has gone through such treatment is referred to as B-staged (an A-staged resin is not crosslinked at all and a C-staged resin fully crosslinked). If stored at a temperature below –18°C, the shelf life of thermoset prepregs is usually 6–12 months. At room temperature, the shelf life of a B-staged prepreg is on the order of days and the desirable stickiness, or tack, is gradually lost as crosslinking progresses. Upon moderate heating, i.e. at final processing, B-staged resins melt before they on further heating completely crosslink (see *Figure 2-29*). The techniques of prepregging and B-staging of resins add a significant advantage to the use of thermoset prepregs in that the prepreg manufacturer has taken care of resin formulation and the processor therefore can disregard such concerns and many of the worker health aspects (see further *Chapter 8*). On the other hand, disadvantages of B-staged prepregs include that they must be stored in freezers, that they normally must be crosslinked under closely controlled conditions, and high price.

Prepregs are available in several forms depending on reinforcement and matrix as well as intended use. With continuous and aligned reinforcement the major impregnation forms are solvent impregnation, melt impregnation, powder impregnation, and commingling. Preimpregnated discontinuous and randomly arranged reinforcement is common in compression molding processes. Such molding compounds, which may be liquid, melt, or powder impregnated, are not referred to as prepregs although they conceptually belong to the same category; they are further discussed in *Section 2.3.6*.

2.3.1 Solvent Impregnation

Recalling earlier sections of this chapter, non-crosslinked thermosets and amorphous thermoplastics are not resistant to all solvents and thus may be solvent impregnated. The polymer is dissolved to significantly lower its viscosity and thus facilitate wetting and impregnation. The reinforcement is led into the solvent bath where the combined efforts of surface tension and the fact that the reinforcement is guided over rollers or bars ensures impregnation, see *Figure 2-48*. Emerging from the bath, the impregnated reinforcement goes through nip rollers that carefully meter the reinforcement-to-solution ratio, whereupon the impregnated reinforcement goes into a drying oven, where the solvent is driven off and recovered. Correctly performed solvent impregnation produces intimately impregnated reinforcement, see *Figure 2-49a*. Both yarns and fabrics may be impregnated in this fashion.

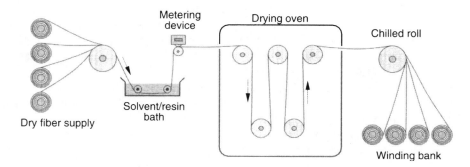

Figure 2-48 Schematic of solvent impregnation. Redrawn from reference [25]

Since B-staged thermosets emerge tacky, or sticky, from the impregnation process they are sandwiched between backing papers or films to prevent layers from adhering to each other (not shown in *Figure 2-48*). After being rolled up, thermoset prepregs are stored in freezers. Thermoplastic prepregs are not sticky and no backing paper is used; rolled-up prepregs are stored at room temperature. Regardless of resin type, residue solvent in the matrix presents a problem since it is often seriously detrimental to the properties of the composite.

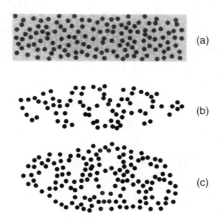

Figure 2-49 Cross sections of different prepreg types (a) solvent- or melt-impregnated; (b) powder-impregnated with polymer sleeve; (c) commingled. Reinforcing fibers are black and matrix is gray. Matrix powder particles and fibers typically have significantly larger diameter than reinforcing fibers

2.3.2 Melt Impregnation

Ideally, melt impregnation is the preferred impregnation process, since it completes the ultimately desired intimate wetting without introduction of a solvent. Unfortunately it is also the most difficult impregnation technique due to the high viscosities involved. With thermosets, an increase in temperature may prove sufficient to achieve a low enough viscosity to complete impregnation (cf. *Figure 2-16*). With thermoplastics the difficulty is even greater since the resin has the consistency of chewing gum and heating does not improve the situation as strongly as with thermosets (cf. *Figures 2-16* and *2-17*). In this case surface tension is of little help, and the first impulse may be to somehow increase the transverse pressure to force the matrix into the reinforcement. However, since an increased pressure also compacts the fiber bed further and thus makes it even less permeable, this tactic has the opposite effect.

In thermoset prepreg manufacture, layers of molten resin of carefully metered thickness are deposited into two carrier films and the reinforcement is sandwiched in between. The resin is worked into the reinforcement through a "kneading" motion achieved by leading the film–resin–reinforcement sandwich over and under rollers (conceptually similar to the impregnation process illustrated in *Figure 2-53*, but in this case with longitudinally oriented and continuous fibers), thus intimately impregnating the reinforcement (see *Figure 2-49a*) to produce a boardy prepreg, see *Figure 2-50*.

Figure 2-50 Different forms of preimpregnated reinforcement. From foreground to background: Comingled carbon/PEEK yarn, melt-impregnated glass/PPS, melt-impregnated carbon/PPS, and fabric woven from commingled carbon/PEEK yarn. Any melt- or solvent impregnated prepreg—thermoset as well as thermoplastic—looks the same on a photograph. Photograph courtesy of Center for Composite Materials, University of Delaware, Newark, DE, USA

Some melt-impregnation techniques for thermoplastics involve the application of molten matrix onto the reinforcement right before the nip between the spread-out reinforcement and a rotating roller. With this approach the pressure forces the matrix through the reinforcement towards the lower pressure on the other side of the reinforcement, i.e. along the direction of the negative pressure gradient. Following several rollers like these the process is finished off by some kind of metering device to ensure that the reinforcement–matrix ratio is correct. The newly fabricated prepregs are cooled and the matrix solidified. Melt-impregnation techniques of this type, which are carefully guarded by patents, are probably only used to impregnate parallel-fiber yarns, since attempts at impregnating fabrics likely would prove extremely difficult due to the restrictions of warp-weft crossovers.

An alternative melt-impregnation technique capable of impregnating fabrics employs a double-belt press (DBP), see *Figures 2-51* and *2-52*. A DBP essentially consists of two steel belts that feed the material through the machine while applying lateral pressure to the material. In the first section of the machine the steel belts are heated, while in the latter section the belts are cooled. A reinforcement fabric sandwiched between two polymer films is fed into the machine. The heat within the press melts the polymer films and the pressure gradient applied by the belts causes the molten polymer to impregnate the fabric, which finally is cooled before exiting the press. Since the residence time within the double-belt press is long, even highly viscous thermoplastics are capable of percolating the reinforcement.

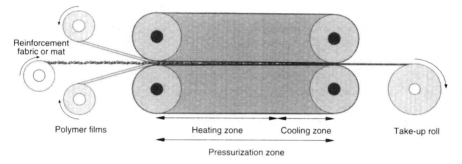

Figure 2-51 Schematic of DBP prepregging of continuous fabric

Figure 2-52 DBP with material feed from right to left. Photograph courtesy of Institut für Verbundwerkstoffe, Kaiserslautern, Germany

An interesting concept is offered by the thermoplastic long discontinuous-fiber reinforced (LDF) material. In this melt-impregnated prepreg product, the fibers are aligned as in an ordinary melt-impregnated prepreg, but the fibers are discontinuous (cf. lower left illustration in *Figure 1-2*), allowing forming to a degree not possible with continuous-fiber reinforced prepregs.

2.3.3 Powder Impregnation

By grinding the solid matrix, usually a thermoplastic, into a fine powder, the reinforcement can be impregnated using a slurry or a fluidized powder bed. With a slurry a liquid, often water, is used to disperse the powder. The reinforcement is then impregnated with this low-viscosity aqueous solution, whereupon the water is driven off. The slurry may contain some kind of agent to promote (temporary) adhesion between matrix powder and fibers, or the powder-impregnated reinforcement may be pulled through an oven to slightly melt the powder onto the fibers.

A fluidized powder bed contains a powder cloud that is kept fluidized by circulating air. Correctly performed, electrostatic attraction and/or friction ensure that the reinforcement passing through the bed is properly filled with resin particles. The use of an oven to fuse the matrix onto the fibers is common. Another option to prevent the powder from being shaken out of the reinforcement involves enclosure of the powder-filled yarn in a simultaneously extruded polymer sleeve of the thermoplastic matrix material, see *Figure 2-49b.*

The main advantages of powder-impregnated reinforcements is that they are flexible (unless subsequently excessively heated) and that they are cheaper than solvent- and melt-impregnated prepregs. The main drawbacks are that the reinforcement is not melt-impregnated, which thus has to be undertaken during manufacturing, and that the powder rarely is evenly dispersed within the reinforcement. The final melt impregnation is often difficult to achieve satisfactorily and the results are liable to be less good than if melt-impregnated prepregs had been used. Another drawback specific to slurry-impregnated materials is that composite properties may suffer from residual adhesive agent, for reasons similar to those encountered with solvent-impregnated prepregs. Powder-impregnated reinforcement, which tends to be unidirectional, is rarely referred to as prepreg, but rather powder-impregnated yarn.

However, not only yarns but also fabrics may be powder-impregnated; resin powder is first distributed onto one side of the horizontal fabric and the powder thermally fused in place, whereupon the other side of the fabric receives the same treatment. Such a powder-impregnation process may be followed by a DBP to produce a fully melt-impregnated fabric or multi-ply laminate.

2.3.4 Commingling

Commingled reinforcement consists of mechanically commingled (combined) reinforcing fibers and fibers spun from a thermoplastic resin, see *Figures 2-49c* and *2-50*. The advantages and limitations of commingled reinforcement are the same as for powder-impregnated yarns (with the difference that there is no worry of residues from a slurry impregnation). Commingled reinforcement is unidirectional and is usually referred to as commingled yarn.

2.3.5 Comparison of Preimpregnated Reinforcement Types

A qualitative comparison between the products of the aforementioned impregnation methods would consider drapeability and flexibility; tack; quality of impregnation; and cost. In general, drape and flexibility is highly desirable to allow conformation to curved shapes and to allow for use in textile processes. Tack is desired to ensure that the material stays where it has been placed without special procedures. High quality of impregnation is of course always desirable and likewise cost should be kept as low as possible.

Drape and flexibility is typically poor for solvent- and melt-impregnated materials, whereas tack is only achievable with thermosets. Not surprisingly, high quality of impregnation goes hand in hand with high cost in solvent- and melt-impregnated materials. The main advantages of thermoset prepregs thus are their impregnation quality and tack. They are also less expensive than their solvent- and melt-impregnated thermoplastic counterparts, which offer equally good impregnation quality. The main advantages of powder-impregnated and commingled reinforcements is that they are flexible enough to be used in textile manufacturing processes such as weaving (cf. *Figure 2-50*) and braiding and that they are, on a relative scale, inexpensive. Their main drawback is that the reinforcement is not melt-impregnated, which creates processing difficulties at a later stage. It deserves to be pointed out that narrow tapes of solvent- and melt-impregnated prepregs may also be used in very specialized weaving processes; *Figure 2-38b* illustrates this feasibility.

The most common prepregs are epoxy-based and usually reinforced with carbon fibers. However, prepregs based on all conceivable combinations of unsaturated polyester, epoxy, and phenolic reinforced with glass, carbon, and aramid are commercially available. Unidirectionally reinforced prepregs are more common than fabric-based prepregs. There are much fewer thermoplastic prepregs on the market, but available combinations include glass-reinforced PP, PAs, and thermoplastic polyesters, as well as carbon-reinforced high-performance thermoplastics.

Applications

Prepregs are mainly found in high-performance composite applications, where the improved properties achieved can motivate their high price. In almost all cases thermoset prepregs dominate usage. Applications include:

- High-performance racing vehicles and yachts (see *Figure 1-19*)
- Aircraft control surfaces and skins (see *Figures 1-22* through *1-26*)
- Spacecraft (see *Figures 1-27* and *1-29*)
- Sporting goods (see *Figure 1-40*)

2.3.6 Molding Compounds

Many techniques for manufacturing composites with more or less randomly oriented and often discontinuous reinforcement also use some form of preimpregnated reinforcement as raw material. However, when preimpregnated reinforcement does not contain oriented and essentially continuous fibers, it is usually not referred to as prepreg but rather as molding compound. With molding compounds it may be more fitting to talk about reinforced resin rather than impregnated reinforcement, since the fiber content tends to be considerably lower than in prepregs and some of the matrix is replaced by inexpensive fillers. With continuous-fiber reinforced prepregs it is appropriate to consider conformation of the prepregs into the desired shape, since the fibers essentially are inextensible from a manufacturing point of view (see in addition *Section 4.1.2*). In contrast, molding compounds have discontinuous or, alternatively, continuous but randomly arranged fibers; the lack of restraining straight fibers together with the lower fiber content allows the material to flow into the desired shape.

The most common thermoset-based molding compound is sheet molding compound (SMC), which is manufactured in a process schematically illustrated in *Figure 2-53*. A paste of fully formulated resin, filler, thickener, and other additives is doctored onto two carrier films and chopped glass fibers are deposited in between, see *Figure 2-54*. The two layers are brought together and the sandwich is worked by rolls to impregnate the reinforcement, see *Figure 2-53*. The resin is then aged, or matured, at a temperature of around 30°C for up to a week before it is ready to be used. During maturing the SMC thickens dramatically from an original resin viscosity of order 10^1 Pa·s to a final viscosity in the range 10^3–10^5 Pa·s. This viscosity increase is necessary to ensure that the resin will not drain out of the reinforcement during molding. The SMC rolls are stored sealed in plastic film to keep volatiles from evaporating.

CSMs (both chopped and continuous types) may be used to manufacture SMC as an alternative to chopping fibers as part of the process. To achieve improved properties in a given direction it is also possible to include longitudinally oriented continuous yarns. SMC with random reinforcement is sometimes denoted SMC-R, SMC with continuous and aligned reinforcement

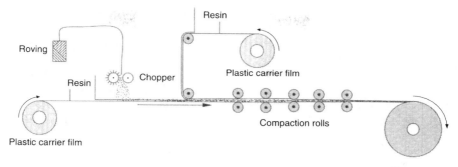

Figure 2-53 Schematic of manufacturing of SMC

Figure 2-54 Chopped glass fibers falling onto the lower resin film in SMC manufacture. Photograph courtesy of Lear Corporation Sweden, Ljungby, Sweden

SMC-C, and SMC with a blend of random and continuous reinforcement SMC-C/R. SMC-R contains fibers 13–25 mm in length (although lengths may vary between 6 and 75 mm) and normally has a fiber volume fraction in the range 0.1–0.3. SMCs may also be characterized by the degree of shrinkage during molding. General purpose grades shrink more than 0.3 percent, low shrink grades between 0.3 and 0.05 percent, and low profile grades less than 0.05 percent. Unsaturated polyesters clearly dominate as SMC resin, although other resins are available for special applications. A typical SMC composition is given in *Table 2-3*.

Table 2-3 Typical SMC and BMC compositions (weight fractions)

Constituent	SMC	BMC
Glass fiber	0.28	0.20
Unsaturated polyester	0.35	0.30
Filler, thickener, pigment, mold release, initiator, etc.	0.37	0.50

Another thermoset molding compound is bulk molding compound (BMC), which differs from SMC in that the feedstock is manufactured by mixing it in bulk, i.e. in a "dough-like" process (which has given rise to the alternative name dough molding compound, DMC). BMC is in the form of a log or a rope and is sometimes called premix. The constituents are basically the same as for SMC, but the filler content is higher and the fiber content lower, see *Table 2-3*, so BMC is likely to flow more easily than SMC and may thus be used to manufacture components of greater geometric complexity. Fiber are generally less than 13 mm long and fiber volume fractions are in the range 0.1–0.2. Unsaturated polyester-based BMC is normally used directly after manufacture without maturing, but with other resins maturing or evaporation of solvents may be necessary. By far the most common constituents in BMC are glass and unsaturated polyester, although molding compounds based on epoxies and phenolics reinforced with glass and high-performance fibers are commercially available as well. While almost all prepregs are manufactured by someone other than the final user, SMC and BMC is commonly manufactured by the molder himself.

The thermoplastic cousin of SMC is called glass-mat reinforced thermoplastic (GMT) and is available in sheet form reinforced with randomly oriented fibers, which may be continuous or discontinuous. The most common technique for manufacturing GMT is to use a DBP (cf. *Figures 2-51* and *2-52*), where two CSMs and three polymer films or layers of extruded molten polymer are sandwiched before entering the press. Since this type of more or less completely melt-impregnated GMT ends up being a few millimeters thick and thus stiff, it is stored flat. In another method of manufacturing GMT, similar to paper-making, chopped fibers, resin powder, and additives are dispersed in an aqueous slurry, which is deposited onto a moving belt where the water is driven off. This type of GMT thus consists of a porous fiber structure containing matrix powder, which, if desired, may be more or less completely melt-impregnated in a DBP. Both these GMT types are manufactured by dedicated raw material suppliers, but there is another type where the molder mixes chopped fibers with the resin and extrudes an exactly metered amount of material just prior to molding without allowing intermediate solidification (the front end in *Figure 1-16* is manufactured with this type of material and process).

Also GMT is available with part of the reinforcement continuous and oriented; alternatively, randomly reinforced GMT may be combined with continuous-fiber reinforced prepregs right before molding. Fiber volume fractions are in the range 0.1–0.3 and fiber lengths in the range 10–30 mm, unless the reinforcement is continuous which is more common. The commercial incarnation of GMT is massively dominated by glass-reinforced PP. In contrast to thermoset molding compounds, the resins in the thermoplastic relatives are usually not filled, although they contain pigments, mold releases, etc. In North America GMT is often referred to by the common Azdel trade name.

Applications

Since molding compounds are significantly cheaper and much simpler to process than prepregs they are the standard raw material for almost all compression molding operations (see *Sections 4.2.4* and *4.3.3.1*). BMC is also widely used in injection moulding (see *Section 4.2.3.1*). Although used for a wide range of large-volume commodity products, the most common applications are found in the automobile industry. Thermoset molding compounds are still more common than their thermoplastic relations, although this is gradually changing primarily due to thermoplastics offering faster processing, elimination of health problems, and simplified recyclability. However, thermoset molding compounds yield superior surface finish and thus may be used for visible parts of an automobile, whereas thermoplastic parts tend to be hidden from immediate view. Applications include:

- Automobile and truck components, such as body panels (thermoset-based compounds only, see *Figures 1-11* and *1-12*), underbody panels, front ends, seat backs, etc. (see *Figures 1-13* and *1-16*)
- Containers and housings
- Electrical and machinery components

2.4 Core Materials

When the flexural rigidity of a composite is not sufficient, the use of the sandwich concept offers an alternative without a significant weight penalty, see *Figure 2-55*. The core supports lateral loads experienced by the sandwich component through shear, for which reason shear strength and modulus are the most relevant properties of a core material. The sandwich concept further relies on the faces being at constant distance from one another, thus stipulating a reasonably high compressive modulus for the core; already a small degree of compression of the core causes a significant decrease in flexural rigidity. Since composites often are weight-critical a low core density is desirable. Moreover, some core materials provide very good thermal and acoustic insulation properties to the sandwich. The most common sandwich core types are wood, honeycomb, corrugated, and expanded polymer foams, see *Figure 2-56*.

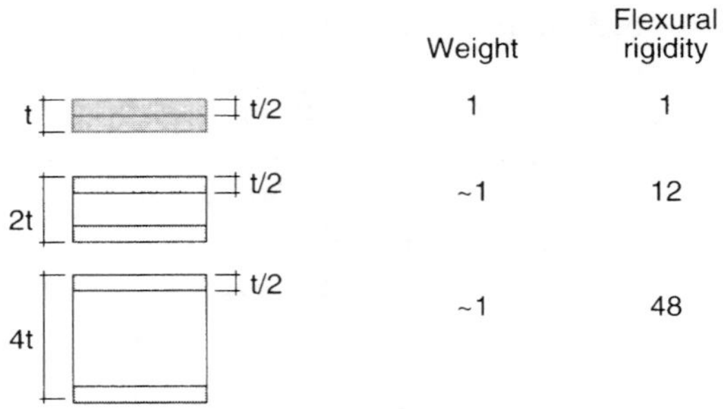

	Weight	Flexural rigidity
t ▭ t/2	1	1
2t ▭ t/2	~1	12
4t ▭ t/2	~1	48

Figure 2-55 The sandwich concept

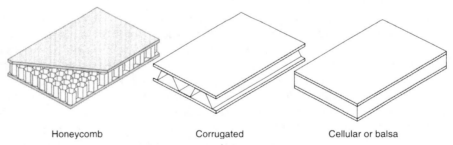

Honeycomb   Corrugated   Cellular or balsa

Figure 2-56 Sandwich core types

2.4.1 Wood Cores

The first core material used in structural sandwich applications was balsa, which found its way into early aircraft applications, such as in the mass-produced British World War II Mosquito fighter. Balsa clearly dominates the wood core category.

Manufacture

The balsa wood is sawed into slices transverse to the grain direction and is then machined into rectangles which are adhesively bonded together side to side to form a larger core block. Since the grain direction of the core blocks is perpendicular to the surfaces this material is called end-grain balsa. It is noteworthy that balsa is a cellular material where the cells, which typically are 0.5–1 mm long and have a diameter of 0.05 mm, are aligned with the grain direction.

Types and Forms

End-grain balsa is available in densities ranging from 100 to 300 kg/m³.

Characteristics

The main advantages of balsa cores are good mechanical properties, that they are simple to machine, that faces can be laminated directly onto the core, and low price. Disadvantages include severe moisture sensitivity, poor conformability, and that the properties are significantly non-uniform within the block since it has been bonded together from smaller sections. The problem of moisture sensitivity is reduced if water-impermeable faces are laminated or bonded onto the core. Since the grain (and cell) direction is perpendicular to the faces and moisture only diffuses easily in the grain direction, moisture damage is typically localized.

Applications

End-grain balsa is in extensive use in for example leisure boats.

2.4.2 Honeycomb and Corrugated Cores

The name honeycomb core stems from this core type's normally hexagonal cellular structure, which strongly resembles the beeswax honeycomb that honeybees use to store honey and eggs. Honeycomb cores are widely used in the aerospace industry where their high price is acceptable to achieve the high specific stiffness and strength uniquely offered by this core type. Among the many materials that may be used to manufacture structural honeycomb cores are unreinforced and fiber-reinforced polymers, metals, and paper.

Manufacture

Most honeycomb cores are manufactured according to the expansion method illustrated in *Figure 2-57*. A thermoset adhesive is printed in strips onto the sheet material, whereupon numerous sheets are stacked, heated, and placed in a press to crosslink the adhesive. After crosslinking has been completed the stack is expanded by laterally pulling on it. Metals yields plastically and retain the expanded shape. In contrast, non-metallic materials do not retain their shape after expansion and are therefore supported in the extended shape while they are dipped in a thermoset resin, which subsequently is crosslinked at elevated temperature to set the expanded shape. Cores may be sliced into desired thickness either before or after expansion.

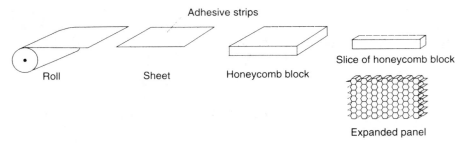

Figure 2-57 Expansion method of honeycomb core production. Redrawn from reference [26]

High-density and/or thick materials may be made into honeycomb cores with the corrugation method illustrated in *Figure 2-58*. The sheets are first corrugated, the adhesive applied, the sheets stacked, and the temperature elevated to crosslink the adhesive.

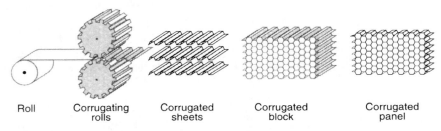

Figure 2-58 Corrugation method of honeycomb core production. Redrawn from reference [26]

Unreinforced thermoplastic honeycomb cores may be produced through extrusion of core sections of, say, 100x100 mm, that subsequently are assembled into larger blocks. The assembly obviously is similar to that used for balsa cores, but with the difference that thermoplastic cores can be fused together through localized melting thus eliminating the need for an adhesive.

Types and Forms

While numerous different materials and material combinations are used in honeycombs, the most common are aluminum and aramid-fiber paper impregnated with a phenolic resin, the latter having the (Du Pont) trade name Nomex. Other reinforcing materials include glass, carbon, and aramid fabrics, usually impregnated with a phenolic, although unsaturated polyesters and polyimides have been used as well. Even honeycombs made from Kraft paper sometimes are used in building applications. Recent developments have added unreinforced and fiber-reinforced thermoplastics to the honeycomb family; candidate polymers include PP and various elastomers as well as extruded thermoplastic tubes thermally fused into cores with circular cells. Although the conventional hexagonal cell shape clearly dominates, several other cell shapes are available, see *Figure 2-59*.

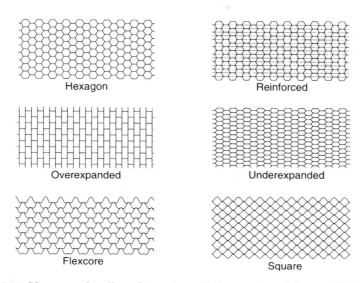

Figure 2-59 Honeycomb cell configurations. Redrawn from reference [26]

Honeycomb cores are, among other things, characterized by the final core density, which depends on material, cell configuration, cell size, wall thickness, etc. Densities are normally in the range 20–200 kg/m³.

Corrugated cores may be fabricated from the same kind of materials and processes (cf. *Figure 2-58*) as honeycomb cores, but are often made from Kraft paper. They are less expensive, difficult to conform to doubly curved surfaces, and have highly orthotropic properties. Corrugated cores are not common in structural composite applications.

Characteristics

Naturally the characteristics of a honeycomb core significantly depend on its constituents, but also on cell configuration and wall thickness. The predominant advantage of honeycomb cores is their excellent specific mechanical properties since their density is so low. Disadvantages include moisture sensitivity, difficult machining, difficult bonding of faces, and high price. Temperature sensitivity and drapeability may be good or poor depending on core material and cell configuration. Cores manufactured through the expansion and corrugation methods have two perpendicular in-plane directions with significantly different properties, since the bonding procedure creates double cell walls along the bond lines. The direction of the core containing the double cell walls is called the length direction and the direction perpendicular to it is consequently the width direction.

Applications

Due to excellent specific properties, high price, and rather difficult processing requirements, honeycomb cores are mostly limited to applications where cost is not the prime concern or where weight savings are highly valued:

- Aircraft control surfaces and skins (see *Figures 1-22* through *1-26*)
- Spacecraft (see *Figures 1-27* through *1-29*)
- Sporting goods

2.4.3 Expanded Foam Cores

In sheer consumption, cellular plastics in the form of expanded foams dominate as sandwich core material. Any polymer, thermoset and thermoplastic alike, may be expanded in any of several different manners and the density ranges available mean that a nearly limitless scope of properties are achievable to suit almost any application.

Manufacture

Polymers are foamed using foaming agents, which may be physical or chemical in nature. Physical blowing agents are gases that are dispersed in the liquid polymer and that expand to form voids when the temperature increases or the pressure decreases. Chlorofluorocarbons (known as CFCs or freones) used to be the most popular physical blowing agents, but are being phased out due to their detrimental effect on the earth's ozone layer; replacements are under development and several are already in use. Chemical blowing agents are mixed into the polymer and decompose into gases, often nitrogen or carbon dioxide, when the processing temperature reaches the decomposition temperature of the blowing agent.

The foam expansion may be of batch-character or it may be continuous, typically meaning extrusion into ambient atmosphere employing a physical blowing agent. In the former case a sheet is cast from the polymer which already contains a chemical blowing agent. Following solidification or gelation of the sheet it is expanded while free-hanging in a hot air oven or lying in a heated water bath. The expanded structure thus obtained is finally set through cooling or steam treatment. In an in-between situation thermoplastic beads (with diameters of the order of a few millimeters) are first partially expanded in a semi-continuous process and are later placed in a closed mold into which steam is injected to further expand and fuse the beads together. In either case, the surface of the foamed material tends to have markedly higher density than the bulk. This is desired in applications such as automobile dashboards, but for structural applications the surface is often machined off, thus obtaining nominal dimensions and reasonably uniform density.

Types and Forms

Foams for structural sandwich applications usually have densities in the range 40–400 kg/m³, although it usually varies noticeably within a block and between blocks. Foams may have open cells, where the walls between cells have ruptured, or closed cells, where each cell is completely enclosed, see *Figure 2-60*. In most foams both open and closed cells are present. Although all polymers may be expanded, relatively few are used in structural sandwich applications.

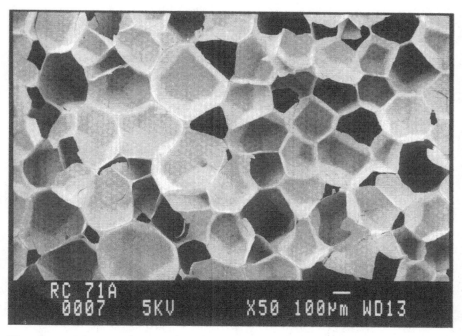

Figure 2-60 Micrograph of machined surface of expanded polymer foam with closed cells. Note 100 μm scale bar in lower right corner. Photograph courtesy of Röhm, Darmstadt, Germany

PURs may be foamed into a large variety of both thermoplastic and thermoset foams with varying degrees of closed cells. The disadvantage of PURs are the relatively poor mechanical properties, while the major advantages include low price and the fact that PURs may be in-situ foamed between two faces to form a finished sandwich, although in-situ foaming is likely to be of limited use in structural applications.

PS foams are thermoplastic, have closed cells, good mechanical properties, and are inexpensive. However, PS cores do not tolerate many solvents,
including styrene, thus limiting their use with composite faces.

The most common foam cores used in structural composite applications
are clearly the polyvinyl chlorides (PVC), which are available in both thermoplastic and thermoset versions. The mechanical properties and the
temperature tolerance are significantly better than for PURs and PSs, but
PVC cores cost more. The major difference between thermoplastic and thermoset PVC cores are that the latter generally have better mechanical
properties and temperature tolerance, but are less ductile. Low-density
PVCs have some open cells, whereas the higher densities have closed cells
only.

Even better properties are offered by polymethacrylimides (PMI; shown
in *Figure 2-60*), which are lightly crosslinked and have closed cells. With the
exception of the standard grades quite brittle, the mechanical properties are
very good and the temperature tolerance allows PMI cores to be used with
some high-temperature crosslinking epoxies. Not surprisingly, PMI cores
are very expensive.

Just like PMI cores, PEI cores offer very good mechanical properties and
high-temperature tolerance at a high price. The major difference from PMI
cores is that PEIs are very ductile.

Syntactic cellular plastics are sometimes used as sandwich cores and are
at times confused with expanded polymers, but owe their cellular nature
and low density to a totally different mechanism. Syntactic cellular plastics
consist of a mixture of polymer and microspheres, which are very small, hollow glass or polymer spheres.

Characteristics

Just as with honeycombs, the characteristics of foam cores significantly
depend on the polymer in question, but also on wall thickness, cell size,
whether cells are open or closed, etc. All in all, mechanical properties, thermal properties, and moisture tolerance may be anywhere between poor and
very good. In general cores are simple to machine, are amenable to direct
lamination of faces, while their price ranges from low to very high. Despite
removal of core block surfaces, their density, cell size, and thus properties
vary noticeably both through the thickness and in the plane. All thermoplastic cores may be thermoformed (i.e. moderate heating to soften the core
followed by forming) to almost any geometry although properties are modified in the process. However, some thermoset cores, most notably PMIs and
to some degree PVCs, have such low crosslink densities that they may be
thermoformed. All foam cores provide excellent thermal insulation and
acoustic damping.

Applications

Foam cores are used in a wide range of applications, from low-performance products utilizing PURs and PSs, via highly stressed structures with PVCs, to high-performance applications with PMIs and PEIs. Application examples include:

- Automobile, truck, and bus parts (see *Figures 1-14* and *1-15*)
- Refrigerated truck and railroad containers
- Leisure boats (see *Figure 1-18*)
- High-performance racing vehicles and yachts (see *Figure 1-19*)
- Mine counter-measure vessels and high-speed passenger ships (see *Figures 1-20* and *1-21*)
- Aircraft propellers and helicopter rotor blades (see *Figure 1-26*)
- Building panels
- Sporting goods (see *Figure 1-39*)

2.5 Adhesives

Adhesives are used in many composite applications to join composite parts and to join composites to other materials, commonly metals; such joining issues are further discussed in *Section 5.2.2*. However, an adhesive is probably only considered a true constituent when it is used to join faces and core into a sandwich structure. Whatever the application, the adhesive should have good shear strength to enable efficient load transfer between components. Though shear typically is the preferred mode of load transfer, adhesive joints also tend be subject to some degree of peeling, i.e. a tensile load transverse to the plane of the joint. In sandwich applications the rule of thumb is that the tensile strength of the adhesive should be higher than that of the core. An appropriate adhesive should have good wetting properties, but an adhesive for sandwich applications also must be capable of completely filling the surface cells of balsa and foam cores, while for honeycomb applications it should only partially enter the cells and wet the cell walls to create a fillet and thus increase the bond area. Since all commonly used adhesives are polymers, most of what was said about polymers in *Section 2.1* applies also to adhesives. In composite applications thermoset adhesives clearly dominate over thermoplastics.

The most common high-performance adhesives are of the epoxy family and offer good mechanical properties, high temperature tolerance, and little shrinkage. For adhesive purposes, epoxies are available in paste, powder, and film form. Since high toughness is required of an adhesive to withstand peeling and general impact loads, and epoxies are inherently brittle, they are often modified to improve toughness. Such modified, or toughened, epoxies have been copolymerized with an elastomer, which apart from improved toughness unfortunately also imparts decreased temperature tolerance due to the reduced crosslink density. Toughened epoxies dominate in aerospace applications, where they are usually used in the form of a film which is mat-reinforced to simplify handling and reduce flow during processing. The reinforcing fibers may be glass or organic in origin. Although epoxies offer very good properties in most applications, the requirements on the preparation of the surfaces to be bonded are stringent lest the bond strength be significantly reduced. Room-temperature crosslinked epoxy adhesives have T_gs in the range 50–100°C, whereas epoxy adhesives crosslinked at elevated temperatures may have T_gs in the range 120–180°C.

PURs adhere well to most materials and are quite common in less demanding applications. The main advantages are decent mechanical properties and an extreme flexibility in formulation to suit almost any process and application. For some PUR adhesives the crosslinking is initiated by moisture, meaning that it may be sufficient to spray the surfaces to be bonded with water prior to application of the adhesive. Compared to the alternatives the temperature tolerance of PUR adhesives is poor.

In less demanding applications various acrylic compounds, such as urethaneacrylates and cyanoacrylates, are also used as adhesives and offer good mechanical properties that are not very dependent on crosslinking conditions. Acrylics also dissolve grease, meaning that they bond surfaces that have not been perfectly prepared. T_gs are up to 100°C.

Phenolic adhesives have poor peel strength but offer very high temperature tolerance. Elastomer-toughened phenolics have better peel strength but lower temperature tolerance; T_gs are in the range 150–200°C. Even higher temperatures may be reached using BMI and PI adhesives.

In applications where faces are laminated directly onto the core, the adhesive is usually the matrix resin itself; examples are unsaturated polyesters and vinylesters laminated directly onto balsa or foam cores. In these situations, the core is often primed with neat resin, which is allowed to partially or fully crosslink before lamination commences (see further *Section 4.2.1.3*). Since solvents evaporate during crosslinking of unsaturated polyesters and vinylesters they shrink to such a degree that significant residual stresses may build up, consequently impairing performance.

When thermoplastics are used as adhesives they are usually quite logically called hot-melt adhesives. Due to the temperatures required to melt thermoplastics, they are in most cases incompatible with thermoset faces and cores, but are used to bond temperature-tolerant materials in highly automated processes; common hot-melt adhesives include PAs and thermoplastic polyesters. Thermoplastics are likely to be the adhesive of choice if both faces and core are thermoplastic. Indeed, if faces and core are based on the same thermoplastic, no additional adhesive may be required if there is excess matrix in the faces or if the core surface is melted. If there is not enough matrix, bonding is usually enhanced by addition of a neat polymer film of the same polymer as used in faces and core.

2.6 Composites Based on Inorganic Matrices

Among the alternatives to polymer matrices in fiber-reinforced composites are carbon, metals, and ceramics. The interest in alternative matrices is primarily due to the desire to reach higher service temperatures than polymers can deliver. While metal matrices and to some degree carbon matrices already are in use, ceramic matrices, albeit attracting intense research efforts, largely tend to be in the developmental stage.

2.6.1 Carbon–Carbon Composites

The promise of reaching service temperatures up to 2,000°C is one of the most attractive aspect of carbon matrices reinforced with carbon fibers. Carbon–carbon composites tend to have continuous, three-dimensional reinforcement. The carbon reinforcement is impregnated with a phenolic or petroleum pitch, which is reduced to more or less pure carbon in a repeated impregnation-pyrolysis sequence. Alternatively, the carbon matrix may be deposited through CVD of hydrocarbon gas which also is carbonized in a repeated pyrolysis process. These slow and iterative processes make carbon–carbon composites very expensive.

Among the advantages of carbon–carbon composites are high strength, high modulus, low creep, high thermal conductivity, high specific heat, low CTE, low density, high wear-resistance, and biocompatibility. The major disadvantages are high cost and the fact that carbon starts to oxidize around 400°C. Really high service temperatures may thus only be realized in inert atmosphere (e.g. space) or if an oxygen-impermeable surface coating is applied to the composite.

High-temperature applications are in rocket engines and in "re-entry conditions", i.e. as heat shields or leading edges on spacecraft and rockets reentering the earth's atmosphere. A reasonably common low-temperature application is as brake disks in aircraft, race cars, etc., where thermal and wear properties are major advantages that allow substantial improvements in wear resistance and weight savings.

2.6.2 Metal–Matrix Composites

Also metal–matrix composites (MMC) offer improved service temperatures over polymer–matrix composites. MMCs further provide good mechanical properties, high conductivity, low thermal expansion, but a high density. Two different developments may be discerned for MMCs; as a means of improving the properties over unreinforced metals or as a way to increase the temperature tolerance of continuous-fiber reinforced composites.

In the former and more common case, discontinuous fibers, whiskers, or particles are used and the MMC basically is viewed as an improved metal with macroscopically isotropic properties. In this case manufacturing methods typically are modified versions of conventional metal forming operations, e.g. casting and sintering. When comparing MMCs to metals, the addition of reinforcement yields improved high-temperature properties, such as strength, stiffness, fatigue, wear, etc. In the latter case, continuous and aligned fibers are used and the resulting properties are anything but isotropic. In this case fiber preforms are impregnated with the liquid metal.

Reinforcements used in MMCs include carbon, boron, SiC, Al_2O_3, and metal wires such as tungsten. Metal matrices include aluminum, titanium, magnesium, copper, and alloys thereof. Applications may be found in satellites, rocket engines, and jet engine turbines.

2.6.3 Ceramic–Matrix Composites

Ceramics are attractive mainly due to their high-temperature tolerance, whereas the most prominent disadvantage is the extreme sensitivity to flaws, which leads to unpredictable and brittle failures at very low strains. When it comes to ceramic–matrix composites (CMC) the reinforcement is seen as a way to decrease this flaw sensitivity and to produce a less brittle and more predictable failure. Reinforcement fibers or whiskers toughen the neat ceramic through crack deflection, crack bridging, debonding, fiber pullout, etc. Analogous to the two-faced approach to MMCs, CMCs may also be seen as a way to improve the toughness of a ceramic or as a way to improve the temperature tolerance of continuous-fiber reinforced composites over what polymer matrices permit.

The reinforcements tried include SiC, carbon, and metal wires. Since CMCs are likely to contain microcracks, the oxidation problem of carbon reinforcement at temperatures above 400°C must be considered. Some CMCs perform well up to 1,000°C. There are currently very few commercial applications of CMCs and they must still be considered as being in the early stages of development. Potential applications include rocket nose cones and engine parts.

2.7 Summary

The constituent materials of a composite are normally considered to be matrix and reinforcement, but may also include fillers and additives. While the reinforcement carries the bulk of the structural loads imposed on a composite, the support of the matrix is crucial. The matrix transfers external loads to the reinforcement, redistributes loads to surrounding fibers when an individual fiber fractures, supports fibers to prevent buckling in compression, and protects the reinforcement from the environment. Polymers are by far the most common composite matrices. A polymer is a high molecular-weight organic compound primarily consisting of carbon and hydrogen atoms bound to each other by covalent, or primary, bonds. There are two major polymer types: thermoplastics and thermosets.

Thermoplastics consist of long molecules with covalent bonds acting within molecules, whereas secondary bonds act between molecules. Since secondary bonds are an order of magnitude weaker than covalent bonds, thermoplastics may be melted largely without affecting the intramolecular structure. In thermosets covalent bonds may form between molecules resulting in solidification of the resin due to a drastic increase in molecular weight. Since these covalent bonds between molecules cannot be broken without simultaneously breaking the covalent bonds within molecules, thermosets cannot be melted. The intermolecular covalent bonds in thermosets translate into higher stiffness and strength than for thermoplastics, but also into a lower strain to failure. All polymers have a characteristic transition temperature where molecular movement is rapidly regained resulting in a notable reduction in stiffness; this temperature is referred to as glass transition temperature and in practice defines the maximum temperature the polymer can withstand. Once again, courtesy of the intermolecular covalent bonds, thermosets tend to have higher temperature tolerance than thermoplastics. From a manufacturing point of view thermosets are essentially Newtonian fluids, whereas thermoplastics are shear-thinning; both polymer types have viscosities that are notably temperature dependent. When a thermoset is fully formulated, intermolecular bonds start to form and the viscosity gradually increases, thus defining the allowable processing time. If thermoplastics are kept molten for too long they start to degrade, which thus determines the allowable time to complete processing.

Thermoplastics dominate in unreinforced applications and have also seen a breakthrough in certain composite applications, where PP and PAs are the most common matrices, but thermosets nevertheless strongly dominate as matrices in structural composites. The main reasons for the dominance of thermosets are their good mechanical and thermal properties, relatively low cost, and low viscosity, where the latter greatly facilitates reinforcement impregnation. By far the most common thermosets are unsaturated polyesters, followed by epoxies. Unsaturated polyesters have advantages in terms of an attractive balance between cost and performance as well as uncomplicated crosslinking requirements, but are limited in their temperature tolerance. Epoxies offer improved mechanical and thermal properties, but cost more and are demanding in terms of crosslinking requirements.

Commonly available fibrous reinforcement types are glass, carbon, and aramid, where glass is by far the most common and aramid the least common. Carbon and aramid fibers are manufactured in a manner that results in alignment of molecules in the fiber direction, meaning that they are transversely isotropic, whereas glass fibers are isotropic. Among the reasons for the dominance of glass fibers are low cost and good mechanical, thermal, and electrical properties. Carbon fibers, and to a lesser degree aramid fibers, offer significant improvements in mechanical properties at notably higher cost.

In most cases a good reinforcement–matrix interface is crucial to a well-functioning composite. To this end, a coupling agent is applied to glass fibers to promote covalent bonds to the intended matrix. Carbon fibers are in contrast chemically modified to promote chemical reactivity of the fiber surface. Both glass and carbon fibers are brittle and abrasive and are therefore given a protective size to facilitate subsequent handling. With glass the size is often applied simultaneously with the coupling agent, whereas carbon fibers generally receive their size following surface treatment. Aramid fibers, which are neither brittle nor abrasive, are not sized and no commercially viable way of promoting the fiber–matrix interface appears to be available.

Only in a few composite manufacturing techniques are reinforcing fibers used in the yarn form in which they are produced. Apart from parallel-fiber yarns, the most common reinforcement forms are woven fabrics, non-crimp fabrics, knitted fabrics, braids, and random mats. The main attraction of using fabrics and mats is that they are easy to handle and that reinforcement orientation is controlled. It is common practice to further enhance handleability through use of combination fabrics and preforms, where the latter are net-shape reinforcement arrangements.

Prepregs are widely used to enhance handleability, eliminate resin formulation, eliminate reinforcement impregnation, and improve component properties. Prepregs, which are generally unidirectionally carbon-fiber reinforced epoxies, are most common in high-performance applications where their high cost can be justified. Molding compounds, which generally consist of randomly oriented glass fibers in a highly filled unsaturated polyester resin, are the conceptual commodity equivalent of prepregs.

When dealing with sandwich components there are two additional constituents—core and adhesive. Common core types are balsa, honeycomb, and expanded polymer foams. Most important are aluminum and Nomex honeycomb cores in aerospace applications and expanded foams in most other areas. Just as the reinforcement–matrix interface is crucial to a well-functioning composite, a strong face–core interface is crucial to a sandwich. In many cases this requires a separate adhesive bond and by far the most common adhesives come from the epoxy family.

Suggested Further Reading

Since this chapter covers several disciplines it is impossible to find any one source covering it all. However, for an excellent and recent in-depth treatment of polymer physics, the reader is encouraged to obtain:

U. W. Gedde, *Polymer Physics*, Chapman & Hall, London, UK, 1995.

while a text covering organic polymer chemistry is:

K. J. Saunders, *Organic Polymer Chemistry*, Chapman & Hall, London, UK, 1988.

Encyclopedias that provide useful information on most aspects relevant to the topics of this chapter include:

International Encyclopedia of Composites, Ed. S. M. Lee, VCH Publishers, New York, NY, USA, 1990-1991 (6 volumes).

Engineered Materials Handbook, Volume 1, Composites, ASM International, Metals Park, OH, USA, 1987.

Engineered Materials Handbook, Volume 2, Engineering Plastics, ASM International, Metals Park, OH, USA, 1988.

Engineered Materials Handbook, Volume 3, Adhesives and Sealants, ASM International, Metals Park, OH, USA, 1990.

References

1. J. M. Rego Lopez and U. W. Gedde, "Morphology of Binary Linear Polyethylene Blends," *Polymer*, **29**, 1037-1044, 1988.

2. F. N. Cogswell, *Thermoplastic Aromatic Polymer Composites*, Butterworth-Heinemann, Oxford, UK, 1992.

3. A. V. Tobolsky, *Properties and Structure of Polymers*, John Wiley & Sons, New York, NY, USA, 1960.

4. "Matrix Systems, Araldite LY 556, Hardener HY 917, Accelerator DY 070," Ciba-Geigy, Switzerland.

5. "Fibredux 6376," Ciba-Geigy Plastics, Cambridge, UK, 1991.

6. L. E. Taske II, Personal communication, BASF, Charlotte, NC, USA, 1990.

7. "The Thermal and Oxidative Thermal Degradation of APC-2," ICI Composites.

8. S. S. Voyutskii, *Autohesion and Adhesion of High Polymers*, John Wiley & Sons, New York, NY, USA, 1963.

9. K. Jud, H. H. Kausch and J. G. Williams, "Fracture Mechanics Studies of Crack Healing and Welding of Polymers," *Journal of Material Science*, **16**, 204-210, 1981.

10. A. B. Strong, *Fundamentals of Composites Manufacturing: Materials, Methods, and Applications*, Society of Manufacturing Engineers, Dearborn, MI, USA, 1989.

11. J. F. Jansson, K. A. Olsson and S. E. Sörelius, *Fiber Reinforced Plastics. Thermosets, Materials — Methods — Environment*, Swedish Tech Books, Solna, Sweden, 1989.

12. K. J. Saunders, *Organic Polymer Chemistry*, Chapman & Hall, London, UK, 1988.

13. C. A. May, "Epoxy Resins," in *Engineered Materials Handbook, Volume 1, Composites*, ASM International, Metals Park, OH, USA, 66-77, 1987.

14. J. C. Watson and N. Raghupati, "Glass Fibers," in *Engineered Materials Handbook, Volume 1, Composites*, ASM International, Metals Park, OH, USA, 107-111, 1987.

15. R. J. Diefendorf, "Carbon/Graphite Fibers," in *Engineered Materials Handbook, Volume 1, Composites*, ASM International, Metals Park, OH, USA, 49-53, 1987.

16. R. J. Morgan and R. E. Allred, "Aramid Fiber Composites," in *Handbook of Composite Reinforcements*, Ed. S. E. Lee, VCH Publishers, New York, NY, USA, 5-24, 1993.

17. B. Z. Jang, *Advanced Polymer Composites: Principles and Applications*, ASM International, Materials Park, OH, USA, 1994.

18. D. L. Caldwell, "Interfacial Analysis," in *Handbook of Composite Reinforcements*, Ed. S. M. Lee, VCH Publishers, New York, NY, USA, 283-298, 1993.

19. W. D. Bascom, "Fiber Sizing," in *Engineered Materials Handbook, Volume 1, Composites*, ASM International, Metals Park, OH, USA, 122-124, 1987.

20. W. D. Bascom, "Interphase in Fiber Reinforced Composites," in *Handbook of Composite Reinforcements*, Ed. S. M. Lee, VCH Publishers, New York, NY, USA, 298-310, 1993.

21. F. K. Ko, "Braiding," in *Engineered Materials Handbook, Volume 1, Composites*, ASM International, Metals Park, OH, USA, 519-528, 1987.

22. D. S. Brookstein, "Textile Structures, Interlocked Fibers," in *Handbook of Composite Reinforcements*, Ed. S. E. Lee, VCH Publishers, New York, NY, USA, 630-634, 1993.

23. Promotional material, Devold AMT, Langevåg, Norway, 1997.

24. V. M. Karbhari, "Progressive Crush Response of Multi-Element Foam-Filled Preform RTM Structures, I: Architecture and Rate Effects," *Journal of Composite Materials*, **29**, 734-750, 1995.

25. F. S. Dominguez, "Prepreg Tow," in *Engineered Materials Handbook, Volume 1, Composites*, ASM International, Metals Park, OH, USA, 151-152, 1987.

26. J. Corden, "Honeycomb Structure," in *Engineered Materials Handbook, Volume 1, Composites*, ASM International, Metals Park, OH, USA, 721-728, 1987.

CHAPTER 3

PROPERTIES

The topic of this chapter represents one of the major disadvantages of polymer composites. For most traditional engineering materials, such as metals, wood, and concrete, there are centuries' worth of experience and compiled property data for the designer to draw upon. In contrast, for polymers and even more so for fiber-reinforced polymers there is a remarkable scarcity of comprehensive information on properties of both constituents and composites. The lack of reliable data bases partly reflects a reasonably new engineering field, but reasons can also be sought in the multitude of suppliers of nominally identical materials and in suppliers constantly improving their products, thus quickly making data bases outdated. A perhaps even more pertinent reason for the lack of comprehensive information is the extreme design flexibility available to the designer, meaning that few applications—including reinforcement architecture and resin formulation—are like any previously encountered. Further, laboriously determined property data often remain proprietary since extensive characterization is expensive. Finally, composite properties strongly depend upon how the component was manufactured, i.e. through which technique and under what time, temperature, and pressure conditions.

The emphasis of this chapter is on thermal and basic mechanical properties, but it intentionally refrains from delving into the quagmire of mechanical properties such as impact, fracture, fatigue, creep, etc. due to their complexity and vastness. Naturally numerous other properties, such as electrical, optical, moisture and corrosion resistance, etc. are also of interest to the designer, but any comprehensive information on these are even harder to come by than on mechanical and thermal properties and are considered beyond the scope of this chapter. The properties quoted herein are from several different sources and thus may be slightly contradictory, which illustrates the significant differences existing between material formulations, processing conditions, and test methods. Unless otherwise noted, all properties are for room-temperature (RT) conditions.

The intention of this chapter is merely to give the reader a feeling for typical properties of constituent materials on their own and in composite form. For more comprehensive information on properties of both constituents and composites, the reader may want to consult the sources recommended at the end of this chapter. In an actual design situation, the raw material supplier may very well possess the most relevant property data available. For some manufacturing techniques, e.g. pultrusion (see *Section 4.2.6*), many products are standard off-the-shelf items for which the composites producer is likely to have already determined the basic properties. Many companies nevertheless see the need to determine the properties of material candidates on their own, either because the material supplier's information is considered too optimistic, because the load case in question does not correspond to the test conditions used to determine this information, because manufacturing techniques or conditions significantly differ, or due to legal or contractual requirements.

3.1 Matrices

For a matrix to be able to perform its tasks of supporting and protecting the reinforcement a number of properties are of relevance. The most pertinent mechanical properties are usually moduli and strengths in tension, compression, and shear, while ultimate strain and fracture toughness may also be important. The properties of the matrix usually determine the environmental tolerance of a composite; tolerance to elevated temperature and aggressive environments, such as UV light, oxygen, water, chemicals, fuels, etc., thus are of paramount importance. A basic overview of the matrices treated in *Chapter 2* is given in *Tables 3-1* and *3-2*. As with all property comparisons of such broad material families—as well as with the quantitative information of this chapter—the information should be seen as indicative only.

3.1.1 Thermal Properties

For thermosets and amorphous thermoplastics the maximum continuous use temperature usually is slightly below the glass-transition temperature, T_g, whereas for thermoplastics with appreciable degree of crystallinity temperatures in excess of T_g and even close to the crystalline melt temperature, T_m, may be permissible for limited periods of time (cf. *Section 2.1.3.1*). *Table 3-3* illustrates indicative transitions temperatures for some unreinforced, or neat, polymers as well as common or recommended processing temperatures, T_{proc}. An alternative way of expressing the upper use temperature is the heat deflection temperature (HDT), see *Section 6.2.3*.

Table 3-3 further gives typical zero shear-rate shear viscosities, η_0, for a few polymers only, since viscosity data usually are quite difficult to come by.

Table 3-1 Overview of properties of unreinforced,] or neat, polymers [1]

	Tensile strength	High temperature capability	Flame spread	Chemical resistance	Mold shrinkage	Cost
Thermosets						
Polyester	**	**	***	**	high	low
Vinylester	**	*	**	***	low	medium
Epoxy	**	**	**	**	low	medium
Phenolic	*	***	**	**	low	low
PUR	***	*	**	***	medium	low
BMI	***	***	**	**	low	high
PI	***	***	**	**	low	high
Thermoplastics						
PP	*	*	***	***	high	low
PA	*	**	***	***	high	low
PPS	**	***	*	***	medium	medium
PEEK	**	**	**	***	high	high
PEI	**	**	**	***	medium	medium
PES	*	***	**	***	medium	medium
PAI	***	***	**	***	medium	high

Table 3-2 Ratings used in *Table 3-1* [1]

Property	Rating	Thermosets	Thermoplastics
Tensile strength [MPa]	*	<50	30 – 90
	**	50 – 70	90 – 140
	***	>70	>140
High temperature capability [°C]	*	<120	
	**	120 – 180	
	***	180 – 300	
Rate of flame spread (UL 94†)	*	Low (5V)	
	**	Medium (V-0, V-1, V-2)	
	***	High (HB)	
Chemical resistance (to weak alkali)	*	Poor	
	**	Fair	
	***	Good	
Mold shrinkage [%]	Low	<0.5	<0.5
	Medium	0.5 – 1.0	0.5 – 1.0
	High	>1.0	1.0 – 1.5
Cost [$/kg, 1992 bulk prices]	Low	<1.4	
	Medium	1.4 – 3.5	
	High	>3.5	

† UL94 is a test standard for rate of flame spread along the surface of a specimen. The lowest rate of flame spread, i.e. the most desirable property, is designated 5V and the highest rate of flame spread HB.

Table 3-3 Transition temperatures, processing temperatures, and viscosities of selected neat polymers. The shear viscosities quoted are the zero shear-rate values at the given temperatures. The lower epoxy viscosities refer to a not-crosslinked resin while the higher epoxy viscosities refer to a B-staged prepreg resin (cf. *Figure 2-16* and *Section 2.3*) [2-14]

	T_g °C	T_m °C	T_{proc} °C	η_0 Pa·s
Thermosets				
Polyester	70 –	—	20 –	0.2 – 1 (RT)
Vinylester	70 –	—	20 –	0.2 – 1 (RT)
Epoxy	65 – 175	—	25 – 175	0.75 (60°C) – 1 (RT) 3 (150°C) – 100 (80°C)
Phenolic	300	—	130 – 160	
PUR	135	—	145 – 205	
BMI	230 – 345	—	230 – 245	0.3 (165°C) – 100 (50°C)
PI	315 – 370	—	315	
Thermoplastics				
PP	-20 – -5	165 – 175	>185	$10^1 – 10^2$ (230°C)
PA 6	50 – 70	225	225 – 290	
PA 12	45	180	180 – 270	175 (220°C)
PA 6,6	55 – 80	265	270 – 325	
PET	80	245 – 265	260 – 310	
PPS	85	285	300 – 355	
PEEK	145	345	360 – 400	380 (380°C)
PEI	215	—	350 – 425	
PES	225	—	340 – 380	10^3 (360°C)
PAI	245 – 275	—	<400	$>10^5$ (340°C)

As discussed in *Section 2.1.3.4*, the viscosity of all polymers decreases with increasing temperature, while thermoplastics are shear-thinning. The behaviors exhibited in *Figures 2-16* and *2-17* thus can be assumed indicative for the respective matrix families although the magnitudes and degrees of temperature and shear-rate dependencies vary from polymer to polymer. An alternative measure of the viscosity of thermoplastics common in the plastics industry is the melt flow rate, which denotes the number of grams of material that in a time frame of 10 minutes flows through a die of given geometry and under a given pressure drop (see references [15] and [16] for further details). (With given capillary geometry and pressure drop, it is possible to correlate the melt flow rate to the shear viscosity through basic fluid-mechanics calculations.)

Table 3-4 similarly gives typical thermal properties of some neat polymers. The table illustrates that the thermal properties do not vary drastically between polymers, although a comparison with other materials may put

Table 3-4 Thermal properties of selected neat polymers [1,4,5,17–19]

	k W/m°C	C_p kJ/kg°C	α 10^{-6} °C^{-1}
Thermosets			
Polyester	0.17 – 0.22	1.3 – 2.3	55 – 100
Epoxy	0.17 – 0.20	1.05	45 – 65
Phenolic	0.12 – 0.24	1.4 – 1.8	25 – 60
PUR	0.17 – 0.21	1.3 – 2.3	70 – 100
PI	0.10 – 0.34	1.05 – 1.5	25 – 80
Thermoplastics			
PP	0.11 – 0.17	1.8 – 2.4	80 – 100
PA 6	0.24	1.67	80 – 83
PA 12	0.21 – 0.31	1.26	61 – 100
PA 6,6	0.24	1.67	80
PPS	0.29	1.09	49
PEEK	0.25	1.34	40 – 47
PEI	0.07		47 – 56
PES	0.26	1.0	55
References (bulk properties)			
Aerospace-grade aluminum	130	0.96	23.6
Carbon steel	54	0.46	11

these properties in perspective; *Figure 3-1* consequently graphically illustrates the vast differences in coefficient of thermal conductivity (CTC), k, of various materials. *Table 3-4* further illustrates that the specific heat, or heat capacity, C_p, is similar for all polymers and that it is slightly higher than for metals; in fact, C_p does not vary significantly for most solid materials. The variation of the coefficient of thermal expansion (CTE), α, of polymers is an order of magnitude and it is noteworthy that the CTE of polymers is significantly higher than for metals. Since the data of *Table 3-4* apply at room temperature it is important to recall from the discussion of *Section 2.1.3.1* what happens as the temperature increases. CTC may increase or decrease with temperature depending on polymer, while C_p and CTE both increase with temperature—especially after T_g is passed. Unfortunately, these temperature dependencies are notoriously difficult to find in the literature.

3.1.2 Mechanical Properties

Given that the selected matrix is capable of withstanding the environmental conditions of an intended application, the mechanical properties usually are the most important ones from a design point of view. While *Figure 3-2* attempts to illustrate approximate relationships between common engineering materials in terms of modulus and strength, it should be borne in mind

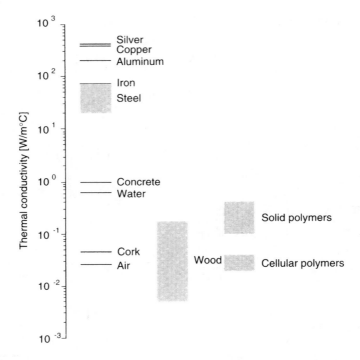

Figure 3-1 Thermal conductivities for polymers and other materials. Note the logarithmic scale. Data from *Tables 3-4* and *3-27* and reference [19]

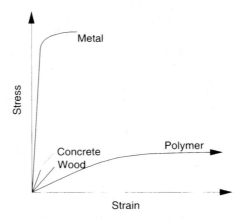

Figure 3-2 Schematic stress-strain relationships for common construction materials

Table 3-5 Mechanical properties of selected neat polymers [1,18]. ρ denotes density, E elastic (Young's) modulus, σ* strength, and ε* strain to failure

	ρ kg/m^3	E GPa	σ* MPa	ε* %
Thermosets				
Polyester	1100 – 1230	3.1 – 4.6	50 – 75	1.0 – 6.5
Vinylester	1120 – 1130	3.1 – 3.3	70 – 81	3.0 – 8.0
Epoxy	1100 – 1200	2.6 – 3.8	60 – 85	1.5 – 8.0
Phenolic	1000 – 1250	3.0 – 4.0	60 – 80	1.8
PUR	1200	0.7	30 – 40	400 – 450
BMI	1200 – 1320	3.2 – 5.0	48 – 110	1.5 – 3.3
PI	1430 – 1890	3.1 – 4.9	100 – 110	1.5 – 3.0
Thermoplastics				
PP	900	1.1 – 1.6	31 – 42	100 – 600
PA	1100	2.0	70 – 84	150 – 300
PPS	1360	3.3	84	4.0
PEEK	1260 – 1320	3.2	93	50
PEI	1270	3.0	105	60
PES	1370	3.2	84	40 – 80
PAI	1400	3.7 – 4.8	93 – 147	12 – 17
References (bulk properties)				
Aerospace-grade aluminum	2800	72	>540	
Carbon steel	7790	205	640	

that large variations in properties and stress–strain behavior occur within each material category. In the unreinforced state, polymers obviously have far lower modulus and strength than all other materials depicted. Whereas most materials essentially are elastic with some (metals) or virtually no (wood, concrete, ceramics) plastic behavior, polymers and in particular thermoplastics exhibit a very high degree of plastic elongation, cf. *Table 3-5* which gives indicative mechanical properties of neat polymers. The table illustrates that the differences in density, modulus, and strength between polymers are not as significant as one might have expected, whereas thermoplastics—with a couple of exceptions—have failure strains that are an order of magnitude higher than thermosets.

3.2 Reinforcements

While the composite matrix normally is responsible for properties such as temperature and environmental tolerance, the reinforcement primarily determines a composite's mechanical properties, since the reinforcement typically has strength and modulus one to two orders of magnitude greater than polymer matrices. While it is relatively straightforward to determine thermal and mechanical properties of neat polymers since they can be cast into blocks, it is—with the exception of plain tensile testing—significantly more difficult to perform such measurements on fibers with diameters of order 10 μm. Thus, most properties available are from tests on composites and the properties of the reinforcement have been calculated using micromechanics relations (see *Sections 3.3.1* and *3.3.2*). A general overview of some of the reinforcements treated in *Chapter 2* is given in *Table 3-6*.

Table 3-6 Overview of reinforcement properties. HM denotes high modulus and HS high strength [1]

	Strength	Modulus	Ease of bonding	Abrasion resistance	Flame spread	Ultimate strain	Cost
PE	**	**	*	*	***	***	High
E glass	**	**	***	***	**	**	Low
R glass	***	**	**	***	**	**	Medium
Aramid (HM)	***	**	*	*	***	**	High
Carbon (HS)	***	***	***	*	***	*	High

Ratings used in table

Ease of bonding:	availability of coupling agents
Abrasion:	resistance to wear
Rate of flame spread:	UL 94 rating
Cost [$/kg, 1992 bulk prices]:	Low <2.2
	Medium 2.2 – 17.5
	High >17.5
Ranking:	* Low
	** Medium
	*** High

3.2.1 Thermal Properties

Table 3-7 gives representative thermal properties of some reinforcement types. The table illustrates a couple of interesting differences in reinforcement properties. First, glass fibers are isotropic, as discussed in *Section 2.2.1*. However, perhaps the most relevant property in this context is the negative longitudinal CTE of aramid and carbon fibers, while another is the high longitudinal CTC of carbon, which for some carbon fiber types even exceeds the CTC of the best metal conductor, silver. The table also gives approximate maximum use temperatures, T_{max}.

3.2.2 Mechanical Properties

Table 3-8 gives representative mechanical properties of most common reinforcements, while *Figure 3-3* graphically displays the same information. Both table and figure illustrate the vast range of reinforcement properties available to the designer. While *Table 3-5* illustrates that variations in mechanical properties of matrices (with the exception of failure strain) are small, there are certainly most significant differences in the mechanical properties of different reinforcement types. It is also noteworthy that the stiffer the fiber, the lower the strength and the strain to failure; it is thus not possible to have both high strength and strain to failure and high modulus.

Table 3-7 Thermal properties of selected reinforcement types. Indices l and t denote properties in longitudinal and transverse fiber directions, respectively [1,18–23]

	k_l	k_t	C_p	α_l	α_t	T_{max}
	W/m °C		kJ/kg °C	10^{-6} °C^{-1}		°C
Spectra 900 (PE)						120
E glass	10 – 13		0.45	5.4		350
S-2 glass	1.1 – 1.4		0.41	1.6		300
Kevlar 49 (aramid)	0.04 – 1.4		0.769	-4.3	41	250
Carbon (PAN)	7 – 70		0.7 – 0.9	-0.5 – -0.7	7 – 10	600
Carbon (pitch)	100 – 520		0.7 – 0.9	-0.9 – -1.6	7.8	500
Nicalon SiC	11.6			3.1		1300
References (bulk properties)						
Aerospace-grade aluminum	130		0.96	23.6		
Carbon steel	54		0.46	11		
Silver	419		0.23			

Table 3-8 Longitudinal tensile properties of selected reinforcement types. HS/S stands for high strength/strain, IM for intermediate modulus, and UHM for ultra-high modulus [1,18,21,23,24]

	ρ	E_l	σ_l^*	ε_l^*
	kg/m^3	GPa	GPa	%
Spectra 900 (PE)	970	117	2.6	3.5
Spectra 1000 (PE)	970	172	2.9 – 3.3	0.7
E glass	2570 – 2600	69 – 72	3.45 – 3.79	4.5 – 4.9
S-2 glass	2460 – 2490	86 – 90	4.59 – 4.83	5.4 – 5.8
Kevlar 49 (aramid)	1440	131	3.6 – 4.1	2.8
Kevlar 149 (aramid)	1470	186	3.4	2.0
Carbon (HS/S)	1700 – 1900	160 – 250	1.4 – 4.93	0.8 – 1.9
Carbon (IM)	1700 – 1830	276 – 317	2.34 – 7.07	0.8 – 2.2
Carbon (HM)	1750 – 2000	338 – 436	1.9 – 5.52	0.5 – 1.4
Carbon (UHM)	1870 – 2000	440 – 827	1.86 – 3.45	0.4 – 0.5
Al$_2$O$_3$	3200 – 3950	300 – 379	1.38 – 2.1	0.4 – 1.5
SiC	2700 – 3300	45 – 480	0.3 – 4.9	0.6 – 1
References (bulk properties)				
Aerospace-grade aluminum	2800	72	>0.54	
Carbon steel	7790	205	0.64	

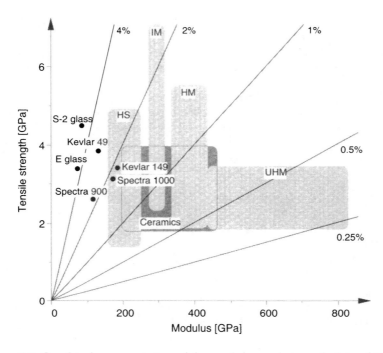

Figure 3-3 Graphical representation of the modulus and strength data of *Table 3-8* with equistrain lines added

Section 2.2 discussed the fact that organic fibers and fibers that are organic in origin (i.e. carbon) have very different properties in the longitudinal and transverse directions since they have been drawn, as partially illustrated by *Table 3-7*. The equivalent mechanical properties are not widely available, but *Table 3-9* illustrates some sample moduli of such anisotropic reinforcement types.

Table 3-9 Moduli of selected reinforcement types. The transverse modulus of Kevlar 49 is the filament compressive modulus. LM denotes low modulus [22,25]

	E_l GPa	E_t GPa
Kevlar 49 (aramid)	122	0.76
Carbon (LM)	230	40
Carbon (HM)	380 – 390	21

Since many high-performance composites are weight-critical and contain 60–70 volume percent reinforcement, comparisons become even more instructive if they are made in terms of specific properties, i.e. normalized with respect to density, see *Table 3-10*. Since most high-performance composites are designed for stiffness it is noteworthy that the specific modulus of reinforcements varies by more than an order of magnitude.

Table 3-10 Specific longitudinal tensile moduli and strengths of selected reinforcements. Data from *Table 3-8*

	E_l/ρ $MPa/(kg/m^3)$	σ_l^*/ρ $MPa/(kg/m^3)$
Spectra 900 (PE)	121	2.7
Spectra 1000 (PE)	177	3.0 – 3.4
E glass	28 – 29	1.3
S-2 glass	35	1.8
Kevlar 49 (aramid)	91	2.5 – 2.8
Kevlar 149 (aramid)	127	2.3
Carbon (HS/S)	84 – 147	0.74 – 2.9
Carbon (IM)	151 – 186	1.28 – 4.16
Carbon (HM)	169 – 249	0.95 – 3.15
Carbon (UHM)	220 – 442	0.93 – 1.84
Al_2O_3	50 – 231	0.45 – 2
SiC	57 – 157	1 – 1.5
References (bulk properties)		
Aerospace-grade aluminum	26	>0.193
Carbon steel	26	0.082

The qualitative categorization of the multitude of carbon fibers of *Tables 3-8* through *3-10* and *Figure 3-3* obviously is somewhat arbitrary and certainly partially contradictory, but it is nevertheless quite common in the composite industry. The development of new carbon fiber types stretches the envelope of ultra-high modulus fibers further to the right in *Figure 3-3*; the 1 TPa (1,000 GPa) limit has already been reached, although such fibers are not yet widely available. The envelope is similarly stretched upwards in *Figure 3-3* to ever higher strengths. Pitch-based fibers extend the envelope in terms of higher moduli, whereas PAN-based fibers are responsible for the strength improvements.

3.3 Composites

All properties discussed above pertain to the constituents on their own. When the constituents are combined into a composite they all lend some degree of their own properties to the composite; this is after all the underlying concept

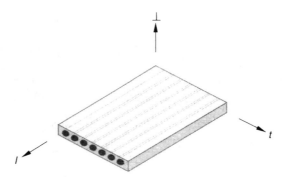

Figure 3-4 A composite lamina

of a composite. The simplest possible composite form is the lamina, which consists of an assembly of unidirectional fibers or a fabric impregnated with a matrix, see *Figure 3-4*.

A composite lamina is (just like any other composite) characterized by the relative amount of constituents it contains. Since the constituents tend to have different densities, two different measures are possible. The volume fraction, V, relates the volume of a constituent to the total lamina volume, whereas the weight fraction, W, relates the weight of a constituent to the lamina weight. It is also quite common to quote the constituent percentage rather than the fraction. The constituents commonly quantified in this fashion are fibers, matrix, fillers, and voids, although voids are for natural reasons only quantified in terms of volume. In most cases it is the fiber content that is of primary interest since it has the greatest influence on a composite's mechanical properties. Partly for reasons of tradition, fiber weight fractions are common with glass-reinforced composites, whereas volume fractions typically are used in most other instances. Another reasons for quoting glass-fiber content in weight fraction is that the glass content of a composite is generally determined through burning off the resin. The weight of the composite before burning and the weight of the glass fibers remaining after burning give the fiber weight fraction. In contrast, composites containing organic fibers or fibers organic in origin cannot have their fiber content determined in the same manner since such reinforcements would not go through the burning procedure unaffected (see *Section 6.2.1* for further discussion).

The fiber weight fraction, W_f, may be expressed in terms of the fiber volume fraction, V_f, as:

$$W_f = \frac{V_f \rho_f}{V_f \rho_f + V_m \rho_m} \tag{3-1}$$

or, correspondingly, the fiber volume fraction may be expressed in terms of the fiber weight fraction as:

$$V_f = \frac{W_f \rho_m}{W_f \rho_m + W_m \rho_f} \tag{3-2}$$

where indices f and m refer to fiber and matrix, respectively. *Figure 3-5* graphically illustrates relationships between weight and volume fractions for five common ratios of constituent densities (ρ_f/ρ_m), see *Table 3-11*, assuming that fiber and matrix are the only constituents. *Equations 3-1* and *3-2* (as well as others introduced later in this chapter) are easily generalized to take more than two constituents into account if necessary.

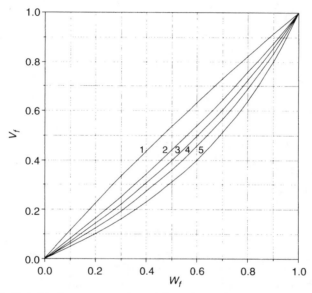

Figure 3-5 Graphical representation of the relationship between weight and volume fractions for the density ratios listed in *Table 3-11*

Table 3-11 Density ratios used in *Figure 3-5* and possible corresponding reinforcement types (cf. *Table 3-8*) given a matrix density of 1,100 kg/m³

Curve number	$\dfrac{\rho_f}{\rho_m}$	Possible reinforcement type	Reinforcement density kg/m³
1	0.88	PE	970
2	1.31	Aramid	1440
3	1.55	Carbon (low-density)	1700
4	1.82	Carbon (high-density)	2000
5	2.27	Glass	2500

Even if one does not take the overall part geometry into account, there are numerous variables in composite design, such as constituent types, fiber content, fiber orientations, etc. It has therefore proved quite convenient to be able to approximately predict composite properties. The following sections briefly look at basic relationships to predict some thermal and mechanical properties of composite laminae. Although the properties of unidirectionally reinforced composites of any cross-section may be predicted using the same relationships, they will likely prove inadequate when trying to predict properties of composites with more complex geometries, such as laminated composites composed of laminae stacked at different angles. Whilst the topic of laminate theory is considered beyond the scope of this text the interested reader should encounter little problem locating one of the many textbooks covering the field of laminate theory, some of which are suggested at the end of this chapter.

Despite laminate theory not being covered in this text it will prove useful to understand laminate stacking sequence notations. A general laminate stacking sequence is written $[\alpha_a/\beta_b]_{cs}$, where α and β are the lamina, or ply, orientations. a and b denote the number of adjacent plies that have the same orientation (the number is omitted if equal to one), while c denotes the number of repeats of the bracketed configuration and finally s denotes symmetry, if applicable, with respect to the laminate midplane to eliminate out-of-plane deformations. The notation $[-60/0_2/60]_{2s}$ thus denotes a sequence of plies with consecutive orientations of $-60°$, $0°$, $0°$, $60°$, $-60°$, $0°$, $0°$, $60°$, $60°$, $0°$, $0°$, $-60°$, $60°$, $0°$, $0°$, $-60°$.

3.3.1 Prediction of Thermal Properties

To enable determination of lamina properties from those of the constituents, several assumptions must be made; among these are that the matrix is isotropic and the fibers transversely isotropic. Whilst it may be questionable whether the matrix really is isotropic, glass fibers certainly are, whereas carbon and aramid fibers clearly are transversely isotropic.

Through micromechanics-based considerations one may derive expressions to determine the thermal properties of laminae. The CTEs may be expressed as [26]:

$$\alpha_l = \frac{V_f \alpha_{fl} E_{fl} + V_m \alpha_m E_m}{V_f E_{fl} + V_m E_m} \tag{3-3}$$

$$\alpha_t = V_f \alpha_{ft}\left(1 + v_{flt}\frac{\alpha_{fl}}{\alpha_{ft}}\right) + V_m \alpha_m (1 + v_m) - (V_f v_{flt} + V_m v_m)\alpha_l \tag{3-4}$$

where v is the Poisson's ratio and α_l is given by *Equation 3-3*. Similar expressions for the CTCs are [27]:

$$k_l = V_f k_{fl} + V_m k_m \tag{3-5}$$

$$k_t = \frac{k_{fl} k_m}{V_f k_m + V_m k_{ft}}$$ (3-6)

while an expression for the heat capacity has been proposed in reference [28]:

$$C_p = \frac{V_f \rho_f C_{pf} + V_m \rho_m C_{pm}}{V_f \rho_f + V_m \rho_m}$$ (3-7)

3.3.2 Prediction of Mechanical Properties

Micromechanics-based predictions of lamina mechanical properties are significantly more common than the thermal relationships of the preceding section. The most basic relationships for prediction of mechanical properties are called rule of mixtures, which is an established and reasonably effective means of roughly estimating the elastic moduli. (While not common, all micromechanics-based lamina property relationships could be referred to as rules of mixture.) Employing the same assumptions as in the previous section, one may estimate the modulus of elasticity (Young's modulus) in the longitudinal direction as:

$$E_l = V_f E_{fl} + V_m E_m$$ (3-8)

Equation 3-8 is sometimes called the parallel model and constitutes an upper bound for the modulus of a composite. Normalizing *Equation 3-8* with respect to the matrix modulus one obtains:

$$\frac{E_l}{E_m} = V_f \frac{E_{fl}}{E_m} + V_m$$ (3-9)

which is plotted to the left in *Figure 3-6* for different fiber volume fractions and ratios of E_f/E_m (recall that this ratio normally is of order 10^1–10^2 for polymer matrix composites). Obviously the rule of mixtures predicts that the lamina modulus in the fiber direction is a volume-weighted average of the constituent moduli.

The modulus in the transverse direction may also be estimated using the rule of mixtures, thus obtaining:

$$E_t = \frac{E_{ft} E_m}{V_f E_m + V_m E_{ft}}$$ (3-10)

Equation 3-10 is sometimes referred to as the serial model and may be regarded as a lower bound for the modulus of a composite. Normalizing *Equation 3-10* with the matrix modulus results in:

$$\frac{E_t}{E_m} = \frac{1}{V_f(E_m/E_{ft}) + V_m}$$ (3-11)

which is plotted to the right in *Figure 3-6* for different fiber volume fractions and ratios of E_f/E_m. The figure indicates that unless the fiber volume fraction is high, the lamina modulus in the transverse direction essentially is the modulus of the matrix.

The parallel model may also be used to estimate the major (in-plane) Poisson's ratio

$$\nu_{lt} = V_f \nu_{flt} + V_m \nu_m \tag{3-12}$$

and the serial model correspondingly may be employed to estimate the (in-plane) shear modulus

$$G_{lt} = \frac{G_{flt} G_m}{V_f G_m + V_m G_{flt}} \tag{3-13}$$

although Poisson's ratios and shear moduli of fibers rarely are known quantities.

Comparisons with experimental data have shown that the parallel model yields reasonably accurate results, whereas the serial model predicts moduli considerably deviating from experimental data, see further *Figure 3-14*.

The rule of mixtures may be used to approximately predict the longitudinal stiffness also of lamina that are not unidirectionally reinforced. By introducing a reinforcement efficiency factor, β, an engineering rule of mixtures may be expressed as

$$E_l = \beta_f V_f E_{fl} + \beta_m V_m E_m \tag{3-14}$$

The reinforcement efficiency factor takes into account the amount of fibers that are effective in the direction of interest and is given in *Table 3-12* for

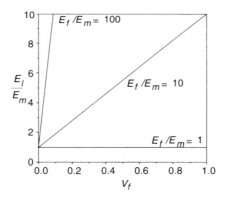
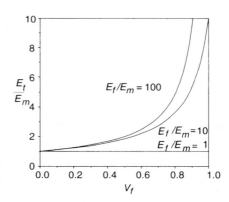

Figure 3-6 Normalized elastic moduli in longitudinal (left) and transverse directions (right; cf. *Figure 3-4*) as functions of fiber volume fraction and ratio of fiber and matrix moduli

different reinforcement types; β_m remains equal to unity. In for example a UD fabric, which may have 90 percent of the reinforcement in the warp direction and 10 percent in the fill direction, the reinforcement efficiency factor is 0.9 and 0.1 in the two directions, respectively. If the fibers are not perfectly aligned in the nominal direction the value of the reinforcement efficiency factor is reduced accordingly.

It is also possible to predict the lamina strengths based on micromechanics considerations although such expressions require extensive assumptions, including that all fibers are perfectly parallel, have exactly the same properties (i.e. fail simultaneously), and are perfectly bonded to an isotropic matrix. In longitudinal tension one may express the strength as [27]:

$$\sigma_l^* = V_f \sigma_{fl}^* + V_m \sigma_m^* \tag{3-15}$$

If the fibers are much stronger than the matrix (which holds true for all common fiber-reinforced polymers) and the fiber volume fraction is significant (which is the case for all structural composites), *Equation 3-15* may be simplified into

$$\sigma_l^* = V_f \sigma_{fl}^* \tag{3-16}$$

Since these strength expressions are based on such highly ideal assumptions they are to be regarded as upper bounds for the strength; indeed they predict strengths significantly higher than experimentally determined ones. A variety of equivalent expressions for transverse, shear, and flexural strengths have been derived and may be found in the pertinent literature, including textbooks on mechanics of composite materials. Such textbooks also discuss the assumptions and accompanying weaknesses of the micromechanics models above, including other and more refined models, and invariably also describe how to determine the effective properties of laminated composites.

Prediction of laminae and laminate properties using the equations above as well as other and more refined models no doubt is convenient in the early design stages. It cannot be overemphasized, however, that such predicted properties rarely are sufficient to finalize a design. For some applications the

Table 3-12 Reinforcement efficiency factors [29]

Reinforcement type	β_f
Aligned unidirectional reinforcement	1
Bidirectionally symmetric reinforcement	0.5
Randomly in-plane arranged reinforcement	0.375
Randomly in-space arranged reinforcement	0.2

more detailed information of raw material suppliers' data sheets may suffice for design purposes, but in many cases the only way to determine reliable composite properties may be to manufacture samples and to perform the relevant tests, as further discussed in *Section 6.2*.

3.3.3 Comparisons of Predicted Mechanical Properties

It was previously mentioned that composites are often used in weight-critical applications and that it is therefore instructive to consider the specific properties. In *Section 3.2.2* the specific mechanical properties of reinforcements were briefly discussed while *Figures 3-7* through *3-9* illustrate the specific properties of epoxy-based composites with a fiber volume fraction of 0.6.* *Figure 3-7* displays the properties in the longitudinal direction of unidi-

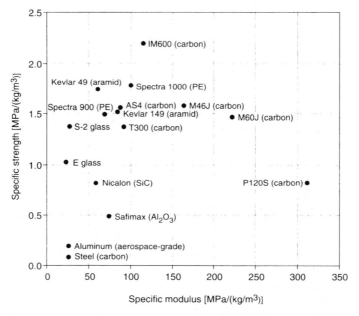

Figure 3-7 Specific longitudinal tensile strengths and moduli of unidirectionally fiber-reinforced epoxy composites with a fiber volume fraction of 0.6 and, for reference, a couple of metals

---

* The data of *Figure 3-7* were computed using *Equations 3-8* and *3-16*, the data of *Figure 3-8* were computed using *Equation 3-10* and a transverse equivalent of *Equation 3-15* from reference [27], while finally the data of *Figure 3-9* were computed using *Equation 3-14* and *Equation 3-15* modified with reinforcement efficiency factors of 0.25 (since the fibers are aligned in four equal directions. Property data of constituents from *Tables 3-5* and *3-8* and reference [1]; bulk properties of metals from reference [18]. Due to the lack of comprehensive transverse fiber properties (cf. *Table 3-9*), the predictions behind *Figure 3-8* employ the longitudinal properties, which is really only appropriate for glass fibers.

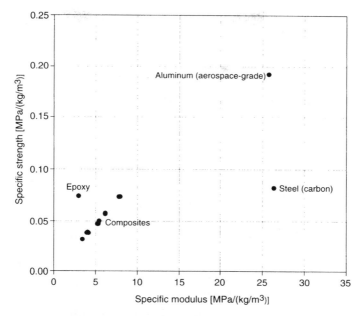

Figure 3-8 Specific transverse tensile strengths and moduli of unidirectionally fiber-reinforced epoxy composites with a fiber volume fraction of 0.6 and, for reference, a couple of metals

rectionally reinforced composites while *Figure 3-8* shows the corresponding transverse properties. *Figure 3-7* clearly illustrates the significant superiority of all composites over metals as long as only the longitudinal direction is considered. However, as soon as one considers the transverse direction of a unidirectionally reinforced composite (*Figure 3-8*), the relationship is quite the opposite and all composites basically have the same properties, which almost entirely are those of the matrix. One thus may suspect that unidirectionally reinforced composites have limited (albeit important) uses.

Since composites are obviously highly anisotropic it is common practice to orient the reinforcement in several directions to lessen this anisotropy. This may be achieved through introduction of random mats for low- and medium-performance composites or by introduction of fabrics for medium- to high-performance applications. However, even more common in high-performance applications is to use unidirectionally reinforced laminae oriented to achieve a quasi-isotropic laminate, which refers to the attempt at approximating isotropy. The simplest example of a quasi-isotropic laminate is $[-60/0/60]_s$ while the next simplest, which is quite common, is $[0/90/+45/-45]_s$. The latter stacking sequence has been assumed for *Figure 3-9*, which illustrates that even in this configuration composites are quite competitive with metals, but the playing field has now been leveled a bit.

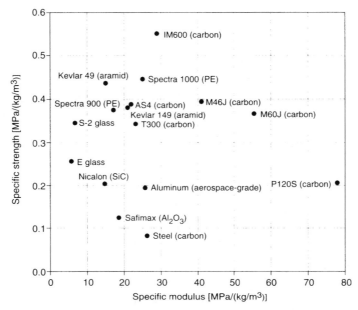

Figure 3-9 Specific tensile strengths and moduli of quasi-isotropic fiber-reinforced epoxy composites with a fiber volume fraction of 0.6 and, for reference, a couple of metals

Having established that composites are anisotropic it is appropriate to consider the properties in any given direction. While any textbook on mechanics of composites treats this topic rigorously, the present treatment will be limited to a qualitative discussion. *Figure 3-10* consequently illustrates the directional nature of one quadrant of some composite types as well as the isotropy of metals. In this figure the radial distance from the origin for a given angle is proportional to a property, such as strength, modulus, CTE, CTC, etc. Note that *Figure 3-10* does not show specific properties. While metals have more or less the same properties in all directions, composites clearly do not. The (0°) laminates of *Figures 3-7* and *3-8* are strongly anisotropic, while the laminates of *Figure 3-9* have three identical directions per quadrant (0°, 45°, and 90°). Similarly a balanced 0°/90° fabric has two directions with the same property per quadrant, whereas a random mat would exhibit approximately isotropic (in-plane) properties. One of the most important lessons to be learnt from *Figure 3-10* is that if the reinforcement of a composite (or the load) is slightly misaligned, the properties may be drastically lower than expected.

In the previous two sections it was implicitly assumed that composites and their constituents are elastic. In the longitudinal direction of a unidirectionally reinforced composite this is generally true; composites are brittle with a linear stress-strain behavior up to an ultimate strain on the order of a

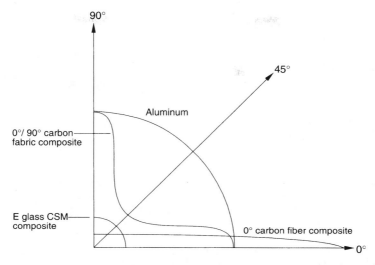

Figure 3-10 The directional nature of one quadrant of some composites and, for reference, aluminum. The relative magnitude of a property for any angle is indicated by the radial distance from the origin

few percent. When loaded in the transverse direction, on the other hand, the behavior is usually elasto-plastic and viscoelastic. This should however not be too surprising since the behavior in the former case is fiber-controlled and in the latter matrix-controlled, cf. *Figure 3-2*. Metals, on the other hand, tend to be elasto-plastic in all directions, with ultimate strains one to two orders of magnitude greater than for composites.

3.3.4 Experimentally Determined Thermal Properties

It was previously mentioned that it is rare for predictions to entirely replace experimentally determined properties. This section and the next therefore attempt to give some examples of experimentally determined properties for use in, for example, simple design tasks as part of a course in mechanics of composites. For more extensive data the reader is referred to some publications at the end of the chapter as well as to product literature from the pertinent raw material suppliers.

Figure 3-11 illustrates the normalized flexural strength of unidirectionally glass-fiber reinforced composites with different matrices as function of temperature. The composites were filament wound into flat laminates. The samples were conditioned at the quoted temperature for 30 minutes prior to testing and the tests were performed at the same temperature. The figure illustrates the obvious fact that properties quickly deteriorate with increasing temperature due to the softening of the matrix. The figure further illustrates the high temperature tolerance of phenolics.

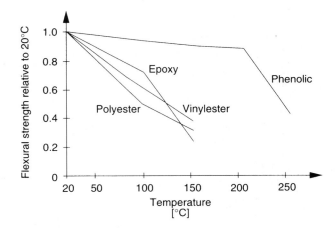

Figure 3-11 The influence of temperature on the flexural strength of unidirectional-ly glass-fiber reinforced composites with different matrices. The strengths have been normalized to the respective strengths at 20°C. Redrawn from reference [1]

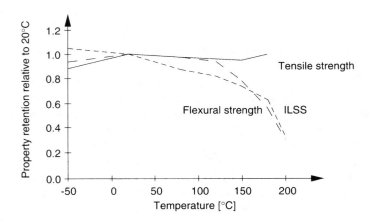

Figure 3-12 The influence of temperature on tensile, flexural, and interlaminar shear strengths (ILSS) of carbon-fabric reinforced epoxy. The strengths have been normalized to the respective strengths at 20°C. Redrawn from reference [1]

Figure 3-12 shows the property retention of hand laid-up aerospace-type carbon-reinforced epoxy prepreg postcured at 190°C to improve high-temperature properties. The prepreg reinforcement was a balanced satin weave and $V_f = 0.60$. The first observation is that the laminate properties displayed in *Figure 3-12* show better retention at high temperatures than the laminate properties shown in *Figure 3-11*, which primarily is due to the high-quality epoxy resin in the prepreg. The second observation is that the tensile strength essentially is unaffected by temperature changes since it is fiber-dominated, whereas flexural and interlaminar shear strengths steadily decrease due to their dependency on resin properties.

Table 3-13 gives some thermal properties of various reinforcement–matrix combinations, while *Figure 3-13* shows the temperature-dependency of the thermal properties of unidirectionally carbon-fiber reinforced PEEK, which probably is the most well-characterized thermoplastic-based composite to be found in the open literature.

Table 3-13 Thermal properties of composites [1,30-32]

Reinforcement	Matrix	V_f	k_l W/m°C	k_t W/m°C	α_l 10^{-6} °C^{-1}	α_t 10^{-6} °C^{-1}
E glass	Polyester	0.3 – 0.6	0.3 – 0.35		5 – 14	
E glass	Epoxy	0.4 – 0.75	0.3 – 0.35		4 – 11	
E glass	PP	0.35			7	9
E glass (SMC)	Polyester	0.18		0.16 – 0.26	20	
E glass (SMC)	Polyester	0.27				8 – 11
E glass (GMT)	PP	0.08			32	
E glass (GMT)	PP	0.13		0.22	29	
E glass (GMT)	PP	0.19		0.27	27	
E glass	Phenolic	0.22	0.22		15	
Kevlar 49	Epoxy	0.64	1.25			
Kevlar 49	Epoxy	0.58 – 0.68		0.41		
Kevlar 49	Epoxy	0.5			-2.1	68.6
T300 (carbon)	Epoxy	0.6	3	0.59	0.3	36.5
Carbon (HS)	PEEK	0.61	4.1	0.47	0.23	28.9
M46J (carbon)	Epoxy	0.6	26	1.1	-0.7	36.5
P75 (carbon)	Epoxy	0.6	114	1.2		
P75 (carbon)	Epoxy	0.65			-1.08	31.7

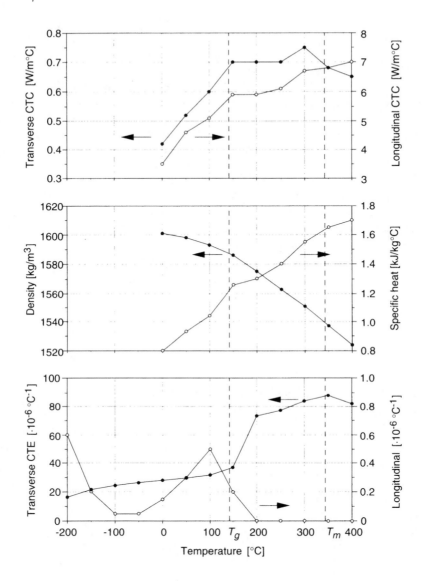

Figure 3-13 Thermal properties of unidirectionally carbon-fiber reinforced PEEK as functions of temperature at $V_f = 0.61$. Values for CTE in excess of 150°C are estimated. T_g and T_m indicated in graphs. Data from reference [30]

3.3.5 Experimentally Determined Mechanical Properties

In analogy with the previous section, this section also attempts to provide some representative property data for composites. *Tables 3-14* through *3-17* give properties of glass-reinforced polyesters, vinylesters, epoxies, and phenolics, while *Tables 3-18* and *3-19* and *Figure 3-14* present properties of aramid- and carbon-reinforced epoxy composites. *Tables 3-20* through *3-23* deal with composites manufactured from thermoplastic prepregs, while *Tables 3-24* through *3-26* present properties of thermoset and thermoplastic molding compounds. In most cases test methods akin to those discussed in *Section 6.2* have been used to determine the data of the following tables and figures.

Table 3-14 gives mechanical properties of glass-fiber reinforced polyester. For $W_f = 0.34$ the reinforcement was a chopped strand mat; for $W_f = 0.47$ and 0.53 the reinforcement was a combination mat consisting of a balanced woven fabric and a random mat; while for $W_f = 0.61$ the reinforcement was the combination mat together with a UD fabric. The resin was a so-called isophthalic NPG-polyester (i.e. formed using neopentyl glycol) and manufacturing was achieved through hand layup. The table clearly illustrates that strengths and stiffnesses increase with fiber fraction in the qualitative fashion that the relationships of *Section 3.3.2* indicate. Further, the tensile strength perpendicular to the laminate is only a fraction of the in-plane strength and a fifth of the tensile strength of the neat matrix, apparently indicating unfavorable fiber–matrix bonding. It is also noteworthy that the high strain to failure of the matrix is not translated into the composite.

Table 3-14 Mechanical properties of glass-reinforced polyester composites. Indices f and $\perp$ denote flexural and out-of-plane tensile properties, respectively, while σ_{lt}^* denotes ILSS [1]

W_f		*0*	*0.34*	*0.47*	*0.53*	*0.61*
Approximate equivalent V_f		*0*	*0.19*	*0.29*	*0.34*	*0.41*
σ_t^*	MPa	73	100	200	240	420
E_t	GPa	3.1	7.5	13.0	14.5	22.5
ε_t^*	%	6.5	1.8	2.2	2.3	2.3
σ_f^*	MPa	135	140	320	330	
E_f	GPa	3.0	6.0	11.0	11.5	
$\sigma_\perp^*$	MPa			15	15	16
σ_{lt}^*	MPa			35		

Table 3-15 provides properties of glass-fiber reinforced polyesters and vinylesters. The reinforcement used was a balanced woven fabric of areal weight 800 g/m² and a chopped strand mat of 100 g/m²; the laminates were hand laid up. The table illustrates that the properties quoted are reasonably independent of the resin type despite resins of quite different properties being used, consequently indicating that these properties are dominated by the reinforcement rather than by the matrix.

Table 3-15 Mechanical properties of glass-reinforced polyester and vinylester composites. Index c denotes compressive properties [1]

Matrix		Polyester			Vinylester	
		Orthophthalic	Isophthalic	Isophthalic NPG	Flexible	Rubber-modified
V_f		0.53	0.53	0.51	0.53	0.52
σ_t^*	MPa	226	209	205	227	237
E_t	GPa	17.3	16.9	16.1	16.8	16.9
ε_t^*	%	1.8	1.4	1.6	1.7	1.7
σ_c^*	MPa	294	209	258	288	259
E_c	GPa	17.5	17.5	16.5	16.6	15.9
ε_c^*	%	1.9	1.4	1.6		1.7

Table 3-16 Mechanical properties of glass-reinforced polyester, vinylester, and epoxy composites. τ^* denotes shear strength [1]

Matrix		Isophthalic polyester		Urethane-modified vinylester		Epoxy	
Reinforcement		Mat	Fabric	Mat	Fabric	Mat	Fabric
V_f		0.20	0.35	0.18	0.38	0.17	0.35
σ_t^*	MPa	129	241	122	243	122	212
E_t	GPa	8.8	16	6.7	16	8.1	16
ε_t^*	%	1.5	1.5	1.8	1.5	1.5	1.3
σ_f^*	MPa	243	366	224	372	199	365
E_f	GPa	7.6	13	7.4	13	6.9	19
ε_f^*	%	3.2	2.9	3.0	2.8	2.9	2.7
τ_{lt}^*	MPa	9	7	10	9	12	11

Table 3-17 Mechanical properties of glass-reinforced phenolic composites [1]

Reinforcement		Continuous strand mat	Chopped strand mat	Woven fabric	Unidirectional reinforcement
V_f		15	24	42	63
σ_t^*	MPa	40 – 60	100 – 150	280 – 350	
E_t	GPa	4.5 – 6.0	5.5 – 7.5	16 – 20	
ε_t^*	%	1.0 – 2.0	1.8 – 2.5	2.0 – 2.5	2.9
σ_f^*	MPa	75 – 125	150 – 230	300 – 450	1340
E_f	GPa	4.0 – 5.5	5.7 – 7.5	17 – 20	46

Table 3-16 gives properties of glass-fiber reinforced polyester, vinylester, and epoxy. For the lower fiber fractions the reinforcement was a chopped strand mat of 450 g/m², while for the higher fiber fractions a balanced woven fabric was used. The composites were manufactured through hand layup. Once again there appears to be no significant difference between matrix types whereas the expected difference between the two reinforcement types is apparent.

Table 3-17 presents properties of glass-fiber reinforced phenolic composites. The composites with unidirectional reinforcement were filament wound while the others were wet hand laid up. The table shows that at least with this phenolic resin the composite properties are comparable to those achieved with polyesters (cf. *Table 3-14*), but, as indicated by *Figure 3-11*, the temperature tolerance should be significantly higher.

Table 3-18 presents comprehensive mechanical data for unidirectionally aramid-fiber reinforced epoxy composites. *Figure 3-14* similarly shows the tensile strength of unidirectionally aramid-fiber reinforced epoxy composites as a function of fiber volume fraction. Judging from the rule of mixtures for longitudinal tensile strength (*Equation 3-15*) one may be tempted to assume that an increase in the amount of fibers always increases the strength as suggested by the line in *Figure 3-14*. However, the figure shows that as the fiber volume fraction increases above approximately 0.5 the strength levels off, which is likely due to the fact that wetting of the fibers becomes ever more difficult. It is also quite obvious that the rule of mixtures highly overestimates the strength, indicating the importance of keeping the assumptions behind *Equation 3-15* in mind.

Table 3-19 gives properties for some unidirectionally carbon-fiber reinforced epoxy composites. The laminates were manufactured through prepreg hand layup and wet hand layup. The table clearly illustrates the very significant differences in properties achievable with different carbon fiber types.

Table 3-18 Mechanical properties of unidirectionally aramid-fiber reinforced epoxy composites [1]

Reinforcement		Kevlar 49	Kevlar 149
V_f		0.60	0.60
ρ	kg/m^3	1360	1370
σ^*_{lt}	MPa	1500	1450
E_{lt}	GPa	79	107
ε^*_{lt}	%	1.71	1.33
σ^*_{tt}	MPa	27	
E_{tt}	GPa	4.1	
ε^*_{tt}	%	0.85	
σ^*_{lf}	MPa	655	634
E_{lf}	GPa	67	79
σ^*_{lc}	MPa	234	193
E_{lc}	GPa	66	73
ε^*_{lc}	%	0.58	
σ^*_{tc}	MPa	93	
E_{tc}	GPa	5.2	
ε^*_{tc}	%	2.83	
ν_{lt}		0.43	
ν_{tl}		0.31	
τ^*_{lt}	MPa	47	
G_{lt}	GPa	1.5	
τ^*_{tl}	MPa	27	
G_{tl}	GPa	1.5	
σ^*_{lt}	MPa	48 – 69	57

Table 3-20 gives properties of composites manufactured from E glass-reinforced PP and E glass-reinforced PA 12 prepreg. The glass-reinforced PP composites were compression molded from unidirectionally reinforced melt-impregnated prepregs, while the glass-reinforced PA 12 composites were compression molded from powder-impregnated fabric prepregs of two different types. The prepreg denoted UD weave was a crowfoot weave with 90 percent of the fibers in the warp direction and 10 percent in the weft direction, whereas the prepreg denoted balanced weave was a satin weave.

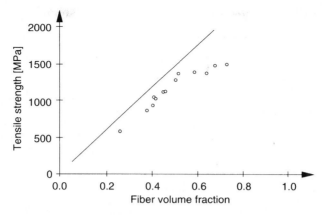

Figure 3-14 Tensile strength of unidirectionally aramid-fiber reinforced epoxy composites as a function of fiber volume fraction. The line represents the rule of mixtures prediction. Redrawn from reference [33]

Table 3-19 Mechanical properties of unidirectionally carbon-fiber reinforced epoxy composites [1]

Reinforcement		AS4	T300	M46J	M60J	P75S	P120S
V_f		0.60	0.60	0.60	0.60	0.62	0.62
ρ	kg/m³	1560	1540	1590	1650		
σ^*_{lt}	MPa	2137	1760	1960	1760	966	1442
E_{lt}	GPa	148	125	260	330	338	524
ε^*_{lt}	%		1.3	0.7	0.5	0.28	0.27
σ^*_{tt}	MPa		80	45	30	33	21
E_{tt}	GPa		7.8	6.9	5.9	6.9	5.7
ε^*_{tt}	%		1.0			0.49	0.4
σ^*_{lf}	MPa	1724					
E_{lf}	GPa	134					
σ^*_{lc}	MPa		1570	880	780	442	270
E_{lc}	GPa		125	245	320	317	559
ε^*_{lc}	%		1.2				
ν_{lt}			0.34				
τ^*_{lt}	MPa		98	59	39		
G_{tl}	GPa		4.4	3.9	3.9	5.9	5.5
σ^*_{lt}	MPa	127	110	80	70	55.2	27.6

Table 3-20 Mechanical properties of E glass-reinforced PP and PA 12 composites [14,34]

Matrix Reinforcement type		PP UD	PA 12 UD weave	PA 12 Balanced weave
V_f		0.35	0.55	0.52
ρ	kg/m^3	1480	1900	1850
σ^*_{lt}	MPa	620	710	350
E_{lt}	GPa	27.5	44	26
ε^*_{lt}	%	2.1	1.5	1.6
σ^*_{tt}	MPa		90	350
E_{tt}	GPa		7.2	26
ε^*_{tt}	%		2.2	1.6
σ^*_{lf}	MPa	570	800	500
E_{lf}	GPa	22	38	22
σ^*_{tf}	MPa		160	500
E_{tf}	GPa		8	22
σ^*_{lt}	MPa		47	42
HDT	°C	156 – 164	178	177

Table 3-21 similarly gives properties of unidirectionally reinforced composites manufactured from high-performance carbon-fiber reinforced thermoplastic prepregs.

Table 3-22 presents properties of unidirectionally reinforced composites made from prepregs consisting of high-strength carbon fibers and PEEK, as well as selected properties of carbon- and aramid-reinforced epoxy composites for comparison.

Table 3-23 illustrates how the increased handleability of commingled materials is also likely to lead to composite properties inferior to those achievable with melt-impregnated prepregs. The drawback of commingled feedstock is that it must be melt-impregnated during processing, which is a non-trivial procedure with such a high-viscosity polymer as PEEK.

Table 3-24 gives representative properties for composites molded from glass-fiber reinforced molding compounds. With the exception of the first BMC listed, which was injection molded, all composites were compression molded. *Table 3-25* then illustrates the significance of fiber length in the molding compound. Although the data of the latter table concerns compression molded BMC, the same trends are to be expected for any raw material form containing discontinuous reinforcement and for any manufacturing technique.

Table 3-21 Mechanical properties of unidirectionally carbon-fiber reinforced, high-performance thermoplastic composites and two high-performance epoxy composites for comparison [35]

Matrix	Reinforcement	Test temp. °C	σ^*_{lt} MPa	E_{lt} GPa	ε^*_{lt} %	σ^*_{tt} MPa	E_{tt} GPa	ε^*_{tt} %	σ^*_{lf} MPa	E_{lf} GPa
PEEK	AS4	RT	2070	142	1.3	78.6	9.7	0.9	1880	121
		100	2010	130	1.3	66.2	8.3	1.1		
	IM7	RT	2890	169	1.6				2040	148
PEKK	AS4	RT							1620	
		93							1390	
PPS	AS4	RT	2090	131		71.7	7.6		1680	103
PAS	AS4	RT	1330	129		59.3	6.9		1480	96.5
PAI	C3000	RT	1380	142					2070	128
PEI	T300	RT		136			8.30			
PI	AS4	RT							1970	97.2
		93							1750	97.2
		149							1600	90.3
		177							1370	91.7
PI†	AS4	RT				37.2–48.3	8.3–10.3		1390	94.5
		177							1000	95.9
	IM6	RT	2340	161	1.46	46.9	8.3		1590	124
Epoxy‡	AS4	RT	2170	145	1.4	53.8	9.70	0.67	2070	128
	IM7	RT	2620	166	1.6					
Epoxy¶	AS4	RT	1830	128	0.8	66.2	63.4	1.3	1700	123
		100	1900	139	1.4	57.9	51.0	1.1		
	IM7	RT	2760	159	1.6	75.8	8.30	1.0		

† PI that permits (relatively) low-viscosity reinforcement impregnation with prepolymer
‡ Epoxy crosslinked at 177°C
¶ Toughened epoxy

Table 3-22 Mechanical properties of carbon-fiber reinforced PEEK composites. Properties of composites manufactured from aerospace-grade carbon-fiber reinforced epoxy prepreg and the corresponding Kevlar 49-fiber reinforced epoxy composite properties of *Table 3-18* are shown for comparison [1,2,30].

Reinforcement		Carbon (HS)	T800H	Kevlar 49
V_f		0.61	≈ 0.6	0.60
σ^*_{lt}	MPa	2130	2740	1500
σ^*_{tt}	MPa	80	79	27
σ^*_{lc}	MPa	1100	1945	234
σ^*_{tc}	MPa	220		93
$\tau^*_{l\perp}$	MPa	120		
τ^*_{lt}	MPa	175		47
$\tau^*_{t\perp}$	MPa	80		
E_{lt}	GPa	137	160	79
E_{tt}	GPa	9.4	8.7	4.1
$E_{\perp t}$	GPa	9.1†		
G_{lt}	GPa	5.1		1.5
$G_{l\perp}$	GPa	4.7†		
G_{tl}	GPa	5.1		1.5
$G_{t\perp}$	GPa	3.2		
$G_{\perp l}$	GPa	4.6†		
$G_{\perp t}$	GPa	3.2†		
ν_{lt}		0.33		0.43
$\nu_{l\perp}$		0.32		
ν_{tl}		0.04		0.31
$\nu_{t\perp}$		0.40		
$\nu_{\perp l}$		0.04†		
$\nu_{\perp t}$		0.40		

† denotes estimated values

Table 3-23 Properties of composites made from melt-impregnated and commingled carbon/PEEK feedstock [30]

Product form	E_{lt} GPa	σ^*_{lt} MPa	E_{lc} GPa	σ^*_{lc} MPa	E_{tt} GPa	σ^*_{tt} MPa
Melt-impregnated	132	>1172	113	1175	8.9	91
Commingled	127	>782	115	865	8.9	64

Table 3-24 Properties of composites manufactured from thermoset molding compounds. The first BMC was injection molded, all others were compression molded [31]

Material	W_f	ρ kg/m^3	σ^*_{lt} MPa	E_{lt} GPa	σ^*_{lf} MPa	E_{lf} GPa	σ^*_{lc} MPa	Mold shrinkage %	HDT °C
Polyester BMC	0.22	1820	33.5	10.5	87.2	10.9		0.4	260
Polyester BMC	0.22	1820	41.4	12.1	88.3	10.9	138	0.1	260
Polyester SMC-R (low glass)	0.2	1780	36.5	11.7	110	9.7	159	0.2	260
Polyester SMC-R (class A)	0.27	1940	70	10.2	152	8.3	165	≥ 0.1	232
Polyester SMC-R (structural)	0.27	1890	77.2	11.7	179	11.7	172	0	260
Epoxy SMC-R (structural)	0.63	1850	269	23.4	476	19.3	228	0.2	302
Epoxy SMC-C (structural)	0.72	1.90	690	44.7	1034	41.4		0.2	302

Table 3-25 The effect of fiber length in compression molded E glass-reinforced epoxy BMC [36]

Fiber length mm	W_f	ρ kg/m^3	σ^*_{lt} MPa	σ^*_{lf} MPa	E_{lf} GPa	σ^*_{lc} MPa
6.4	0.63	1880	120	270	25	190
12.7	0.63	1880	190	470	28	290
31.8	0.63	1880	310	760	32	290

Table 3-26 Properties of composites manufactured from glass/PP GMT. For the last GMT type ($W_f = 0.43$) half of the reinforcement is longitudinally oriented [37,38]

GMT type	W_f	Fiber length	ρ	σ^*_{lt}	E_{lt}	ε_{lt}	σ^*_{lf}	E_{lf}	Mold shrinkage
		mm	kg/m³	MPa	GPa	%	MPa	GPa	%
Powder-impregnated									
	0.25	12	1070	75	4.0	3.3	130	4.8	
	0.30	12	1120	86	4.5	3.0	140	5.5	
	0.35	12	1160	96	5.2	2.8	150	5.9	
	0.40	12	1220	105	5.9	2.7	165	6.4	
Melt-impregnated									
	0.20	continuous	1020	55	3.5	1.8	90	3.5	0.1 – 0.3
	0.30	continuous	1130	70	4.5	1.8	110	4.5	0.1 – 0.3
	0.40	continuous	1190	90	6.0	1.7	140	5.5	0.1 – 0.3
	0.43	continuous	1210	240	12.0	1.7	230	8.5	0.1 – 0.3

Table 3-26 provides properties for components compression molded from GMT consisting of the most common GMT material combination: glass/PP. For all GMTs listed, with the exception of the last one which has half the reinforcement longitudinally oriented, the properties should be approximately isotropic in the plane.

3.4 Core Materials

Although there are also micromechanics-based approaches to predicting the properties of core materials, these naturally require knowledge of the properties of the cell walls, so the common approach in design is to use experimentally determined property data. *Tables 3-27* through *3-29* consequently provide thermal and mechanical properties of selected balsa, Nomex honeycomb, aluminum honeycomb, and expanded foam cores. In terms of CTC, see also *Figure 3-1*.

3.5 Sandwich Components

The effective thermal properties of sandwich components may be calculated from its constituents (i.e. faces, which may be composite laminates, and core) in ways analogous to those discussed in *Section 3.3.1*. However, for transverse heat transfer calculations it is usually more appropriate to treat the sandwich as a structure consisting of three layers and three coupled heat transfer situations.

Table 3-27 Thermal properties of selected cores [39–42]

Core type	ρ kg/m^3	k $W/m°C$	α $10^{-6}\ °C^{-1}$	T_{max} $°C$
Balsa	96	0.0509		
	180	0.0710		
Expanded PUR	30	0.025	100	100
	50	0.025	100	100
Expanded PS	30	0.035	70	80
	60	0.035	70	80
Expanded crosslinked PVC	36	0.022	35	80
	60	0.025	35	80
	100	0.030	35	80
	200	0.043	35	80
Expanded PEI	80	0.035		190
Expanded PMI	52			205
	110			200
	205			200

Table 3-28 Mechanical properties of selected balsa and foam cores [39–42]

Core type	ρ kg/m^3	σ^*_t MPa	σ^*_c MPa	E_c MPa	τ^* MPa	G MPa
Balsa	96	6.9	6.5	2240	1.85	108
	180	15.7	16.0	4930	3.46	188
Expanded PUR	30	0.3	0.2	10	0.2	3
	50	0.4	0.35		0.3	
Expanded PS	30	0.4	0.3	20	0.25	8
	60	1.2	0.9	60	0.6	20
Expanded crosslinked PVC	36	0.9	0.3	20	0.35	13
	60	1.6	0.8	60	0.7	22
	100	3.1	1.7	125	1.4	40
	200	6.4	4.4	310	3.3	85
Expanded PEI	80		0.75	45	0.9	18
Expanded PMI	52	1.6	0.8	75	0.8	24
	110	3.7	3.6	180	2.4	70
	205	6.8	9.0	350	5.0	150

Table 3-29 Mechanical properties of selected Nomex and aluminum honeycomb cores [39]

Core type	ρ kg/m³	d mm	σ^*_c MPa	E_c MPa	τ^*_{length} MPa	G_{length} MPa	τ^*_{width} MPa	G_{width} MPa
Nomex honeycomb	29	3.2	0.62		0.65	27	0.38	16
	64	3.2	4.7		1.6	62	0.98	38
	96	3.2	7.4		2.6	96	1.4	52
	200	3.2	26		4.1	138	2.7	98
	32	6.4	1.0		0.72	32	0.40	18
	32	9.6	1.0		0.66	27	0.38	17
	32	12.8	0.95		0.56	20	0.26	12
Aluminum honeycomb	50	3.2	2.3	660	1.7	310	1.0	130
	98	3.2	9.1	2030	4.9	700	2.7	260
	50	4.8	2.7	660	1.8	310	1.0	130
	37	6.4	1.6	400	1.2	89	0.72	100
	55	6.4	3.0	790	2.0	340	1.2	150

The most common reason for using the sandwich concept is to achieve a structure with high flexural rigidity (cf. *Figure 2-55*). The flexural rigidity of a sandwich structure may be calculated as [43]:

$$D = \frac{E_{face}t^3_{face}}{6} + \frac{E_{face}t_{face}(t_{face} + t_{core})^2}{2} + \frac{E_{core}t^3_{core}}{12} \qquad (3\text{-}17)$$

where t is thickness and indices *face* and *core* refer to face and core, respectively. The first term of *Equation 3-17* corresponds to the flexural rigidity of the faces about their individual neutral axes, the second term the stiffness of the faces associated with bending about the neutral axis of the entire sandwich, and the third term the flexural rigidity of the core. In many cases it is reasonable to assume that $t_{face} \ll t_{core}$ and $E_{core} \ll E_{face}$, in which case *Equation 3-17* reduces to [43]:

$$D = \frac{E_{face}t_{face}(t_{face} + t_{core})^2}{2} \qquad (3\text{-}18)$$

3.6 Summary

One of the major disadvantages of composites is that it is difficult to obtain comprehensive and reliable design data. Reasons for this may be found in the fact that composites are a relatively young engineering field experiencing rapid developments on all fronts, but also in the fact that there is no such thing as a standard material formulation or reinforcement orientation sequence, meaning that each design is different from the last. The fact that the way in which a composite is manufactured significantly affects final component properties (cf. *Chapters 2* and *4*), and that experimentally determined properties depend on how characterization is carried out (cf. *Chapter 6*), further makes compilation of reliable material data bases difficult. In providing data for constituents and composites that are intended to be representative, this chapter aims to complement the discussion of *Chapter 2* with quantitative information.

While the temperature tolerance of candidate matrices varies significantly, heat transfer and thermal expansion properties are similar. Also mechanical properties of polymers are largely comparable, although there are some polymers that stand out from the rest; the only consistent difference is that failure strains of thermoplastics tend to be significantly higher than for thermosets.

Whereas matrices generally exhibit small property differences, the opposite holds true for reinforcements. Thermal properties vary widely and it is particularly notable that glass fibers are isotropic, whereas carbon and aramid fibers are transversely isotropic. Of significance also is that carbon and aramid fibers have negative longitudinal CTEs and that the longitudinal CTCs of carbon fibers are comparable to those of the best metal conductors. Nevertheless, for structural applications the mechanical properties are the most relevant and this is where the largest differences are to be found. For common reinforcement types, tensile moduli span from 70 to over 800 GPa (cf. aluminum at 70 GPa and carbon steel at 200 GPa), tensile strengths from 1 to over 7 GPa (cf. aluminum at 0.5 GPa and carbon steel at 0.6 GPa), and tensile failure strains from 0.5 to 6 percent. If density is taken into account and specific mechanical properties are considered, the difference to metal alternatives is even greater. This immense variation provides the designer with a fantastic variety of properties to suit virtually any need, but also a significant challenge in dealing with the inherent anisotropy of composites.

The rule-of-mixture approach may be used to estimate lamina properties from those of the constituents for use in initial analyses and more refined techniques may be utilized to predict properties of cross-plied laminates, but it is important to realize that such results are essentially never accurate enough to finalize a design. This chapter therefore provides some basic and largely representative thermal and mechanical composite properties, and the following section suggests further and considerably more comprehensive sources of property data. When such data are insufficient, the raw material supplier is likely to possess more accurate data, but in many cases the only reliable way to obtain data for a critical design is to manufacture samples from candidate material systems and then characterize them using methods discussed in *Chapter 6*. With sandwich components properties may often be quite adequately predicted from known properties of faces and core.

Suggested Further Reading

Since the intention of this chapter is primarily to give the reader a feeling for properties of constituents and composites and to provide a rudimentary data base for basic design exercises, the reader no doubt may be left desiring more information than has been given herein. The reference list following this section is naturally a good source of information, but the author would like to suggest some other good sources. Starting with mechanics of composite materials there is a multitude of textbooks available that provide excellent treatments, so it is difficult to single out any particular book. However, good textbooks include:

I. M. Daniel and O. Ishai, *Engineering Mechanics of Composite Materials*, Oxford University Press, New York, NY, USA, 1994.

R. M. Jones, *Mechanics of Composite Materials*, Hemisphere Publishing Corp., New York, NY, USA, 1975.

Quite often raw material suppliers and in some cases composite manufacturers possess the most accurate property information short of actually determining the properties oneself. Supplier-independent data bases for constituents and composites are nevertheless available in software and book form, but they tend to be lacking in comprehensiveness. Among the more easily accessible sources of compiled property data are:

N. L. Hancox and R. M. Mayer, *Design Data for Reinforced Plastics—A Guideline for Engineers and Designers*, Chapman & Hall, London, UK, 1994.

International Encyclopedia of Composites, Ed. S. M. Lee, VCH Publishers, New York, NY, USA, 1990-1991 (6 volumes).

Engineered Materials Handbook, Volume 1, Composites, ASM International, Metals Park, OH, USA, 1987.

Engineered Materials Handbook, Volume 2, Engineering Plastics, ASM International, Metals Park, OH, USA, 1988.

Modern Plastics Encyclopedia, McGraw-Hill, New York, NY, USA (published annually).

The first reference below provides an exhaustive listing of glass-fiber properties and suppliers, while the second reference provides an equally exhaustive listing for carbon and other high-performance fibers as well as prepregs based on these reinforcements:

T. F. Starr, *Glass-Fibre Directory and Databook*, 2nd edn., Chapman & Hall, London, UK, 1997.

T. F. Starr, *Carbon and High Performance Fibres Directory and Databook*, 6th edn., Chapman & Hall, London, UK, 1995.

Composites manufactured through pultrusion (see *Section 4.2.6*) are often available off the shelf in standard configurations and such composites are generally well characterized. Properties of such components, as well as design aids, are available from most pultruders of any size, and the following design manuals are therefore mere examples:

Design Manual, Morrison Molded Fiber Glass Company, Bristol, VA, USA.

Design Guide, Creative Pultrusions Inc., Alum Bank, PA, USA.

Designmanual, Fiberline Composites, Kolding, Denmark, 1995.

For additional information on sandwich-related topics the choices are somewhat limited. Sandwich property data are scattered throughout the scientific literature and in suppliers' data sheets, while on the design side classical and somewhat inaccessible texts have been followed by the following comprehensive treatments:

D. Zenkert, *An introduction to Sandwich Construction*, EMAS, Warley, UK, 1995.

The Handbook of Sandwich Construction, Ed. D. Zenkert, EMAS, Warley, UK, 1997.

References

1. N. L. Hancox and R. M. Mayer, *Design Data for Reinforced Plastics—A Guideline for Engineers and Designers*, Chapman & Hall, London, UK, 1994.

2. "Fibredux 6376," Ciba-Geigy Plastics, Cambridge, UK, 1991.

3. "Matrix Systems, Araldite LY 556, Hardener HY 917, Accelerator DY 070," Ciba-Geigy, Switzerland.

4. R. E. Laramee, "Thermal and Related Properties of Engineering Thermosets," in *Engineered Materials Handbook, Volume 2, Engineering Plastics*, ASM International, Metals Park, OH, USA, 439–444, 1988.

5. S. W. Shalaby and P. Moy, "Thermal and Related Properties of Engineering Thermoplastics," in *Engineered Materials Handbook, Volume 2, Engineering Plastics*, ASM International, Metals Park, OH, USA, 445–459, 1988.

6. P. G. Galanty and J. J. Richardson, "Polyethylene Terephtalates (PET)," in *Engineered Materials Handbook, Volume 2, Engineering Plastics*, ASM International, Metals Park, OH, USA, 172–176, 1988.

7. D. G. Brady, "Polyphenylene Sulfides (PPS)," in *Engineered Materials Handbook, Volume 2, Engineering Plastics*, ASM International, Metals Park, OH, USA, 186–191, 1988.

8. "Data Sheet 2: Making Consolidated Sheet from Aromatic Polymer Composite, APC-2," ICI Fiberite Corporation, 1987.

9. L. E. Taske II, Personal communication, BASF, Charlotte, NC, USA, 1990.

10. R. E. Fines and J. P. Bartolomucci, "Polyether-imides (PEI)," in *Engineered Materials Handbook, Volume 2, Engineering Plastics*, ASM International, Metals Park, OH, USA, 156–158, 1988.

11. E. C. Watterson, "Polyether Sulfones (PES, PESV)," in *Engineered Materials Handbook, Volume 2, Engineering Plastics*, ASM International, Metals Park, OH, USA, 159–162, 1988.

12. J. E. Fitzpatrick, "Polyamide-imides (PAI)," in *Engineered Materials Handbook, Volume 2, Engineering Plastics*, ASM International, Metals Park, OH, USA, 128–137, 1988.

13. M. M. Konarski, "Curing BMI Resins," in *Engineered Materials Handbook, Volume 1, Composites*, ASM International, Metals Park, OH, USA, 657–661, 1987.

14. "Product Information Vestopreg," Hüls, Marl, Germany, 1994.

15. ISO 1133, "Determination of the Melt-Mass Flow Rate (MFR) and the Melt Volume-Flow Rate (MVR) of Thermoplastics".

16. ASTM D1238, "Standard Test Method for Flow Rates of Thermoplastics by Extrusion Plastometer".

17. *Modern Plastics Encyclopedia '95*, McGraw-Hill, New York, NY, USA, 1995.

18. *Formelsamling i hållfasthetslära*, 9th edn., Department of Solid Mechanics, Royal Institute of Technology, Stockholm, Sweden, 1986.

19. J. P. Holman, *Heat Transfer*, 4th edn., McGraw-Hill, New York, NY, USA, 1976.

20. Anon., "Fibers," in *Engineered Materials Handbook, Volume 1, Composites*, ASM International, Metals Park, OH, USA, 360–362, 1987.

21. A. R. Bunsell, "Fiber Reinforcement," in *Handbook of Composite Reinforcements*, Ed. S. M. Lee, VCH Publishers, New York, NY, USA, 199–217, 1993.

22. R. J. Diefendorf, "Carbon/Graphite Fibers," in *Engineered Materials Handbook, Volume 1, Composites*, ASM International, Metals Park, OH, USA, 49–53, 1987.

23. "S-2 Glass Fiber, Owens-Corning," Toledo, OH, USA, 1993.

24. J. J. Pigliacampi, "Organic Fibers," in *Engineered Materials Handbook, Volume 1, Composites*, ASM International, Metals Park, OH, USA, 54–57, 1987.

25. R. J. Morgan and R. E. Allred, "Aramid Fiber Composites," in *Handbook of Composite Reinforcements*, Ed. S. M. Lee, VCH Publishers, New York, NY, USA, 5–24, 1993.

26. I. M. Daniel and O. Ishai, *Engineering Mechanics of Composite Materials*, Oxford University Press, New York, NY, USA, 1994.

27. G. Eckold, *Design and Manufacture of Composite Structures*, Woodhead Publishing, Cambridge, UK, 1994.

28. C. C. Chamis, "Simplified Composite Micromechanics Equations for Hygral Thermal and Mechanical Properties," SPI Composites Institute Annual Conference, **38**, 21-C, 1983.

29. J. A. Quinn, *Design Data Fibreglass Composites*, 2nd edn., Fibreglass Limited Reinforcements Division, St. Helens, UK, 1981.

30. F. N. Cogswell, *Thermoplastic Aromatic Polymer Composites*, Butterworth-Heinemann, Oxford, UK, 1992.

31. R. D. Pistole, "Compression Molding and Stamping," in *Engineered Materials Handbook, Volume 2, Engineering Plastics*, ASM International, Metals Park, OH, USA, 324–337, 1988.

32. J. J. C. Jansz, "GMT: Ein Werkstoff stellt sich vor," Symalit AG, Lenzburg, Switzerland, 1995.

33. A. Mittelman and I. Roman, "Tensile Properties of Real Unidirectional Kevlar/Epoxy Composites," *Composites*, **21**, 63–69, 1990.

34. "Plytron GN 638 T, Unidirectional Glass-Fibre/Polypropylene Composite," Borealis, Stathelle, Norway, 1995.

35. N. J. Johnston, T. W. Towell, and P. M. Hergenrother, "Physical and Mechanical Properties of High-Performance Thermoplastic Polymers and Their Composites," in *Thermoplastic Composite Materials*, Ed. L. A. Carlsson, Elsevier, Amsterdam, The Netherlands, 1991.

36. W. G. Colclough, Jr. and D. P. Dalenberg, "Bulk Molding Compounds," in *Engineered Materials Handbook, Volume 1, Composites*, ASM International, Metals Park, OH, USA, 161–163, 1987.

37. L. A. Berglund and M. L. Ericson, "Glass Mat Reinforced Polypropylene," in *Polypropylene: Structure, Blends and Composites, Volume 3, Composites*, Ed. J. Karger-Kocsis, Chapman & Hall, London, UK, 202–227, 1995.

38. "GMT Breaks Through in Automotive Structural Parts," *Reinforced Plastics*, April 1992.

39. D. Zenkert, Personal communication, 1992.

40. "Divinycell Technical Manual H-Grade," Barracuda Technologies, Laholm, Sweden, 1991.

41. "Airex R82 High Performance Foam," Airex, Sins, Switzerland.

42. "Rohacell A, Rohacell WF," Röhm GmbH, Darmstadt, Germany.

43. D. Zenkert, *An Introduction to Sandwich Construction*, EMAS, Warley, UK, 1995.

CHAPTER 4

MANUFACTURING TECHNIQUES

Metal and wood are usually obtained in block, rod, plank, or sheet form for use as raw material in machining, molding, and joining operations to produce the final component, although there are some metal-working exceptions such as casting, sintering, and extrusion without intermediate semifinished product forms. For the most part the situation is quite different with fiber-reinforced composites. With composites both raw material and final component are normally made in the same manufacturing operation, though the use of prepregs and molding compounds (see *Section 2.3*) represents an in-between situation. Moreover, with thermoplastic matrices it is possible to first manufacture composite bar or sheet stock for subsequent forming—but rarely extensive machining—although such a procedure may eliminate part of the potential economic advantage of using composites. Whether making raw material and component in the same operation or using prepreg or molding compound, "net-shape" and "near–net-shape" are adjectives often used to describe composite manufacturing techniques. The expressions refer to the extent of post-manufacturing machining and trimming needed, which from both economical and property viewpoints naturally should be minimized.

Not surprisingly the use of composites presents a whole new array of challenges for the designer. Not only must he deal with an anisotropic material in component design, but he must also have a great deal of understanding of the processing requirements of candidate raw materials and the specifics of potential manufacturing techniques in order to be able to estimate the final properties of the part. First of all, for a manufacturing technique to come into consideration at all it must obviously be capable of producing a part of the required geometry. However, the technique used also determines surface finish, possible reinforcement configurations, available resins and their properties as matrices, as well as a host of other properties—not least cost. To illustrate the challenges facing the composite component designer, *Table 4-1* outlines some of the many issues that must be considered when trying to specify raw materials and manufacturing technique for a particular design.

With metals, which are essentially isotropic, homogenous, and have predictable properties, material selection and component design normally precede manufacturing considerations. If attempting the same methodology with composites (which unfortunately is not uncommon), the end product is unlikely to be a very good solution. This is partly the reason why composites rarely are successful as direct replacements for metal parts—to be successful composite components must be designed with composites in mind from the onset of the design process.

Given that an appropriate manufacturing technique has been selected and a mold manufactured, success in composite manufacturing in the end always boils down to a matter of closely controlling temperature and pressure throughout the process and throughout the component. The basic requirements for any manufacturing technique are to use sufficient pressure to maintain the liquid reinforcement–matrix mass in the desired shape at a specified temperature for the time required to allow it to become dimensionally stable. If the reinforcement must be impregnated as part of the process, this difficult operation adds another requirement in that a pressure gradient to drive the matrix through the fiber bed is necessary. For thermoset processes, the time frame from (preimpregnated) molding compound to demoldable part is on the order of a few minutes, whereas when the raw material is prepreg or fibers and matrix separate, the corresponding time frame may be hours or even days. For thermoplastic processes employing prepregs and molding compounds, the time to demolding (assuming that the raw material already is molten) is from less than a minute to a few minutes, since the part only needs to be cooled to achieve dimensional stability.

While this chapter concentrates on commercially common and emerging manufacturing techniques, there is really no limitation to what means may be employed to manufacture a composite. Thus, any process in which the liquid reinforcement–matrix mass is shaped and subjected to sufficient heat and pressure is likely to result in a composite; whether the resulting composite properties are acceptable and whether the technique is economically feasible are entirely different issues. Thus, when attempting to understand the essential aspects of the techniques discussed in this chapter (as well as any other), considering answers to the following questions should prove helpful:

- What is used as the mold to give the composite its shape?
- How is the raw material made to conform to the mold?
- How is the impregnation pressure gradient applied (if applicable)?
- How is the consolidation pressure applied?
- How is the heat applied (and removed)?

Table 4-1 Manufacturing-related issues to consider in design

Requirements due to component performance specifications
- Reinforcement
  - type
  - continuous or discontinuous
  - random or oriented
  - size, coupling agent, and surface treatment
  - volume fraction
- Matrix
  - thermoset or thermoplastic
  - temperature tolerance
  - environmental tolerance
  - compatibility with reinforcement
  - additives and fillers
  - void fraction
- Surface finish
- Dimensional tolerance
- Holes, undercuts, bosses, and ribs
- Inserts and fasteners

Requirements on manufacturing technique imposed by raw material
- Reinforcement
  - continuous or discontinuous
  - random or oriented
  - volume fraction
- Matrix
  - viscosity
  - temperature, pressure, and time requirements
  - toxicity

Requirements on manufacturing technique imposed by geometry
- Overall size
- Thickness
- Hollow or solid
- Constant or varying cross-sectional shape
- Single or double curvature

Requirements on manufacturing technique imposed by economy
- Production rate and total series length
- Degree of automation
- Post-molding machining, trimming, and surface preparation
- Cost of
  - capital equipment
  - mold
  - raw material
  - labor
  - energy
- Mold life

Further, the many cross-references throughout this chapter ought to convince the reader that there is no reason why different manufacturing techniques cannot be combined to create a "new" technique. The absolute freedom in selecting, or even inventing, a manufacturing technique to suit any condition and end use reinforces the fact that with composites the only limitation is the imagination. Not surprisingly, this imagination also applies to the terminology used in manufacturing. There is therefore rarely a unique name for a technique and the ones used herein are merely believed to be the most common ones.

In an attempt to aid the reader in comparing the predominant characteristics of different manufacturing techniques and the resulting composites, the commercially most relevant techniques (including the emerging technique of diaphragm forming) are qualitatively compared in an admittedly crude fashion. Where possible, the characteristics have been categorized into two or three levels as shown below:

Technique Characteristics

Equipment cost:	Low, intermediate, high
Mold cost:	Low, intermediate, high
Labor cost:	Low, intermediate, high
Raw material cost:	Low, intermediate, high
Feasible series length:	Short, intermediate, long
Cycle time:	Short, intermediate, long
Crosslinking/consolidation requirements:	Simple, complex
Health concerns:	Low, intermediate, high

Component Characteristics

Geometry:	—
Size:	—
Holes and inserts:	Possible, difficult
Bosses and ribs:	Possible, difficult, not possible
Undercuts:	Possible, difficult, not possible
Surfaces:	—
Fiber arrangement:	—
Typical fiber volume fractions:	—
Void fractions:	Low, intermediate, high
Mechanical properties:	Poor, intermediate, good
Quality consistency:	Poor, intermediate, good

"—" indicates where two- or three-level categorization has been deemed inappropriate or impossible. The categorizations no doubt involve some major generalizations, but they attempt to portray the "mainstream" version of each technique and thus do not take uncommon or fancy process alterations into account. With two exceptions the techniques for thermoplastic composite manufacturing have been characterized with reference to their thermoset relations since the similarities are extensive and reiteration of the full characteristics deemed unnecessary.

4.1 Generic Manufacturing-related Issues

Several issues related to composite manufacturing apply to more than one manufacturing technique. A few of the more notable issues—mold fabrication, reinforcement conformability, prepreg conformability, and heat transfer—are therefore treated in the following to avoid unnecessary repetition.

4.1.1 Mold Fabrication

Possibly the most critical aspect in composite manufacturing is to have a well-functioning mold and all technique descriptions discussed in the remainder of this chapter assume that such a mold already exists. However, mold design and fabrication are not only critical to the success of a manufacturing operation, they also represent some of the most difficult issues in composite manufacturing and take years to master. While mold design is beyond the scope of this treatment, mold materials and fabrication techniques will be briefly discussed.

Molds are also known as tools, tooling, mandrels, and dies; the term used is largely determined by convention and by the manufacturing technique in which it is used. In this context, a mold is any object that somehow helps maintain the liquid reinforcement–matrix mass in an exact predetermined shape until a dimensionally stable composite is obtained. While molds to a limited degree can be used in more than one manufacturing technique, they are normally very technique-specific.

In most cases the material used for the mold is critical. Materials that have been successfully used for molds include metals, composites, expanded polymer foams, syntactic foams, neat polymers, elastomers, wood, plywood, chipboard, plaster, ceramics, concrete, and monolithic carbon. Arguably the most important issue to consider when selecting mold material is that of thermal expansion, since most composite manufacturing techniques involve significant changes in temperature. To avoid inaccuracies in composite component dimensions and inclusion of unnecessary residual stresses, the coefficient of thermal expansion (CTE) of the mold should be as closely matched to the CTE of the component as possible, cf. *Table 4-2*.

The way in which a mold is fabricated is greatly influenced by precision requirements of the application, the manufacturing technique in which it will be used, what raw materials will be used, series length, and mold material. Particularly the temperatures and pressures that the mold will

Table 4-2 Representative CTEs for different mold materials [1]

Material	α $10^{-6}\ °C^{-1}$
Aluminum	22.5
Steel	12
Electroplated nickel	13
Cast ceramic	0.7
Silicone rubber	1.8 – 3.8
Glass-reinforced epoxy prepreg	20
Glass-reinforced epoxy, wet layup	22 – 25
Carbon-reinforced epoxy prepreg	5 – 7
Carbon-reinforced epoxy, wet layup	9 – 11

experience during composite manufacturing largely determine what mold materials and fabrication concepts are feasible. There are three conceptually different ways in which a mold of a given shape can be fabricated: direct manual fabrication, use of a master model, and direct machining from a block of material.

4.1.1.1 Direct Manual Mold Fabrication

In direct manual fabrication, the mold shape is built up from contoured material sheet, as illustrated in *Figure 4-1*. In the simplest case the contours are cut out of cardboard, whereas sheet metal, wood, plywood, or chipboard are more common. The geometries of these contours would likely be extracted from a computer-aided design (CAD) drawing or a finite element (FE) model, but may in the less critical case be directly conceived by the designer. Once the contours are properly aligned and fixed to each other or to a common base (see *Figure 4-1*), the mold surface is attached, see *Figure 4-2*; in this case plywood planks are nailed to the contours. Nail heads and gaps between planks are then spackled and sanded to provide a smooth surface. In many cases the surface is then coated with lacquers, sealants, and release agents to further enhance surface finish and release properties. The surface smoothness is most important since an as-molded component will never have a better surface finish than the mold, while release agents ensure that the consolidated component can be separated from the mold without damaging either component or mold.

Direct manual fabrication is common in boat and ship building as well as in applications where exact shape is not absolutely critical. The predominant advantages of this mold fabrication approach are low cost and comparatively short lead times, meaning that it is economically feasible to use such molds to manufacture prototypes, one-of-a-kind components, and short

Figure 4-1 Contours lined-up to fabricate a sailboat mold. Contours have been sawed from chipboard and plywood. Photograph courtesy of Marten Marine Industries, Auckland, New Zealand

Figure 4-2 Finished sailboat mold. Mold surface has been constructed from plywood planks. Photograph courtesy of Marten Marine Industries, Auckland, New Zealand

series. Another advantage is that molds of virtually any size can be economically fabricated. Among the disadvantages are that it is difficult to achieve high geometric precision and that molds often end up having limited durability, i.e. they are likely not to be feasible for long series; indeed, they may very well disintegrate or require dismantling during demolding of the crosslinked component.

4.1.1.2 Mold Fabrication Using a Master Model

With this mold fabricating approach, the first step is to somehow manufacture an exact copy, called master pattern, master model, or just master, of the component that one wants to manufacture. There are numerous ways to manufacture a master model, but most of them employ a block of material that gradually is machined to the desired shape, see *Figures 4-3a* and *4-3b*. This block may be high-quality wood, expanded polymer foam, or it may be a near-net-shape block cast from for example polymer, syntactic foam, plaster, or ceramic; following casting, the near-net-shape block is then machined to exact dimensions. While this machining may be carried out by hand— and certainly is in prototype work—it is more common that a computer-controlled grinding machine executes the machining according to a CAD or FE model. After machining of the master model is finished, its surface is treated in much the same way as a mold fabricated through the direct manual procedure.

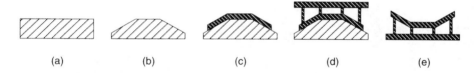

(a) (b) (c) (d) (e)

Figure 4-3 Fabrication of mold using master model (a) block of material for master; (b) finished master; (c) finished mold shell; (d) mold shell with backup structure; (e) finished mold

Once the master model is finished there are a number of ways in which a mold shell can be fabricated onto the master, see *Figure 4-3c*. One of the most common ways is to laminate a composite onto the master model. While it is possible to laminate such a mold shell using wet hand layup techniques (see *Section 4.2.1.1*) and materials, it is more common to layup epoxy-based prepregs. The reason for the preference of prepregs is that they make it easier to achieve high-quality results. The way in which to layup and consolidate prepregs is described in *Section 4.2.2*; the manufacturing technique is the same regardless of whether the component being made is a composite mold shell or a final component. The difference lies in the materials used; there are prepreg types specifically formulated for mold fabrication. Another way to build up the mold shell is to electroplate nickel onto the surface of the master model (electroforming); although very time-consuming, this technique yields high-quality tooling.

Once a mold shell has been fabricated, it must be strengthened with a backup structure to ensure dimensional stability and strength, see *Figure 4-3d* and *4-3e*. It is common that this backup structure is manufactured from the same type of material as used in the mold shell to eliminate distortions due to differences in CTE. With composite mold shells there are basically two ways to strengthen the shell, see *Figure 4-4*.

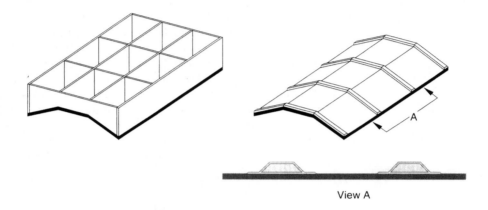

View A

Figure 4-4 "Egg-crate" (left) and integrally (right) stiffened molds. With the traditional egg-crate solution, the backup structure is generally built up from sheet material that is machined to size. In contrast, integral stiffeners are manufactured and crosslinked at the same time as the mold shell. Stiffness may be increased and manufacturing simplified using foam cores to build up sandwich stiffeners as illustrated at the bottom of the figure. Redrawn from reference [2]

Fabrication of molds using a master model has advantages in that a high degree of precision may be maintained and that several identical molds can be produced. Disadvantages include that fabrication is time-consuming and expensive due to the number of steps involved. The technique is common in aerospace applications.

Another possible way to fabricate a mold from a master model is to use castable materials, such as neat polymers, syntactic foams, ceramics, and concrete, which offer a rapid means of mold fabrication. For some applications sands, salts, and plaster may be used to cast disposable internal molds that following composite manufacturing are dissolved or chipped out to leave a hollow component. Castable materials may be reinforced and also permit integration of heaters and cooling channels.

4.1.1.3 Mold Fabrication through Direct Machining

Perhaps the most obvious way to fabricate a mold is to machine a solid block of material into the desired shape. Materials used with this approach include wood, expanded polymer foams, syntactic foams, metals, monolithic carbon, and many others. This mold fabrication technique is frequently used to produce prototype molds, in which case a block of wood or polymer foam is machined by hand to arrive at the final shape. For production molds the common approach would be to use a computer-controlled grinding facility to machine the shape out of a solid metal block, usually high-grade tool steel, according to a CAD or FE model. In the manufacturing techniques where solid steel molds are used there is usually a need for closed molds, meaning that (at least) two matching mold sections have to be fabricated. While matched metal molds are very expensive, they are durable and tolerate high temperatures and pressures and are therefore the norm in compression molding, injection molding, and pultrusion.

4.1.1.4 Flexible Molds

In some techniques employing open molds, vacuum is used to apply consolidation pressure onto the laid-up laminate as it solidifies. This is achieved by placing a vacuum bag over the laminate and then sealing the bag around the perimeter, see *Figure 4-5*. Air is then expelled from under the bag, which results in atmospheric pressure compacting the contents of the bag against the rigid mold. With this kind of consolidation procedure, which is discussed in more detail in *Sections 4.2.1* and *4.2.2*, one surface of the composite obtains the shape of the rigid mold, while the shape of the other surface is a function of the amount of material and the applied pressure. In order to improve control over the surface that is not in contact with the mold and also to facilitate compaction in difficult areas where bridging or tapering may otherwise occur, caul plates, pressure pads, and dams are used, see *Figure 4-5*. Caul plates and pressure pads, which may be rigid or flexible, permit better control of the pressure applied to the laminate. Flexible caul plates and pressure pads are normally elastomeric in origin, whereas the rigid ones tend to be of metal. Since the CTEs of elastomers are much larger than the CTEs of all common mold materials (cf. *Table 4-2*) and composite material systems (cf. *Table 3-13*), elastomeric pressure pads may also compact a laminate due to thermal expansion if processing takes place at elevated temperature and the pressure pad is restrained. Elastomer molds may be cast, injection molded, or extruded to size.

Another type of flexible mold is where an elastomer tube or bladder is used to compact a laminate against a one-sided mold. In this case consolidation pressure is achieved through pressurization of the tube or bladder with air and sometimes water. This type of mold is frequently used to consolidate hollow components against a rigid external mold.

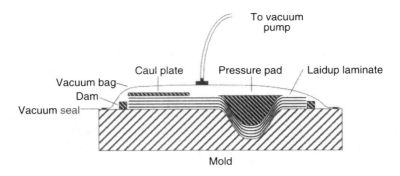

Figure 4-5 Schematic of vacuum bagging and use of caul plate, pressure pad, and dams

4.1.2 Reinforcement Conformability

The degree of conformability of the reinforcement to the intended component shape is of crucial importance. The practical limits of formability are set by reinforcement form and fiber strength, whereas fiber elongation may be neglected in terms of formability. Fiber fracture from tension and fiber buckling from compression should naturally be avoided. The main reinforcement deformation modes are transverse compression, out-of-plane bending, in-plane tension, and in-plane shear, see *Figure 4-6*.

All reinforcement forms may be transversely compressed until interfiber contacts eventually cause fiber fracture. Similarly, all reinforcement forms may be bent to some degree, although bending also requires simultaneous deformation through another mode, fiber fracture, fiber buckling, or mat rupture. A reinforcement allows in-plane tension only if there are no continuous and straight fibers in that direction. Finally, all reinforcement forms may be sheared, but shearing involves changes in the original angle of yarn intersection. In general the degree of conformability is increased by a reduced number of yarn crossover points that may resist movement; i.e. the looser the reinforcement, the greater the conformability, cf. *Section 2.2.7*.

Consequently, all reinforcement forms can be made to conform to a singly curved surface (as long as the radius of curvature is not too small), and all forms, with the exception of the unidirectional, can be draped over a doubly curved surface of moderate curvature. However, such conformation involves changes in the original angle of yarn intersection in weaves and braids and decreases in thickness or even rupture (or buckling) in mats. *Figure 4-7* illustrates how a fabric may be draped over a half-sphere through a combination of out-of-plane bending and in-plane shear; note the large changes in the original right angles of the fabric.

To manufacture more complex composite shapes from flat woven fabrics requires that the reinforcement is cut and several pieces used to compose the

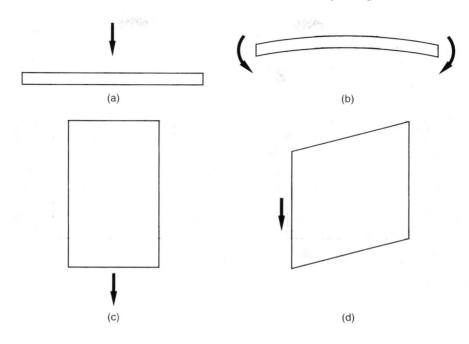

Figure 4-6 Fabric deformation modes; (a) Transverse compression (b) Out-of-plane bending (c) In-plane tension (d) In-plane shear. Redrawn from reference [3]

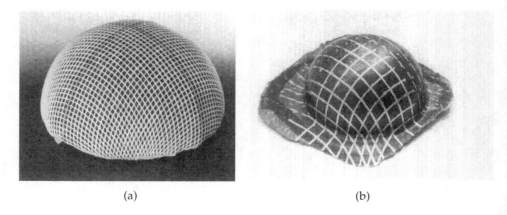

Figure 4-7 Draping of fabric over half-sphere; (a) Simulation with single layer of woven fabric (b) Consolidated multiply composite. The lines drawn onto the top prepreg ply illustrate the fiber orientations that prior to forming were orthogonal. Photographs reprinted from reference [4] with kind permission from the author

shape; *Figure 4-8a* illustrates how a corner may be draped requiring cutting of the fabric along one edge, while *Figure 4-8b* illustrates how a cut can be avoided assuming that the change in reinforcement orientation is acceptable. Alternatively, custom-made preforms, as discussed in *Section 2.2.7.6*, may be used for complex shapes.

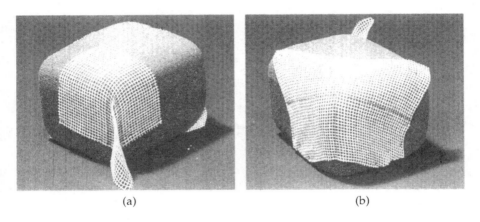

(a)   (b)

Figure 4-8 Draping of woven fabric over corner; (a) Principal axes of fabric aligned with sides of cube causing buckling (b) Principal axes at 45° with sides of cube preventing buckling. Photographs reprinted from reference [4] with kind permission from the author

4.1.3 Prepreg Conformability

In manufacturing of thermoset composites employing in-process impregnation it is generally the conformability of individual reinforcement layers (as discussed in *Section 4.1.2*) that is of the greatest importance. However, in most processes employing prepregs the situation is a little different. Since most prepregs contain continuous and oriented inextensible fibers (which may be in woven fabric form) saturated with lubricating liquid resin, the issue of conformability of a laid-up stack of flat prepregs may be addressed in an alternate fashion, as illustrated in *Figure 4-9*. The figure illustrates possible conformation modes for the prepreg stack and the minimum requirements in terms of flow mechanisms to achieve the respective mode of conformation.

Following the figure, the simplest conformation mode is consolidation using a flexible mold (or vacuum bag), since only resin percolation (transverse or parallel to the fibers) is required. Consolidation between rigid matching mold halves also requires transverse, or squeeze, flow. To shape a singly curved component requires interply shear (between plies) and shaping of a doubly curved component further requires intraply shear (within a ply). In shaping of doubly curved components there is a pronounced risk of wrinkling of the prepreg. The wrinkling tendency is significantly reduced if the prepreg stack to be formed has shape and size resulting in as little excess material around the edges as possible (cf. *Figure 4-7b*) and if it is kept under slight tension during molding. Even if wrinkling does not occur, there are limits to what shapes a prepreg stack realistically can be conformed into. With more complex shapes it is inevitable that thickness variations result, see *Figure 4-10*. Since such variations are often not acceptable, it becomes necessary to use a varying number of prepreg plies over the component even if a uniform thickness is desired.

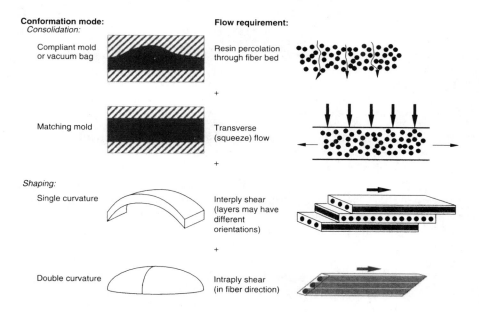

Figure 4-9 Hierarchy of deformation modes for stacked prepreg plies and corresponding required flow mechanisms. Redrawn from reference [5]

4.1.4 Heat Transfer

Heat transfer issues are of great importance in composite manufacturing since they govern crosslinking, melting, solidification, and crystallization, but naturally also in how they influence process efficiency. While far from irrelevant in thermoset composite manufacture, heat transfer issues are

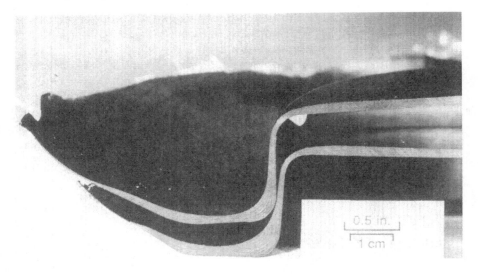

Figure 4-10 Hat section cut to reveal significant thickness variations. Photograph reprinted from reference [6] with kind permission from SAMPE, Covina, CA, USA

especially crucial in manufacturing of thermoplastic composites, for two basic reasons. First, higher processing temperatures are generally required; second, to realize the potential advantage of rapid manufacturing the processing temperature must be reached as quickly as possible and the component likewise must be cooled to enable demolding as quickly as possible. These issues place extraordinary requirements on heating and cooling facilities, not least due to the poor heat transfer capabilities of polymers. It should therefore come as no surprise that almost every conceivable heating method has been tried in manufacturing of thermoplastic composites.

The most commonly used heating methods employ radiative heat transfer and include various types of infra-red (IR) lamps and lasers. IR heaters are often categorized into long, medium, and short wavelengths, where the former tend to be ceramic heaters and the latter quartz lamps. The longer wavelengths essentially heat the surface only, whereas the shorter wavelengths penetrate some distance into the material. The type of heating desired very much depends on the application; in most techniques through-the-thickness heating clearly is desirable, whereas in others it may be preferable to only melt the surface of the material. Major advantages of IR heating include rapid heat transfer and that high heat capacity can be concentrated on a small area, especially if focused spot heaters are used. Disadvantages include that the emitted spectrum must be within the absorption spectrum of the material to be heated, that it is difficult to control the heating in dynamic situations, that it is relatively easy to accidentally overheat the material, that it is difficult to heat non-flat materials evenly, and that it generally results in a large temperature gradient through the thickness of the material. Even higher power may be very accurately applied to small areas using lasers, which in contrast to IR heaters may also be very accurately controlled in dynamic situations. The problem of wavelength compatibility naturally also applies to lasers. Whereas the capital costs associated with IR heating are low, lasers are quite expensive.

Another common heating principle employs heated gas, often air, to heat the material through convection. The advantages of gas heating include relatively low cost, uniform heating, little risk of accidental overheating, and—if an inert gas (such as nitrogen) is used—no risk of polymer oxidation or ignition. Disadvantages include low rate of heat transfer and possible problems with fiber and matrix movement if gas velocity is high. For materials that have large surface area-to-volume ratios, gas heating is ideal since the gas may penetrate the gaps and thus provide fairly even through-the-thickness heating. Raw materials with large surface area-to-volume ratios include GMT manufactured through the paper-making process (cf. *Section 2.3.6*), powder-impregnated and commingled prepregs (cf. *Sections 2.3.3 and 2.3.4*), but also stacks of unconsolidated melt-impregnated prepregs.

Open flame heating represents an interesting combination of radiative and convective heat transfer. Advantages of flame heating include low cost, that high heat capacity may be concentrated on a small area, and that a relatively oxygen-poor environment is created. The main drawbacks are the risk of ignition and difficulty in controlling the rate of heat transfer. Although open flame heating has proved feasible in some applications, it remains rather exotic in composite manufacturing.

Conductive heat transfer using heated bodies, usually metal, in direct contact with the material offers an effective means of closely controlling the heat transfer to the material. However, since polymers tend to adhere to most other materials there are obvious problems associated with this heating concept, which therefore is rarely used except in the form of heated (and cooled) molds where release agents lessen this concern.

Among the more unusual, albeit technically feasible, heating methods are microwave, induction, radio-frequency, ultrasonic, and resistance heating. In the latter case large currents are run through the continuous conductive reinforcing fibers (carbon) which consequently are heated. All these heating techniques allow through-the-thickness heating, but the first four techniques, and to some degree also the fifth, bring on substantial capital costs.

The options for cooling of the molded component are unfortunately not as plentiful as the available heating methods and are essentially limited to convective and conductive modes of heat transfer. Probably the most common cooling situation is conduction to a cooled mold followed by free or forced convective cooling to the surrounding air and sometimes water. While some heating methods allow a certain degree of through-the-thickness heating there is no equivalent option to cool the consolidated monolithic component. The cooling of the interior of the component is thus entirely dependent upon conduction. Indicative surface heat transfer coefficients are given in *Table 4-3*.

Table 4-3 Representative surface heat transfer coefficients for different contacting media [5]

Contacting surface	Surface heat transfer coefficient $W/m^2 {}^\circ C$
Water	≈ 2000
Metal	≈ 400
Air at 10 m/s	≈ 50
Still air	≈ 10

While many of the aforementioned heating techniques are primarily relevant to thermoplastic composite manufacturing, they are also used with thermosets. The most common heating method is conduction from a heated mold, with convective heating running a distant second. Some manufacturing techniques further employ induction and radio-frequency heating to heat through the thickness to ensure more even crosslinking. With specialty resins (of the same family that dentists use to fill in cavities in teeth) ultraviolet (UV) radiation offers a means of initiating crosslinking. While uncommon, such resins nevertheless have been tried in composite applications. The cooling issues with thermoset composites are no different than with thermoplastics.

4.2 Thermoset–Matrix Techniques

In terms of commercial usage, thermosets clearly dominate in structural composite applications. In fact, the first incarnations of fiber-reinforced polymers were based on phenolics and unsaturated polyesters (glass-reinforced unsaturated polyester was made in Germany in the 1940s [7]), whereas thermoplastics entered the scene much later. It should therefore come as no surprise that thermoset manufacturing techniques are more plentiful and with few exceptions more mature than their thermoplastic counterparts.

From a manufacturing viewpoint the main advantages of thermosets over thermoplastics are low viscosity, which makes impregnation comparatively easy, and often modest temperature and pressure requirements during crosslinking meaning that relatively simple molds may be used in many techniques. The main disadvantage is the active chemistry which translates into long processing times and a potentially unhealthy work environment; most discussions of health and safety issues in composite manufacturing are deferred until *Chapter 8*.

4.2.1 Wet Layup

The wet layup manufacturing techniques refer to processes where the resin is applied in liquid form and the reinforcement is impregnated as part of the layup. The layup may either be performed by hand, herein referred to as hand layup (but also known as manual layup, contact molding, and open molding), or in a partially automated procedure, referred to as spray-up.

4.2.1.1 Hand Layup

Wet hand layup no doubt is the simplest and most versatile of all composite manufacturing techniques. Despite being one of the first techniques used, wet hand layup is still widely used since it is extremely versatile, requires little capital investment and prior knowledge, and is highly economical for prototypes and short production series. Whereas little capital investment and prior knowledge is required to get started with hand layup, significant experience and dedicated facilities are required to manufacture high-quality components.

Process Description

Hand layup employs a one-sided mold, male or female, which is treated with a mold-release agent, e.g. layers of polished wax, to facilitate component removal following crosslinking. Normally a neat resin layer, called gelcoat, is sprayed or sometimes painted directly onto the mold before lamination commences, see *Figure 4-11*. The gelcoat resin is selected so as to

Figure 4-11 Spraying of pigmented gelcoat onto release-treated mold. Photograph courtesy of Ryds Båtindustri, Ryd, Sweden

have good environmental resistance. The gelcoat, which usually has a thickness of about 0.5 mm, produces a smooth, cosmetically appealing, and often pigmented surface that hides the reinforcement structure which otherwise may be visible on the composite surface. In, for example, bathroom interiors thermoformed polymer films are often used instead of gelcoat.

To start building up the laminate, an appropriate amount of resin is applied and distributed on top of the gelcoat whereupon the dry reinforcement, typically in mat and fabric form, is placed on top, see *Figure 4-12*. The resin is worked upwards through the reinforcement with a hand-held roller, which also compacts the laminate and works out voids, see *Figure 4-13*.

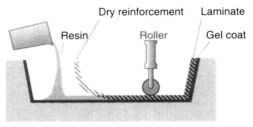

Figure 4-12 Schematic of the wet hand layup

Figure 4-13 Rolling of laid-up laminate. The component is a mold shell for an outrigger of a rowing skiff

After one reinforcement layer has been satisfactorily impregnated and compacted, more resin is applied, another reinforcement layer is placed on top, and the impregnation and compaction step is repeated. This procedure is repeated until the desired number of reinforcement layers have been applied or, less commonly, the desired laminate thickness has been reached. The lamination is sometimes finished with a topcoat that is similar to the gelcoat in function and composition, but since there is no mold on this side of the laminate its surface will not be as good as that of the gelcoat.

If reinforcement with a coarse structure is used close to either surface its structure may be visible on the surface. To improve the surface appearance it is therefore common to use fine fabrics, mats, or veils close to the surface even if gel- and topcoats are used. *Figure 4-14* illustrates a common layup sequence. In some applications it is nevertheless necessary to apply some post-molding coating or surface treatment to get the required finish. Since the resins used in wet hand layup usually crosslink without externally applied pressure (see further the *Crosslinking* section below), it is of utmost importance that the rolling is thoroughly performed since it determines fiber fraction, void fraction, and part thickness.

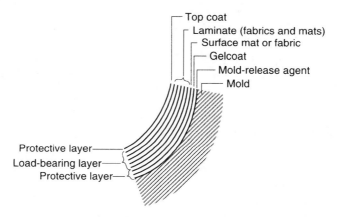

Figure 4-14 Illustration of layup sequence common in wet hand layup

While the previous paragraphs describe the most basic incarnation of wet hand layup there are several ways in which the process may be made more efficient. While pouring of the resin onto the mold indeed is quite common, rollers (akin to those used in painting) saturated with resin may also be used. Other techniques include rollers that distribute resin from within the roller at the push of a trigger (see *Figure 4-15*) and spraying of resin onto the mold (same procedure as shown in *Figure 4-11*). The most refined way of applying the resin is to first impregnate the reinforcement in one processing step and then immediately place the prepreg onto the mold in another. The reinforcement impregnation may either be performed manually on a table or using dedicated impregnation machinery, see *Figure 4-16*. Since impregnated mats easily rip when handled, this procedure is likely to be limited to fabrics. The advantages of such in-house prepregging are more precisely controlled fiber volume fraction and simpler and more rapid laminating on vertical or complex-shaped surfaces.

Figure 4-15 Semi-automated resin application using dispensing roller. Photograph courtesy of Rolf Andersson, Bildbolaget, Västervik, Sweden

Raw Materials and Molds

While wet hand layup is almost entirely synonymous with E-glass-fiber reinforced unsaturated polyester, some applications employ vinylesters and epoxies. In the former case an orthophthalic polyester is normally used as laminating resin and an isophthalic polyester as gel- and topcoat. It is desirable for the resin to have a long pot life to ensure sufficient time to finish laminating a structure without appreciable changes in viscosity, although this naturally has the unwanted side effect of increasing the time required for completion of crosslinking.

By far the most common reinforcement form is CSMs, but also fabrics, often in combination with CSMs, are also used. There is no significant added processing difficulty in using other reinforcements than glass, but it is not common. Since wet hand layup involves short series the reinforcement is normally cut by hand using a knife and steel ruler or scissors.

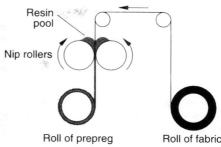

Resin pool

Nip rollers

Roll of prepreg   Roll of fabric

Figure 4-16 Photograph and schematic of in-house prepregging facility. The roll of carbon fabric is placed at the back of the prepregger and is then led into a nip between two counter-rotating steel rollers. The epoxy resin is manually poured into the nip region and the fixed distance between the rollers determine the reinforcement–matrix ratio in the prepreg, which is wound onto a cardboard core at the front of the prepregger

One of the significant advantages of hand layup is that very simple molds may suffice since only low temperatures and low pressures are encountered. It is thus possible to use almost any mold material for prototype and one-of-a-kind manufacturing, including wood, whereas longer production series are likely to require more durable materials, such as composite molds stiffened by wood or aluminum frames; composite molds may also be used for short production series.

Crosslinking

Once lamination has been completed, the laminated structure is normally left to crosslink at room temperature. Depending on resin formulation, part thickness, ambient temperature, and mold heat transfer properties, crosslinking will have progressed far enough for the component to be ready for demolding in a few hours. The crosslinking time may be shortened by increasing the temperature during crosslinking since this stimulates the reaction rate. It may be feasible to place small components in an oven, while larger structures can be heated by blowing warm air under a tarpaulin enclosure containing the structure.

Too high processing temperatures may be caused by overheating (if heating is used) and by the resin exotherm, especially if the laminate is thick (recall the poor heat transfer capabilities of polymers). In extreme cases too high temperatures may even lead to polymer degradation. However, already moderately high processing temperatures may result in residual stresses due to the matrix shrinkage from crosslinking and the fact that the CTE of the matrix is significantly higher than that of the reinforcement. A component thus should not be allowed to crosslink at excessively high temperatures, which unfortunately contradicts the cost-driven desire for rapid crosslinking. To ensure a reasonable crosslinking temperature one may formulate the resin for slow and thus lower-temperature crosslinking (cf. *Figure 2-24*) or one can allow intermediate crosslinking of a fraction of the laminate thickness before lamination resumes. In the latter case the time between laminations should not be too long, usually less than a day or two (although weeks may be acceptable in some cases), lest the interface strength between the laminate sections suffer significantly. The procedure of intermediate crosslinking is not only used to build up thick laminates, but also for very large structures that cannot reasonably be laminated within the pot life of the resin or within a workday.

Technique Characteristics

Equipment cost:	Low
Mold cost:	Low
Labor cost:	High
Raw material cost:	Low
Feasible series length:	Short to intermediate
Cycle time:	Long
Crosslinking requirements:	Simple
Health concerns:	High

Component Characteristics

Geometry:	Any
Size:	Any
Holes and inserts:	Possible
Bosses and ribs:	Possible
Undercuts:	Possible
Surfaces:	One good surface
Fiber arrangement:	Planar arrangement; random or oriented, continuous or discontinuous
Typical fiber volume fractions:	0.2–0.3 with mats, up to 0.5 with fabrics
Void fractions:	Intermediate
Mechanical properties:	Poor to intermediate
Quality consistency:	Poor

Applications

Due to low capital costs and the possibility of making structurally very capable components of significant geometric complexity, wet hand layup is widely used for all kinds of prototypes as well as for products manufactured in short series. However, although it is possible to use reinforcement fabrics, a great many applications only use CSMs and the potential for manufacturing of structurally high-performing components is thus often not exploited. Applications include:

- Vehicle components, such as utility vehicle hardtops
- Leisure boats
- Construction members
- Windmill blades
- Pipes, storage tanks, and vats
- Electrical equipment
- Bathroom interiors and pools

4.2.1.2 Spray-up

The main disadvantage of wet hand layup is the labor intensity which may be reduced by simultaneously spraying chopped reinforcement and matrix onto the mold. Examples of sprayed-up composites are small boats, truck body panels, and bathroom interiors.

Process Description

In spray-up a dedicated spray gun is used to deposit a mixture of chopped reinforcement and matrix onto the mold, see *Figure 4-17*. One or more continuous rovings are fed into the gun, where they are chopped to a predetermined length. Resin and initiator are fed into the gun through separate systems. Mixing may take place in a static or a dynamic mixer within the gun or the components may be separately sprayed to complete mixing right in front of the gun on the way to the mold. In-gun mixing provides more thorough mixing and is in some countries legally required to minimize operator exposure to the initiator. However, in-gun mixing requires cleaning of the gun and the type of gun that sprays resin and initiator separately is therefore often preferred. The latter gun type may be designed in such a way that the resin spray envelopes the initiator spray to lessen the problem of operator exposure to the initiator.

After the gelcoat has been applied onto the release-treated mold or the thermoformed polymer surface film has been placed on the mold, the spray gun is used to deposit a predetermined ratio of fibers and matrix, see *Figure 4-18*. The spray gun is normally hand-held, but for longer production series it may prove economically feasible to mount it on a robot to reduce labor costs. If spraying is performed manually, the skill of the operator has large impact on laminate properties since the thickness is a direct function of the spraying pattern. Just as in hand layup, laminate compaction is achieved with hand-held rollers. The spray gun normally has a resin-only setting for deposition of sufficient amounts of matrix to allow inclusion of unimpregnated fabrics should structural requirements so dictate.

Figure 4-17 Spray gun with roving chopper attached. Photograph courtesy of Aplicator System, Mölnlycke, Sweden

Figure 4-18 Spray-up of chopped fibers and resin. Photograph courtesy of Ryds Båtindustri, Ryd, Sweden

Raw Materials and Molds

Raw material types and molds used in spray-up are similar or identical to those used in wet hand layup, although the dominance of E-glass-fiber reinforced unsaturated polyester is even greater in spray-up. Due to the nature of the process, only rovings are used and the chopper on the spray gun chops them to lengths in the range 10–40 mm. For improved structural properties fabrics may be placed directly onto the mold following spraying of a layer of pure resin. Since final resin mixing occurs continuously in or in front of the spray gun, fast-reacting resins may be used (pot life only needs to allow for spraying and rolling). The resin often contains significant amounts of fillers. The use of thermoformed thermoplastic films instead of gelcoats is more common in spray-up and ensures good surface finish in, for example, sanitary applications.

Crosslinking

Crosslinking requirements and conditions are virtually identical to those in wet hand layup, with the difference that ambient-temperature crosslinking is more common. Since resins may be formulated for more rapid crosslinking than in hand layup, demolding (see *Figure 4-19*) can be performed after a couple of hours.

Figure 4-19 Demolding of sprayed-up component; the deck structure of the boat in *Figure 1-17*. Photograph courtesy of Ryds Båtindustri, Ryd, Sweden

Technique Characteristics

Equipment cost:	Low
Mold cost:	Low
Labor cost:	High
Raw material cost:	Low
Feasible series length:	Intermediate
Cycle time:	Long
Crosslinking requirements:	Simple
Health concerns:	High

Component Characteristics

Geometry:	Any
Size:	Any
Holes and inserts:	Possible
Bosses and ribs:	Possible
Undercuts:	Possible
Surfaces:	One good surface
Fiber arrangement:	Planar arrangement; random and discontinuous
Typical fiber volume fractions:	0.2–0.3
Void fractions:	Intermediate
Mechanical properties:	Poor
Quality consistency:	Intermediate

Applications

Spray-up may also be used to manufacture geometrically complex components and application are similar to those for hand layup. However, with spray-up the resin–reinforcement ratio is well controlled by the spray gun, which leads to more uniform component properties than obtained in hand layup using CSMs only. Since spray-up is less labor intensive than hand layup, longer series are feasible. Applications include:

- Vehicle components, such as utility vehicle hardtops
- Leisure boats (see *Figure 1-17*)
- Construction members
- Pipes, storage tanks, and vats
- Electrical equipment
- Bathroom interiors and pools (see *Figures 1-37* and *1-38*)

4.2.1.3 Hand Layup of Sandwich Components

Particularly in composite boat and ship building it is quite common to use wet hand layup to produce sandwich structures, since stiffness is of essence (cf. *Figure 2-55*) and series are short. Prime examples of its use include one-of-a-kind yachts and passenger ships as well as boats manufactured in short series.

Process Description

Both hand layup and spray-up may be used to manufacture sandwich components, but since the sandwich concept is mainly used to achieve high stiffness, hand layup of fabric-reinforced faces is likely to be the favored technique. There are countless ways in which to hand layup a sandwich structure, three of which are described below.

In the first one, a mold such as that shown in *Figure 4-2* is used and a laminate is built up and crosslinked in the manner described in *Section 4.2.1.1*. Following crosslinking, core blocks and planks are fitted and adhesively bonded to the first laminate. For plane and low-curvature surfaces large and flat core blocks can be used, while sharply curved surfaces may give rise to difficulties, see *Figure 4-20*. To enable core conformation to curved surfaces, small core blocks, or cores with grooves machined in one or several directions to allow for bending, may be used. A product specifically aimed for draping over highly curved surfaces is manufactured by bonding a lightweight fabric onto one side of the core, which subsequently is sawed from the opposite side all the way through the core (but not through the fabric) leaving a drapeable structure with wedge-shaped gaps between small blocks. It is also noteworthy that many cores may be thermoformed into complex shapes (cf. *Section 2.4.3*), although this possibility is rarely used in hand layup.

Figure 4-20 Conformation of core planks to curved surfaces may prove difficult. Photograph courtesy of Marten Marine Industries, Auckland, New Zealand

Gaps between core sections are then filled with a putty that is typically designed to closely mimic core properties. When the putty is fully crosslinked the structure is carefully sanded to provide a smooth surface for subsequent lamination. Prior to lamination of the second face the core structure is often primed to improve adhesion to the laminate, normally using pure resin of the same type as that used in the faces or the aforementioned putty; the primer is often allowed to gel or fully crosslink before lamination commences using the core as mold. Once the second laminate has crosslinked, a lot of manual sanding is usually required to achieve a smooth surface that can be topcoated or painted. Once the hull is finished it must be turned, which is certainly not a trivial exercise if the hull is large. Several cranes are usually required to rotate the hull and very carefully place it in custom-made supports; this exercise may have to take place outdoors to give room for the cranes. Once turned, the mold is removed, which possibly requires mold disassembly first, and work on the inside of the hull may commence, see *Figure 4-21*. Bulkheads, deck, and superstructure are normally prefabricated (often in the same fashion and with the same materials as used for the hull) and are laminated in place using reinforcement strips, see *Figure 4-22*. This basic technique to manufacture sandwich hulls is for example used for yachts of the type shown in *Figure 1-19*.

Figure 4-21 Lamination on the inside of the hull of a sailing yacht

Figure 4-22 Lamination of prefabricated bulkheads onto hull of sailing yacht. Photograph courtesy of Marten Marine Industries, Auckland, New Zealand

Smaller boats are often manufactured in a slightly different fashion. For example the boat in *Figure 1-18* is laminated onto a female mold (the inverse of that shown in *Figure 4-2*). Gelcoat is first sprayed onto the release-treated mold and allowed to crosslink. The laminate is then hand laid up starting with a fine CSM followed by coarser CSMs and some fabrics. When the laminate has crosslinked a filled putty is applied onto the laminate, and precut foam core with the aforementioned grooves to allow for conformation is placed onto the laminate. A vacuum bag is then placed over the core and the vacuum pulls the putty into the core grooves before it crosslinks. The second sandwich face is then laminated directly onto the core and is crosslinked, stiffeners are laminated in place, and finally a topcoat is applied.

Larger boats and ships, particularly if they are one-of-a-kind, are sometimes built according to a third technique. Also in this case contours are set up to create a mold as shown in *Figure 4-1*, but instead of using plywood planks to cover the contours as shown in *Figure 4-2*, foam or balsa core planks are used to create the mold surface. The core is then spackled, sanded, and primed in the previously described fashion before lamination commences. When the first laminate has crosslinked, the semifinished hull is turned and the contours removed to expose the core planks. The inside surface of the core is then also spackled, sanded, and primed and the second laminate laid up and crosslinked before bulkheads, deck, and superstructure are laminated in place. The SES shown in *Figure 1-20* was built using this technique. Also with this technique the outside of the hull ends up unfinished since it is not formed by a mold and extensive manual work is required to reach a high-quality finish.

Raw Materials

Face materials used in hand layup of sandwich components are the same as for conventional hand layup, although glass-reinforced unsaturated polyester has more competition from aramid- and carbon-reinforced vinylesters and epoxies. For extreme boats, such as competition yachts and powerboats, it is even feasible to use (commercially available) prepregs to complement wet layup or to completely replace it, see further *Section 4.2.2*. Since structures are often large, it is common to use in-house prepregging and resin-dispensing rollers to rationalize the lamination work, but also to improve laminate properties and consistency.

Commonly used core materials are expanded polymer foams and balsa cores and in some rare cases honeycomb cores; PVC cores clearly dominate. Sandwich components allow loading points to be integrated into the core prior to lamination; these may be in the form of high-strength (i.e. high-density) core, marine plywood, metal, etc.

Crosslinking

Crosslinking requirements and conditions are similar to those in conventional wet hand layup, although for large and thick ship hulls crosslinking times may run into days. Most sandwich components and structures are crosslinked at room temperature and without externally applied pressure, but it is becoming increasingly common to use elevated temperature and vacuum bagging to improve properties in for example competition yachts and powerboats. In the latter case, complete hulls may be simultaneously vacuum bagged and part of the factory temporarily turned into a crude oven using temporary walls and thermal insulation.

Technique Characteristics

Equipment cost:	Low
Mold cost:	Low
Labor cost:	High
Raw material cost:	Low
Feasible series length:	Short
Cycle time:	Long
Crosslinking requirements:	Simple
Health concerns:	High

Component Characteristics

Geometry:	Any
Size:	Any
Holes and inserts:	Possible
Bosses and ribs:	Possible
Undercuts:	Possible
Surfaces:	One or no good surface (depending on technique)
Fiber arrangement:	Planar arrangement; random or oriented, continuous or discontinuous
Typical fiber volume fractions:	0.2–0.3 with mats, up to 0.5 with fabrics
Void fractions:	Intermediate
Mechanical properties:	Intermediate
Quality consistency:	Poor

Applications

Wet layup is eminently suitable for structurally very capable components manufactured in short series or one-of-a-kind products, mainly due to low capital and mold costs. Applications include:

- Vehicle components and even entire buses (see *Figure 1-14*)
- Refrigerated truck and railroad containers
- Leisure boats (see *Figure 1-18*)
- High-performance racing vehicles and yachts (see *Figure 1-19*)
- Mine counter-measure vessels and high-speed passenger ships (see *Figures 1-20* and *1-21*)
- Helicopter rotor blades
- Building panels

4.2.2 Prepreg Layup

Layup of prepregs is in some respects a refined form of wet hand layup. Use of prepregs instead of impregnating the reinforcement as part of the layup yields higher-performance parts, avoids resin formulation, and dramatically reduces health concerns. Virtually all high-performance composites are manufactured from prepregs due to the superiority in properties thus achievable. Just as with wet hand layup there are varying degrees of automation that may come into question in prepreg cutting and layup.

Process Description

First of all the prepreg must be taken out of the freezer where it has been stored and be allowed to slowly reach room temperature; since thawing should take place in the original packaging to avoid condensation, this may take a day or two. Prior to hand layup the prepreg must then be cut. For prototypes and short series the prepreg may very well be cut by hand using knife and steel ruler, but for longer series automated procedures are used to cut patterns out of wide rolls of prepreg, or broadgoods, see *Figure 4-23*. Machinery intended for cutting of prepregs usually employs reciprocating knives, although ultrasonically vibrating knives, giant "cookie cutters", water jets, and lasers are used as well. Optimization software is used to determine how to cut the plies from the prepreg roll with a minimum of waste. The cutting-pattern, or nesting, software provides the computerized input for the cutting machine motion. Automated prepreg cutting machines are usually capable of cutting several plies of stacked prepreg simultaneously to further improve efficiency.

Figure 4-23 Computer-controlled machine for automatic cutting of broadgoods. Photograph courtesy of Saab Military Aircraft, Linköping, Sweden

Prepregs tend to be used in short production series, so layup is often performed by hand in a process not unlike wet hand layup. The backing paper or film is first removed from one side of the prepreg plies, which are then placed one at a time onto a one-sided mold treated with release agent, carefully tacking the sticky plies together to ensure that no air pockets or contaminants (such as prepreg backing paper) are entrapped. This procedure is repeated until the desired number of plies have been laid up.

Since prepregs are normally very thin a large number are required to build up a laminate and it is critical that they are all placed in exactly the intended location and orientation. A number of aids are used to ensure that layup is correctly performed, such as automatic labeling of the individual plies during cutting, and aligning of holes cut in the prepreg onto pins protruding out of the mold outside the effective mold area (following crosslinking the "ears" laidup around the pins are machined off). More recently, computerized systems using overhead lasers that track the outline of each ply on the mold have become common in the aerospace industry. In high-performance applications layup is performed in rooms where temperature, humidity, and air contaminants are closely controlled.

While hand layup of prepregs is quite common, computer-controlled machinery is sometimes used for longer series where the high equipment and programming costs can be justified. Automated prepreg layup heads are quite complicated pieces of machinery that normally are mounted on an overhead gantry (see *Figure 4-24a*) while the mold is located on a table beneath the gantry, see *Figure 4-24b*. The automated layup head unwinds the

(a) (b)

Figure 4-24 Tape-laying machine with gantry-mounted prepreg layup head (a) facility overview; (b) close-up of layup head. Photographs courtesy of Boeing Commercial Airplane Group, Seattle, WA, USA

prepreg tape from the roll (see *Figure 4-25*), peels off backing paper or film, places the prepreg on the mold using a pressure shoe or roller, cuts off the prepreg tape at the edge of the laminate, and then starts the process all over again. Automated layup machines have limitations in what degree of mold complexity they are able to accommodate, but for example aircraft wing skins, which have moderate curvatures, are well suited for automated layup.

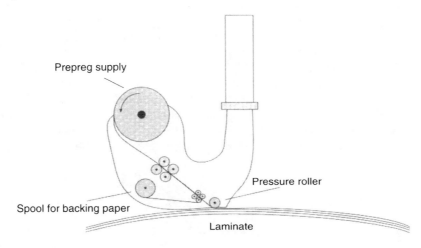

Figure 4-25 Schematic of prepreg layup head

Raw Materials and Molds

In analogy with the mold requirements in wet hand layup, molds for prepreg layup may also be quite simple from a structural point of view, since no substantial loads need to be supported in either case. However, dimensional tolerances of high-performance parts normally manufactured from prepregs are very strict, so molds usually end up being very expensive anyway. The normal aerospace mold consists of a metal or composite mold shell fabricated as outlined in *Section 4.1.1.2*. Caul plates, pressure pads, and dams are commonly used to avoid excessive resin bleeding and thus tapering at edges, holes, and flanges, as well as to apply higher pressure locally to ensure prepreg conformation to the mold in difficult locations, cf. *Figure 4-5*. Although solid molds, such as that shown in *Figure 4-5*, are sometimes used, stiffened mold shells (cf. *Figure 4-4*) allow uniform heating and cooling due to the additional possibility of convective heat transfer from the rear as well as due to the mold's low and uniform thermal inertia.

Laid-up prepreg stacks must be consolidated using a vacuum bag to ensure proper consolidation; a typical vacuum-bag assembly is illustrated in *Figure 4-26*. On top of the laid-up prepreg stack a separator, or release film, is

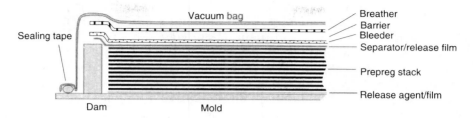

Figure 4-26 Typical vacuum-bagging assembly. Redrawn from reference [8]

placed; the film is either perforated or porous enough to allow entrapped air, volatiles, and resin to escape. The release film is followed by the bleeder, which is a porous fabric or felt whose task is to absorb the excess resin squeezed out of the prepreg stack. The bleeder is followed by the barrier which contains the escaping resin beneath it; it may be of the same material as the separator with the difference that it is not perforated (or porous). The last layer under the vacuum bag is the breather which is intended to ensure even pressure over the part while allowing air and volatiles to escape; the breather is a porous fabric or felt that may be identical to the bleeder. The entire layup is covered by a vacuum bag that is carefully sealed at the edges using sealing tape that sticks to both mold and bag. If there is problem sealing the vacuum bag at the edges or if the mold is porous, it is possible to enclose both the prepreg stack and the entire mold in a vacuum bag (on all sides). Expendable PA film is generally used for the vacuum bag, although reusable elastomer bags are also available. If the consolidated laminate is to be adhesively bonded to at some later processing step, optional peel plies, which are release-treated fabrics, may be placed onto the outermost prepreg ply. Immediately prior to bonding the peel ply is removed to leave a clean and slightly rugged surface suitable for bonding.

Prepreg layup normally employs carbon- and sometimes glass- and aramid-reinforced epoxies that have been B-staged. Less commonly used prepregs are based on unsaturated polyester, phenolic, BMI, and PI resins. Although UD reinforcement clearly dominates, fabric-reinforced prepregs are used to some degree as well. Prepregs come in the form of broadgoods, which may be over a meter wide, and as tape (in widths up to 75 mm primarily for use in tape laying).

Most commercially available UD-reinforced prepregs have fiber volume fractions around 0.6 and the consolidated laminate thus will have the same or higher fiber fraction. Some prepreg types are intended to be bled, i.e. have some of the resin squeezed out in the early stages of the cure cycle; under normal processing conditions some resin is thus absorbed by the bleeder to increase the fiber content of the laminate before the resin gels. Other prepreg types are intended to have a minimal amount of resin squeezed out during processing and it may thus be possible to omit bleeder and barrier.

Prepreg layup is traditionally used in the aerospace industry, but has gradually become very common also for sporting goods. In tube rolling layers of fabric- or UD-reinforced prepregs are wound onto a cylindrical or near-cylindrical mandrel in hoop or helix patterns (see *Figure 4-52* for definitions) in much the same fashion as one would wrap a strip of paper around a pen. Winding may be manual, but dedicated machinery offering various degrees of automation is available. Constant-diameter and tapered tubes of finite lengths are manufacturable; common applications include fishing rods and golf club shafts. In manufacturing of sporting goods, consolidation is usually accomplished without an autoclave; techniques include hoop-wound shrink tape (which contracts at elevated temperature) around cylindrical products, internal and/or external pressure bags, and matched-die molding. Racquets are for example consolidated using an internal pressure tube which compacts the laminate against a rigid external mold.

Through so-called cocuring, several components that traditionally would be separately consolidated and then adhesively bonded can be consolidated into one structure. For example, stringers and stiffeners may be cocured with the wing skin to eliminate a costly secondary processing step. Cocuring naturally requires a more sophisticated and possibly multi-section mold, probably also involving caul plates and very time-consuming layup, but may very well prove cost-effective in the long run. Another version of cocuring is the simultaneous layup of two face laminates, two adhesive films and a core to produce a sandwich structure in a one-step process, see *Figure 4-27*. Since both adhesive and face laminates are often epoxy-based they crosslink more or less simultaneously. However, in some applications where both surfaces of the sandwich must be very good, it may be preferable to pre-consolidate laminates and then bond them to the core in a separate

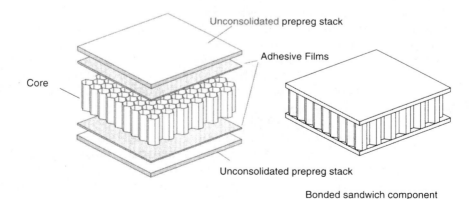

Figure 4-27 Layup for cocuring of sandwich component. Redrawn from reference [9]

manufacturing step to avoid complex and perhaps matching molds or caul plates. In applications such as these, aluminum and honeycomb cores dominate even though high-performance expanded foams are sometimes used.

Crosslinking

Most resins available in prepreg form require increased temperature during crosslinking to provide optimum properties. Heating of the mold or simple heating of the atmosphere outside the vacuum bag similar to practices used in wet layup may be sufficient for low-temperature resins and less critical applications. However, it is significantly more common to use an autoclave to achieve the temperatures and pressures required to properly crosslink and consolidate the composite. An autoclave is a pressure vessel allowing precise control of the temperature and pressure of the internal atmosphere, see *Figure 4-28*. The internal pressure is controlled through injection of pressurized gas into the autoclave. Apart from controlling the internal pressure of the autoclave, i.e. outside vacuum bag and mold, it can also draw vacuum under the vacuum bag. The vacuum inside the bag and the pressure outside the bag thus cooperate in ensuring that the prepreg stack conforms

Figure 4-28 Autoclave with vacuum-bagged component. Photograph courtesy of Kongsberg Aerospace, Kongsberg, Norway

to the mold and remains compacted during crosslinking. The temperature of the consolidating laminate is primarily controlled through heating of the pressurized gas in the autoclave although it may also be possible to heat and cool the mold. The internal autoclave atmosphere may be air for low-temperature crosslinking, but an inert gas such as nitrogen is common at higher temperatures to eliminate the risk of fire.

A typical cure cycle for a high-temperature epoxy is shown in *Figure 4-29*. First vacuum is drawn to compact the prepreg stack and then the temperature is gradually increased to a level that allows the resin to flow (cf. *Figure 2-29*) to eliminate voids and percolate into the bleeder. Then pressure is applied and the temperature increased further to another plateau that is maintained for a couple of hours to allow the resin to fully crosslink. Finally, the temperature is lowered, pressure released, and the part demolded. It deserves to be pointed out that in contrast to the processing conditions shown in *Figure 4-29*, crosslinking specifications for many other resins, including many high-temperature epoxies, often only stipulate one temperature plateau and that vacuum is to be applied initially only. While the prepreg supplier suggests crosslinking conditions, the user often elects to modify them following careful trials. Since autoclaves are expensive and cycle times long, it is common to try to crosslink as many components as possible simultaneously, see *Figure 4-30*.

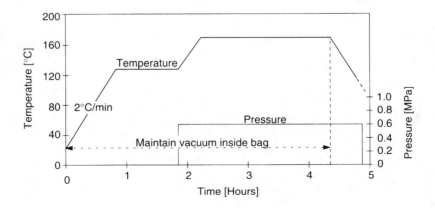

Figure 4-29 Typical crosslinking conditions for epoxy-based prepreg

Figure 4-30 Several separately vacuum-bagged components laid-up on egg-crate stiffened molds stacked together on their way into the autoclave. Photograph courtesy of Saab Military Aircraft, Linköping, Sweden

Technique Characteristics

Equipment cost:	High
Mold cost:	High
Labor cost:	High
Raw material cost:	High
Feasible series length:	Short to intermediate
Cycle time:	Long
Crosslinking requirements:	Complex
Health concerns:	Low

Component Characteristics

Geometry:	Any
Size:	Limited by autoclave size
Holes and inserts:	Possible
Bosses and ribs:	Possible
Undercuts:	Difficult
Surfaces:	One good surface, more with caul plates
Fiber arrangement:	Planar arrangement; oriented and continuous

Typical fiber volume fractions:	Up to 0.7 with UD prepregs, less with fabrics
Void fractions:	Low
Mechanical properties:	Good
Quality consistency:	Intermediate

Applications

Due to high capital costs and the possibility of making structurally very capable components of significant geometric complexity (see *Figure 4-31*), prepreg layup is widely used for high-performance products in a variety of applications, such as:

- High-performance racing vehicles and yachts (see *Figure 1-19*)
- Aircraft control surfaces and skins (see *Figures 1-22* through *1-26*)
- Spacecraft (see *Figures 1-27* through *1-29*)
- Sporting goods

Figure 4-31 Typical aerospace-type components that have been manufactured through hand layup of epoxy prepregs followed by autoclave crosslinking. Photograph courtesy of Saab Military Aircraft, Linköping, Sweden

4.2.3 Liquid Molding

Liquid molding is a generic term encompassing a range of processes by which unreinforced or short-fiber reinforced liquid resin is transferred into a closed mold where it crosslinks before being demolded. All liquid molding processes are capable of producing net or near-net shape components with good dimensional tolerances and are characterized by cost-effectiveness. The major traits distinguishing the liquid molding processes from one another are the degree of resin reactivity and whether the reinforcement is preplaced in the mold or mixed with the resin prior to injection. The most common liquid molding techniques for manufacturing of fiber-reinforced composites are:

- Injection molding
- Resin transfer molding (RTM)
- Vacuum injection molding
- Reinforced reaction injection molding (RRIM)
- Structural reaction injection molding (SRIM)

While only RTM, vacuum injection molding, and SRIM are used to manufacture continuous-fiber reinforced composites, injection molding and RRIM are common techniques to manufacture short-fiber–reinforced composites and are included herein for the sake of logic and completeness. Whereas the manufacturing techniques described in previous sections are quite mature, most liquid molding techniques for composite manufacture are undergoing rapid development in terms of processing technology. This development has resulted in "new" or improved manufacturing techniques that have been awarded additional and confusing acronyms, such as HSRTM (high-speed RTM), SCRIMP (Seeman composites resin infusion molding process), and VARI (vacuum-assisted resin injection/infusion), but they are all variations of the techniques described in the following sections. Nevertheless, the field enjoys much research and development effort and further technique enhancements and application areas are likely.

4.2.3.1 Injection Molding

Injection molding is mainly known as a manufacturing technique for thermoplastics, but it is to a lesser degree also used with thermosets. Injection molded parts often contain no reinforcement at all, but short-fiber reinforcement may be used if stiffness so dictates. Although injection molding is incapable of producing composites with fibers of any appreciable length, it is included herein as an introduction to subsequently described techniques.

Process Description

Powdered or pelletized resin, which optionally may contain short fibers, or BMC is fed into a hopper and then into the barrel of the injection molding machine, see *Figure 4-32*. The screw pushes the material forward in the barrel while at the same time compacting it to remove entrapped gases (mainly air). The friction and the shear induced by the screw heats the resin and transforms it into a liquid before it reaches the injection chamber at the front of the barrel. Due to the fierce action of the screw, any fibers present in the resin are reduced in length. Depending on resin type, the barrel may be cooled to prevent premature crosslinking or carefully heated to obtain the required flow characteristics. It is naturally very important to prevent the resin from starting to crosslink while still in the barrel, since this would require a time-consuming process shutdown to disassemble and clean both barrel and screw.

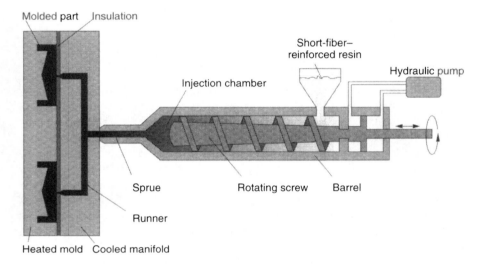

Figure 4-32 Schematic of thermoset injection molding

The material accumulating in the injection chamber gradually forces the screw backward in the barrel. When a carefully metered amount of material has been accumulated, the screw is pushed forward to inject the material into the mold, see the first stage in *Figure 4-33*. A mechanical valve within the barrel (not shown in the figures) prevents material backflow into the barrel. Although the manifold is cooled to prevent gelation in sprue and runners, the mold is heated to initiate the crosslinking reaction; the screw remains in the forward position to pressurize the part until it is dimensionally stable, at which point it is ejected (see stages two to four in *Figure 4-33*). With a cooled manifold such as that shown in *Figure 4-32*, it should be possible to prevent gelation of sprue and runners between injection cycles. However, if they crosslink, for example if a cooled manifold is not used, they are ejected with the part and are discarded since they cannot easily be recycled in the process. The injection pressures used, which range from 10 to over 100 MPa, are considered very high by composite standards and the mechanism clamping the mold halves together (which normally is an integral part of the injection molding machine) obviously must be able to sustain substantial loads.

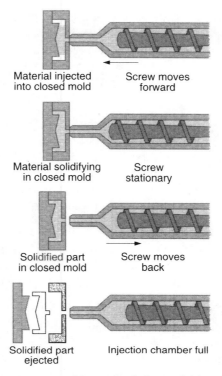

Material injected into closed mold	Screw moves forward
Material solidifying in closed mold	Screw stationary
Solidified part in closed mold	Screw moves back
Solidified part ejected	Injection chamber full

Figure 4-33 Stages in injection molding. Cooled manifold and multiple-cavity mold of *Figure 4-32* omitted for simplicity. Redrawn from reference [10]

Raw Materials and Molds

Virtually any thermoset may be injection molded, but the more common ones are unsaturated polyesters and phenolics. It is common not to use any reinforcement at all in thermoset injection molding, but the resin often contains fillers. When reinforcement is used, it is normally in the form of short glass fibers. The raw material form is commonly pellets or powder, although BMC may used if it is screw-fed into the barrel (since gravity is not sufficient to maintain feed). When using BMC it may very well be manufactured in line with the injection molding operation. Due to the high injection pressures encountered, molds are manufactured from solid tool steel and contain provisions for heating.

Crosslinking

Crosslinking takes place within a couple of minutes in the closed mold. While the part is ejected as soon as it is dimensionally stable, crosslinking usually still continues in the center of the part.

Technique Characteristics

Both injection molding machines and molds are expensive. However, since the technique is highly automated and cycle times fairly short, it is very cost-effective for long series. Thermoset injection molding may become a serious competitor to its much more common thermoplastic relation when tolerance to high temperature is among the requirements for a component.

Component Characteristics

Injection molded parts rarely compete with conventional composite manufacturing techniques since the process does not yield composites with high structural capabilities. The reasons for the poor mechanical properties are that by composite standards the fiber content is low and the fibers are very short and aligned by the flow. Among the advantages are that very complex geometries may be molded and that all surfaces are well controlled.

Applications

Thermoset injection molding is limited to small and moderately sized parts. Components injection molded from BMC include various automobile parts, such as the reflector assemblies for headlights and tail-lights.

4.2.3.2 Resin Transfer Molding

By far the most common liquid molding technique for manufacturing of structurally capable composites is RTM. RTM mainly owes its popularity to its capability for producing large, complex, and highly integrated components, as well as to low capital costs, low mold costs, and good work environment.

Process Description

A schematic of RTM is given in *Figure 4-34*. The process uses a closed mold into which dry reinforcement is placed. In short series, CSMs and fabrics may be cut by hand and manually placed in the lower mold half, but for longer series preforms are used to speed up the process, see *Figure 4-35*. One of the significant advantages of RTM is that it is possible to include inserts and fasteners, as well as cores to produce sandwich components (see *Figure 4-36*), with the reinforcement prior to impregnation. When the reinforcement and possible other constituents have been placed in the lower mold half the mold is closed, see *Figure 4-37*. The mold may be held closed in a variety of ways using hydraulic or pneumatic presses or merely using clamps along the edges of the mold.

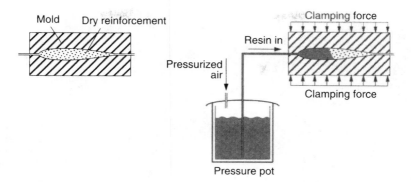

Figure 4-34 Schematic of RTM

Figure 4-35 Thermoformed RTM preforms. Photograph courtesy of Lear Corporation Sweden, Ljungby, Sweden

Following mold closure, liquid resin is injected (i.e. transferred) into the closed mold to impregnate the reinforcement. In the simplest cases it is often sufficient to use a pressure pot containing the preformulated resin, see *Figure 4-34*; pressurized air is injected into this simple pressure vessel forcing the resin out through a tube connected to the mold injection point. However, beyond the prototype stage it is more common to use purpose-built equipment to continuously mix the resin—typically using a static mixer—right before injection, see *Figure 4-38*.

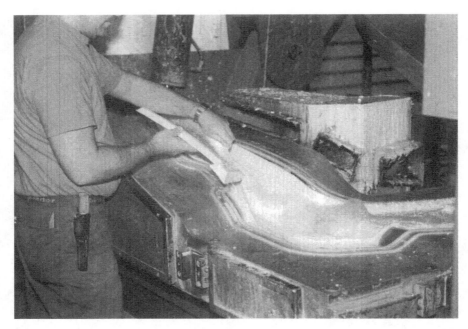

Figure 4-36 Charging of RTM mold with preforms and custom-molded PUR foam blocks containing metal inserts for joining purposes. Photograph courtesy of Lear Corporation Sweden, Ljungby, Sweden

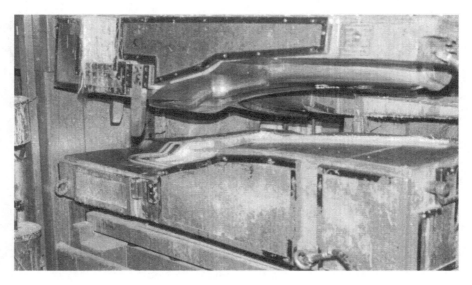

Figure 4-37 Charged RTM mold prior to closing. Photograph courtesy of Lear Corporation Sweden, Ljungby, Sweden

Figure 4-38 Standard RTM pump for normal requirements (left) and computer-controlled pump permitting variable control of pressure or flow rate for more demanding applications (right). Photographs courtesy of Aplicator System, Mölnlycke, Sweden

For small components resin injection may take place through a single injection port, possibly including sprues to distribute the resin to impregnate the reinforcement over a wide front; the sprue may even run all around the part so that impregnation progresses from the part perimeter and toward the center. Larger components may require more than one injection port and sprues to fill the entire mold satisfactorily. The resin is often injected at the lowest point(s) and fills the mold upward to several vents that are usually located at the highest points of the mold to reduce the risk of entrapping air pockets. However, under most conditions gravity effects are secondary to the applied pressure, so air pockets are seldom created through this mechanism. Vents are holes in the mold which allow the entrapped air to escape through clear plastic tubing leading into so-called resin traps (which in the simplest case is a disposable drinking cup). When the flow front impregnating the reinforcement reaches a vent, resin starts to leak into the resin trap, whereupon the tube is clamped shut to minimize resin loss. When resin emerges from the final vent, it is also closed off and infusion halted so as not to overpressurize and possibly deflect the mold. The emerging resin may initially contain voids, which are a sign of processing conditions resulting in air bubbles being entrapped within the reinforcement. In this case the resin should be allowed to escape through the vent until no more air bubbles can be seen in the resin.

During mold filling the resin will follow the path of least resistance meaning that it prefers to penetrate the spaces between reinforcement layers and yarns rather than into tightly compressed yarns. Thus, if the injection rate is too high, only the larger spaces will be filled with resin, the yarns will largely remain unimpregnated, and the advancing resin front will entrap voids. Impregnation of the yarns requires help from capillary forces and the injection rate therefore must be low enough to allow for complete impregnation.

Sometimes injection is aided by drawing vacuum at the vents to displace most of the air within the reinforcement and help draw the resin through the reinforcement. Vacuum also helps seal the mold since it is then pressurized by the external atmosphere. However, in some cases use of vacuum may lead to porosities due to faulty seals between mold halves, or some reinforcing fibers bridging the seal resulting in air leaking in due to the higher external pressure (and not due to styrene boiling and evaporating, as previously suggested).

In contrast to injection molding, RTM does not involve large injection pressures; pressures range from 0.1 MPa to over 1 MPa depending on, among other things, fiber content and resin type. Injection times range from a few minutes for small and simple components to hours for large, complex components with high fiber content.

Raw Materials and Molds

E glass and unsaturated polyesters dominate in most applications, although vinylesters, epoxies, and high-performance reinforcements are used to some degree as well. In particular the recent upsurge in interest from the aerospace industry is centered around carbon-fiber reinforced epoxies. Resin viscosities should be less than 1 Pa·s and preferably below 0.5 Pa·s; heating of the resin is often used to achieve the desired viscosity, but must be performed with care to prevent premature gelation (cf. *Figure 2-18*).

For low-volume applications weaves, braids, and mats are manually placed in the mold, whereas higher-volume applications are likely to employ random-fiber preforms, see *Figure 4-35*. In more demanding applications, fabrics may be cut and then stitched together into a net-shape preform, or 3D fabrics custom-woven to size may be used. There are companies specializing in making RTM preforms for manufacturers that do not have expertise or facilities to produce their own.

It is conceptually quite straightforward to integrate expanded foam and balsa cores as well as inserts and fasteners into the reinforcement prior to mold closure. Although complicating the process, gelcoats may also be used to enhance surface finish.

One of the major advantages of RTM is the modest requirements on the mold since relatively low pressures and temperatures are encountered. It is thus common to use stiffened composite molds for prototypes and short series. In fact, if the crosslinking temperatures are low and injection pressures modest, composite molds may last for a couple of thousand parts. However, for high fiber loadings, resulting in high injection pressures, and long series, solid metal or nickel-plated molds are usually required (see *Figure 4-37*). Mold sections are sealed using elastomer o-ring seals and sometimes shear edges (see also *Figure 4-46*). When considering what mold material to use one must take into account that the injection pressure may cause significant mold deformations unless the mold is stiff enough or fully supported by the mold closing device, such as by the press table of a hydraulic press. Composite molds may also prove inadequate if high-temperature crosslinking epoxies are used, since the mold then must be heated.

Crosslinking

A critical aspect of RTM is to ensure that the onset of crosslinking is kept at bay until the mold is completely filled, which calls for moderation in heating of the resin to achieve low viscosity and thus more rapid mold filling. Once the mold has been filled, it is on the other hand desirable to achieve crosslinking rapidly to allow demolding of the component so that the mold can be used to manufacture the next component. With unsaturated polyesters and vinylesters the crosslinking normally proceeds at ambient temperature, whereas crosslinking of some resins is accomplished by heating of the mold. In either case, vents remain closed and a certain back pressure is maintained at the inlet ports until the resin gels to prevent it from bleeding out of the laminate. The time to demolding (see *Figure 4-39*) is anywhere from a few

Figure 4-39 Demolding of crosslinked RTM component. Injection of this component takes approximately 1 minute and is aided by vacuum, while the total cycle time is such that 3–4 components are molded per hour including mold cleaning and charging. The component is part of the spoiler on top of the driver's cab of a heavy Volvo truck. Photograph courtesy of Lear Corporation Sweden, Ljungby, Sweden

minutes to a few hours depending on component thickness, resin, and processing conditions. With high-temperature epoxies, final crosslinking may be completed after demolding using freestanding postcuring at elevated temperature, assuming that the component is first sufficiently crosslinked within the mold to become dimensionally stable.

Technique Characteristics

Equipment cost:	Low
Mold cost:	Low to intermediate
Labor cost:	Intermediate
Raw material cost:	Low
Feasible series length:	Short to intermediate
Cycle time:	Intermediate
Crosslinking requirements:	Simple
Health concerns:	Low

Component Characteristics

Geometry:	Any
Size:	Any
Holes and inserts:	Possible, but holes are more conveniently made in a secondary step
Bosses and ribs:	Difficult
Undercuts:	Difficult
Surfaces:	All surfaces good
Fiber arrangement:	Planar arrangement (3D arrangement possible); random or oriented, continuous or discontinuous
Typical fiber volume fractions:	Up to 0.6 with fabrics
Void fractions:	Low
Mechanical properties:	Good
Quality consistency:	Good

Applications

While RTM lends itself to both simple prototype manufacturing and relatively long series, it has had the greatest breakthrough in intermediate-length production series. Due to the aforementioned characteristics, RTM is rapidly being accepted for new applications in for example the automotive, aerospace, and marine industries. Applications include:

- Automobile, truck, bus, and train parts, including body panels (see *Figures 1-10* and *4-39*), complete self-supporting bodies (see *Figure 1-15*), and front ends (driver's cabs) for high-speed train engines

- Components for leisure boats, yachts, and ships, including hatches (see *Figure 1-17*) and booms
- Aircraft propellers and interiors
- Construction members

The major reasons for the automobile industry's interest in RTM is that large, complex components with high-class surfaces may be economically manufactured in one step using low-cost molds. Good examples of this are the body panels of exclusive cars and cars manufactured in short series (see *Figure 1-10*). Moreover, the interest in manufacturing large structural parts, such as the entire automobile base plate, is increasing. An early example of this feasibility is the all-terrain vehicle of *Figure 1-15*, where all structural elements (floor, roof, and walls) are manufactured through RTM (in series production since 1981). The sandwich panels, which have PVC and PUR foam cores and glass-reinforced unsaturated polyester faces, are adhesively bonded using an epoxy adhesive to form a self-supporting structure.

4.2.3.3 Vacuum Injection Molding

Although the previous section mentioned the occasional use of vacuum to assist in reinforcement impregnation in RTM, the predominant force driving the impregnation in RTM is the fact that the resin is injected under pressure. However, certain related processes employ vacuum as the sole force driving impregnation. Such a process may employ a closed mold (where the higher external pressure helps seal the mold) or a marginally modified wet layup mold covered by a vacuum bag, see *Figure 4-40*. In the latter and more common case, vacuum draws resin from a container and into the vacuum bag along a line, for example along a symmetry line in the component, as shown in the figure. If fiber content is high, impregnation takes a long time and more rapid results may be obtained by placing a carrier layer between the top reinforcement layer and vacuum bag or sometimes between fabric layers. The function of the porous carrier is to allow the resin to travel rapidly in the plane and then allow for the more time consuming transverse impregnation more or less simultaneously throughout the component. (The key to

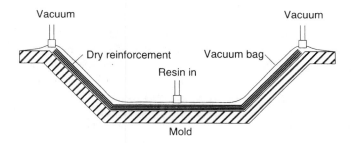

Figure 4-40 Schematic of vacuum injection molding

the SCRIMP process is the use of such a carrier layer directly beneath the vacuum bag.)

The main advantages of vacuum injection molding are that simple molds may be used and that capital costs are very low, essentially only a vacuum pump. It is therefore common that vacuum injection molding is used for laboratory and prototype work. A disadvantage when compared to RTM is that only one surface is well controlled and that thickness and consequently fiber volume fraction are not as closely defined or uniform.

Technique Characteristics

Equipment cost:	Low
Mold cost:	Low
Labor cost:	Intermediate
Raw material cost:	Low
Feasible series length:	Short
Cycle time:	Long
Crosslinking requirements:	Simple
Health concerns:	Low

Component Characteristics

Geometry:	Any
Size:	Any
Holes and inserts:	Possible
Bosses and ribs:	Difficult
Undercuts:	Difficult
Surfaces:	One good surface
Fiber arrangement:	Planar arrangement; random or oriented, continuous or discontinucus
Typical fiber volume fractions:	Up to 0.4 with fabrics
Void fractions:	Low
Mechanical properties:	Good
Quality consistency:	Good

Applications

Vacuum injection molding has chiefly been used to produce large components in short series, such as boats, offshore structures, vehicle body panels, and front ends for high-speed train engines. As the technique is well suited for inclusion of cores, many of these applications are sandwich components.

Among the more spectacular application examples are railroad freight cars for transportation of goods requiring refrigeration. Through use of a thick PUR foam between glass-reinforced vinylester laminates, the insulative capabilities are so good that no refrigeration unit is required for week-long journeys, while at the same time increasing payload by 35 percent over conventional steel freight cars. The 21-m long cars are assembled from two sections, one for the roof while the other integrates floor and walls [11].

4.2.3.4 Reaction Injection Molding

RIM (which is the parent of both reinforced RIM (RRIM) and structural RIM (SRIM)) is similar in concept to conventional injection molding; the significant difference between the techniques lies in the resins used. In injection molding the resins are preformulated and require heating to flow and crosslink in a time frame in the order of minutes. In contrast, RIM utilizes highly reactive two-component resins that are low-viscosity liquids at room temperature; in this case crosslinking is initiated by mixing of the resin components, not by increased temperature.

In a RIM machine, the two resin components constantly recirculate under high pressure in separate systems and are not mixed until the moment of injection (the recirculation loops are not shown in *Figure 4-41*). To initiate mixing and injection, the piston in the mixhead moves up and the two resin components collide in the mixing chamber under high speed and high pressure, see the second stage in *Figure 4-41*. The resin streams collide at linear velocities of 100–200 m/s resulting in pressures of 10–40 MPa. This impingement mixing is completed in 0.1–1 ms and the mixed resin is injected into the mold at low pressure (less than 1 MPa) in a time frame of a few seconds by the continued inflow of more resin into the mixing chamber. Following mold filling, the piston purges the mixhead of any remaining resin, see the last stage in the figure. Since crosslinking is initiated by the mixing and is very rapid, the resin gels within seconds after the mold has been filled, partially aided by heating of the mold. To ensure complete crosslinking, it is

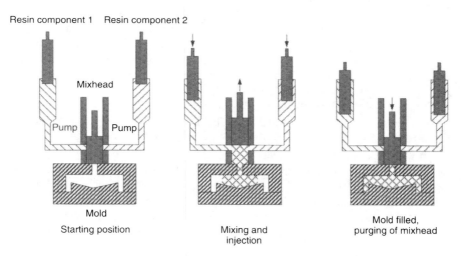

Figure 4-41 Schematic of RIM

important that exactly correct proportions of the resin components are mixed. This requirement, as well as the aforementioned need for constant recirculation and mixing at high velocity and high pressure, means that RIM equipment is quite expensive.

By far the most common RIM resin is PUR, the reactivity of which is easily varied to suit any application, although unsaturated polyesters and epoxies can also be formulated to be very reactive. Ideally, the mixed resin should have a very low viscosity during mold filling and then crosslink as rapidly as possible following mold filling. RIM is used to manufacture a range of automotive and consumer products.

4.2.3.5 Reinforced Reaction Injection Molding

In RRIM short or milled fibers or glass flakes are added to one of the resin components prior to final resin mixing. It has proved difficult to include fibers longer than 0.5 mm, since this leads to a too high viscosity. Although even such short fibers provide significantly increased stiffness, damage tolerance, dimensional tolerance, and lower CTE over what RIM components may offer, articles produced through RRIM nevertheless have poor structural properties from a composite viewpoint, both due to modest fiber length and flow-induced fiber orientation. Also, in RRIM, PURs are almost exclusively used, while the reinforcement is glass.

Provided appropriate molds are used, RRIM may yield class-A* surfaces; the process consequently is popular in the automotive industry, where the short cycle times and the low labor costs are further important advantages. Particularly in North America, RRIM is used for body panels and fascias (covers for bumper beams), see *Figure 1-9*.

4.2.3.6 Structural Reaction Injection Molding

SRIM is the result of conceptually combining RTM with RIM. In SRIM the reinforcement is first placed in the mold (just as in RTM) and following mold closure the highly reactive impingement-mixed resin is injected into the mold (just as in RIM) to impregnate the reinforcement.

While deceptively similar to RTM, SRIM is used for long series where the significantly higher initial cost for the injection equipment (an order of magnitude higher) may be written off. To be feasible in long series, matching metal molds mounted in hydraulic presses for opening and closing are used. The higher resin reactivity means that it is not possible to produce as large components as with RTM, but the cycle times are clearly shorter—down to a minute for small, simple components. Although injection pressures are within the same range as for RTM, the much higher injection rates mean that it is not possible to reach as high fiber fractions and as low void fractions. Further, if large amounts of random reinforcement are used there is also a

---

* Class-A is the automobile industry's notation for the surface quality required for body panels.

greater possibility of fiber washout, i.e. when the reinforcement is moved by the advancing resin front.

To allow for reinforcement impregnation, resins used in SRIM tend to be formulated for lower viscosity and longer inhibition time than those used in RIM; a viscosity in the range 0.1–0.5 Pa·s for 10–20 seconds is typical. Since SRIM requires long series to be economically advantageous, net-shape pre-forms are used to reduce labor intensity. Glass-reinforced PUR clearly dominates in SRIM.

Technique Characteristics

Equipment cost:	High
Mold cost:	High
Labor cost:	Intermediate
Raw material cost:	Low
Feasible series length:	Long
Cycle time:	Short
Crosslinking requirements:	Simple
Health concerns:	Intermediate

The health concerns associated with SRIM are likely to be higher than with RTM due to the predominance of PURs, since one of the resin components used to produce a PUR (isocyanate) is significantly more toxic than for example unsaturated polyesters, see also *Chapter 8*. This concern applies to manual dealing with the unreacted resin components, e.g. during filling of resin tanks, and not to the actual manufacturing.

Component Characteristics

Geometry:	Any
Size:	Intermediate
Holes and inserts:	Possible
Bosses and ribs:	Difficult
Undercuts:	Difficult
Surfaces:	All surfaces good
Fiber arrangement:	Planar arrangement; random or oriented, continuous or discontinuous
Typical fiber volume fractions:	Up to 0.4
Void fractions:	Intermediate
Mechanical properties:	Intermediate
Quality consistency:	Good

Applications

Just like RIM, SRIM is favored by the automobile industry since short cycle time is of such critical importance. However, while RTM is feasible in short series, SRIM is used for long series where the significantly higher capital costs may be written off. Applications include bumper beams and spare wheel wells, as well as consumer products.

4.2.4 Compression Molding

Compression molding is the technique used to manufacture the vast majority of all composite components. The reasons for this volumetric dominance is that compression molding is most cost-effective for long and very long production series and has had the greatest successes in the automobile industry, where the similarity with the familiar technique of sheet metal stamping is a significant advantage. Vehicle body panels are the most common applications.

Several different thermoset-based materials may be compression molded; the most common is SMC, followed by BMC and other material forms. Throughout the following sections the discussion will focus on the use of SMC since it clearly dominates usage, whereupon the differences between the use of SMC and alternative raw material forms will be pointed out. In this context it is noteworthy that through linguistic bastardization the acronyms SMC and BMC have become synonymous not only with the raw material forms (cf. *Section 2.3.6*), but also with compression molding using the respective materials.

Process Description

In compression molding matching male and female mold halves are used, see *Figure 4-42*; this is why an alternate name for the process is matched-die molding. To mold a component, the SMC roll is brought out of storage and pieces cut to both size and carefully controlled weight according to specifications, see *Figure 4-43*. Depending on methodology, the carrier films are either removed prior to or following cutting. The prepared stack of SMC sheets, which is called charge, is placed on the lower mold half, see *Figure 4-44*. The mold is then rapidly and forcefully closed using a hydraulic press to force the charge to flow to fill the mold. The mold is heated, usually with circulating steam or oil or through the use of electrical cartridge heaters, to ensure that crosslinking is initiated and completed. When the part is dimensionally stable it is demolded (see *Figure 4-45*) and the cycle repeated.

The charge covers between 20 and 90 percent of the total surface of the mold cavity and the remaining parts are filled through the forced material flow. Since the amount of raw material determines component thickness and reinforcement orientation it is important that the charge has the intended weight and shape, also known as charge pattern, and that it is placed in the correct position in the mold. When good surface quality is required, the charge covers a smaller portion of the mold since flow generally promotes a good surface. When the structural performance is the most important, the

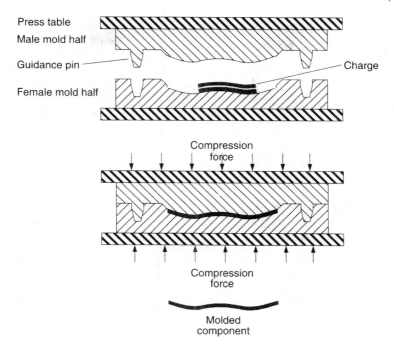

Figure 4-42 Schematic of compression molding

Figure 4-43 Cutting of SMC using steel ruler and knife. Photograph courtesy of Lear Corporation Sweden, Ljungby, Sweden

Figure 4-44 Placing of the SMC charge in the mold. Photograph courtesy of Lear Corporation Sweden, Ljungby, Sweden

Figure 4-45 Demolded SMC component. The component is the roof of a farm tractor. Photograph courtesy of Lear Corporation Sweden, Ljungby, Sweden

charge covers a greater portion of the mold, since this limits flow-induced reinforcement orientation. This normally undesired orientation may cause properties to vary by a factor of two within the same component depending on location and direction.

The main difference in processing SMC and BMC is that the BMC charge typically covers less of the mold cavity and the flow and thus fiber orientation becomes more pronounced. Crosslinking of condensation-type thermosets, such as some phenolics which are not uncommon in BMC, produces a byproduct that must be allowed to escape so as not to entrap voids. With such materials the mold is first closed to fill it, before it is slightly and briefly reopened to allow the condensation products to evaporate, whereupon the mold is closed again to complete crosslinking. This extra processing step is called degassing or breathing.

Although compression molded components may have class-A finish as molded, in-mold coating (IMC) is sometimes used to further improve surface finish (such as on the SMC components on the truck of *Figure 1-12*). In low-pressure IMC the molding pressure is reduced when the component has crosslinked—the mold may even be opened ever so slightly—and a neat resin is injected between component and mold, whereupon molding pressure is applied again to crosslink the surface coating. In high-pressure IMC the resin is injected under high pressure between partially crosslinked component and mold without releasing molding pressure. In contrast to low-pressure IMC, which adds up to a minute to the press cycle, high-pressure IMC does not extend the molding cycle. The reason is that since molding pressure is not released during injection, the molded component does not need to be crosslinked throughout the thickness and the coating resin can thus crosslink at the same time the component completes crosslinking. In some applications thermoformed films are used instead to enhance the surface finish.

In wet compression molding dry reinforcement, possibly in the form of a preform, is first placed in the lower, female mold half whereupon the normally filled resin is poured on top (or vice versa) before the heated mold is closed. While temperatures are similar to those used in molding of SMC and BMC, pressures are lower, sometimes much lower, and cycle times somewhat longer. Wet compression molding may also be carried out with room-temperature resins and without mold heating.

A further process variation employs a spray-up gun (see *Section 4.2.1.2*) to deposit a mixture of chopped fibers and resin onto the lower mold half. In this case it may be sufficient to heat the upper mold half only to allow sufficient time to deposit the required amount of material onto the (lower) mold half without risk of premature gelation. With spray-up it is difficult to ensure a repeatable charge pattern unless the spray gun is robot-mounted.

Yet another process variation combines injection molding and compression molding. An injection molding machine injects a predetermined amount of material into a partially closed mold which following injection is closed to fill the mold cavity completely.

Raw Materials and Molds

SMC and BMC are clearly the most common raw material forms. Heavily filled unsaturated polyesters dominate, but vinylester, phenolics, and epoxies are used as well; in all but very rare cases, E glass is the reinforcement of choice. In components that are not highly stressed it is normal to use BMC and randomly reinforced SMC (SMC-R), whereas for structural applications some degree of oriented and/or continuously reinforced SMC (SMC-C, SMC-D, etc.) is used and the reinforcement content is likely to be higher.

Wet compression molding normally also employs E glass and filled unsaturated polyesters. The reinforcement may be hand-cut mats and possibly fabrics, whereas longer series may call for net-shape preforms.

Steel molds, which are usually chrome- or nickel-plated to improve wear resistance and component surface finish, are used in production. The molds are equipped with guidance devices, as shown in *Figure 4-42* for precise alignment of the mold halves and with mechanical or pneumatic ejection devices to aid in demolding. Through the use of telescopic shear edges, as shown in *Figure 4-46*, around the entire parting line, air and matrix but not the reinforcement are allowed to escape, since the shear edge is designed to close before the flow front reaches it. This edge arrangement also reduces or possibly even eliminates edge trimming of demolded components to allow for near–net-shape manufacturing. Note that the component thickness is determined by the volume of the charge; the function of the stop block in *Figure 4-46* is only to prevent mold damage upon accidental closing of an empty mold. Design of molds for compression molding is far from trivial and hosts of guidelines based on both science and art are employed by mold designers. A basic principle is

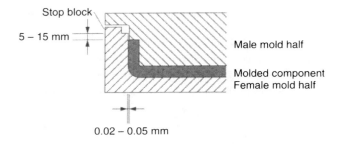

Figure 4-46 Mold detail showing telescopic shear edge

that part thickness should not vary by much (less so for BMC), especially if good surface finish and dimensional stability is required. Further, sharp corners must be avoided and a draft angle should be used on near vertical sections (i.e. surfaces parallel to the press motion) to allow for component removal without damaging its surface (1° is sufficient for short sections, whereas up to 3° may be required for longer sections). Merging flow fronts should be avoided, since they create so-called knitlines where few (if any) fibers bridge the line, which thus ends up very weak. Though stiffening ribs and bosses are common, they must be carefully designed to avoid sink marks on the other surface.

Crosslinking

In molding of SMC and BMC the mold is normally heated to a temperature in the range 120–180°C if the material is unsaturated polyester-based, while the applied pressure varies within a wide range depending on a number of factors; pressures are likely in the range 3–20 MPa. Once again heavily dependent upon a number of factors, such as part size, thickness, intricacy, and resin, the time between mold closing and opening is usually between 1 and 4 minutes; low-pressure IMC adds up to a minute to the molding time. In order to reduce cycle time components may be postcured following demolding, but then a fixture may be needed to maintain dimensional accuracy.

If room-temperature crosslinking resins are used, the process is often called cold compression molding. Cycle times are naturally much longer than with mold heating and pressures tend to be an order of magnitude lower than in conventional compression molding.

Technique Characteristics

Compression molding of SMC and BMC is characterized by:

Equipment cost:	High
Mold cost:	High
Labor cost:	Intermediate
Raw material cost:	Intermediate
Feasible series length:	Long
Cycle time:	Short
Crosslinking requirements:	Simple
Health concerns:	Low

With cold compression molding and to some degree with wet compression molding, pressures are so low that mold requirements are reduced when compared to molding of SMC and BMC. This means that equipment and particularly mold costs may be significantly reduced; in cold molding composite molds may be used for prototypes and short series.

Component Characteristics

Whereas SMC components largely tend to be thin, two-dimensional shell structures, such as vehicle body panels, BMC parts are usually smaller and of more intricate geometry. Characteristics of compression molded SMC composites include:

Geometry:	Any
Size:	Limited by press size
Holes and inserts:	Inserts possible, vertical holes difficult
Bosses and ribs:	Possible
Undercuts:	Difficult
Surfaces:	All surfaces good, but IMC usually required for Class-A finish
Fiber arrangement:	Planar arrangement; random and discontinuous (continuous and partially oriented fibers possible)
Typical fiber volume fractions:	0.2–0.3, up to 0.5 with SMC-C
Void fractions:	Intermediate
Mechanical properties:	Poor; intermediate with SMC-C
Quality consistency:	Good

Components compression molded from BMC are characterized by:

Geometry:	Any
Size:	Limited by press size
Holes and inserts:	Inserts possible, vertical holes difficult
Bosses and ribs:	Possible
Undercuts:	Difficult
Surfaces:	All surfaces good, but IMC usually required for Class-A finish
Fiber arrangement:	Planar arrangement; random and discontinuous
Typical fiber volume fractions:	0.1–0.2
Void fractions:	Intermediate
Mechanical properties:	Poor
Quality consistency:	Good

While it is possible to mold in inserts and holes, the material flow has to diverge before the insert or hole and then converge after it, thus creating a knitline; it may therefore in some cases be preferable to drill holes and put in inserts in a secondary operation (see also *Chapter 5*). Large holes may nevertheless be molded (cf. *Figure 4-45*) if SMC is continuously placed all around the hole (cf. *Figure 4-44*) to eliminate knitlines. Through complex mold designs, including more than two mold sections or moving mold inserts, it is possible to mold components with undercuts, but mold cost and reliability consequently suffer.

Wet compression molding shares most traits with SMC molding, with the overall exceptions that bosses, ribs, and undercuts are not possible with normal procedures and that fabrics may be used to control fiber orientation.

Applications

Compression molding is the most popular in automobile and truck applications. Among the reasons for this are that shell-like structures are common and production series long. However, possibly the biggest reason is the fact that these industries have ample experience of and equipment for sheet metal stamping and the processing techniques, component and mold design procedures, and presses are similar and sometimes identical to composite compression molding. The step from stamping of sheet metal to compression molding of composites is therefore relatively short. When comparing these two techniques it is also important to realize that stamping of a sheet metal component is an incremental process requiring several molds (and presses). Thus, although one molding step in a sheet metal line certainly is much faster than the single molding step in SMC or BMC molding, the overall cycle times are comparable and the capital costs definitely are in favor of the composite alternative. Applications of compression molded composites are numerous; they are mainly found in areas such as:

- Automobile, truck, and bus components, such as body panels (see *Figures 1-11*, *1-12*, and *4-45*), bumper beams, front ends, etc.
- Containers and housings
- Electrical and machinery components
- Bathroom interiors

4.2.5 Filament Winding

Filament winding offers a highly efficient and automated means of precisely placing impregnated reinforcement yarns onto a rotating mold. While the process is cost-effective, the structures that may be filament wound are essentially limited to convex and closed geometries that may be rotated. Common applications includes pressure vessels, pipes, launch tubes, and other rotationally symmetric components.

Process Description

In the simplest incarnation a filament winding facility may be likened to a lathe where the rotating mold, in filament winding usually called mandrel, is mounted between head and tail stocks. In most cases the mandrel is a rotationally symmetric body such as that shown in *Figure 4-47*. The figure schematically illustrates how a continuous supply of reinforcement, normally several rovings in parallel, is impregnated in a resin bath and then wound onto the mandrel to create interlocked and balanced laminae. *Figure 4-48* shows a production facility.

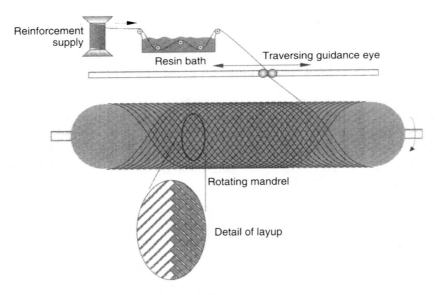

Figure 4-47 Schematic of filament winding

In the name of simplicity the process schematic of *Figure 4-47* overlooks several important aspects of the technique and it is therefore more carefully discussed in the following. To achieve a constant degree of impregnation the rovings must be kept under constant tension, which normally is achieved through some kind of brake on the roving spools. If the winding tension is too low the laminate is not fully compacted and if the tension is too high the rovings gradually migrate toward the mandrel and the laminate may become resin-starved near the inside (and resin-rich near the outside surface); too high tension may also lead to fiber breakage and fuzzing. The roving supply and the tensioning devices may be mounted on the same transversely moving carriage as the impregnation station and guidance eye (or another carriage synchronized with it); alternatively, it may be mounted at a significant distance behind the carriage to minimize the effects of changes in distance between roving spools and carriage due to its translation (see *Figure 4-48*). In the latter case it may be necessary to have a rewind capability for the roving spools to compensate for rapid carriage movements.

Figure 4-48 Filament winding facility. In the background the creels of roving spools are visible. Photograph courtesy of ABB Plast, Piteå, Sweden

Right before and after the impregnation station the rovings are guided by pins to form an essentially homogenous, thin band. The impregnation station may be of two basic types; either a simple dip-through tank as shown in *Figure 4-47* or it may employ a pick-up cylinder as illustrated in *Figure 4-49*. The advantage of the pick-up cylinder is that a well-defined resin thickness is applied onto the cylinder through adjustment of the doctor blade. With the pick-up cylinder solution the resin may instead be continuously pumped onto the cylinder in metered amounts to eliminate aging of a large amount of resin present in a bath. With either impregnation solution the reinforcement–resin ratio depends on band tension and resin viscosity, but may also be controlled with nip rollers. The final fiber fraction also depends on mandrel diameter, since the higher the mandrel curvature (i.e. the smaller the radius), the higher the normal force between mandrel and bands, leading to more resin being radially squeezed out of the laminate.

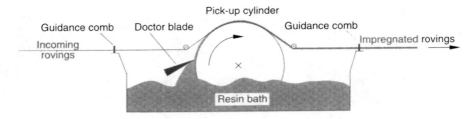

Figure 4-49 Schematic of resin bath with pick-up cylinder

Following impregnation the roving band is precisely placed onto the mandrel using for example a guidance eye or another comb, see *Figure 4-50*. In general terms the layup must follow the geodesic path, which is the shortest distance between two points on a surface. When following the geodesic path no lateral forces act on the band, meaning that it will stay where it has been placed (assuming that the surface is not concave). If one follows the geodesic path, all subsequent band paths are defined once a starting angle has been selected. Since this places a serious restriction on the geometries that may be wound, it is of interest to be able to deviate from the geodesic

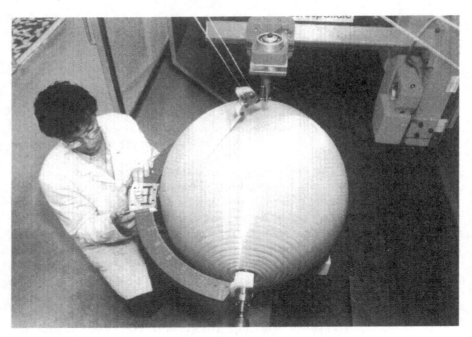

Figure 4-50 Winding of sphere with precise guidance of the band. The component is a fuel tank for space use, such as illustrated in *Figure 1-28*. Photograph courtesy of Aerospatiale, Les Mureaux, France

path. Relying on friction between the band to be placed and the mandrel or previously wound laminae it is possible to deviate from the geodesic path, but since the resin lubricates the band only careful and gradual deviations are permitted to avoid slippage. Both higher band tension and smaller mandrel diameter lead to a higher normal force, meaning that it is easier to deviate from the geodesic path.

The layup pattern essentially may be of three different kinds: helical, hoop, and polar, see *Figure 4-51*. The most common and most versatile pattern is the helical and it can be shown that for a rotationally symmetric body the geodesic path must satisfy:

$$r \sin \alpha = constant \tag{4-1}$$

where r is the local mandrel radius and α the local winding angle. With helical winding the roving bands are not laid side by side for each complete revolution of the mandrel. A given number of revolutions, which depends on angle, band width, and mandrel length, are required before the band is laid directly beside itself. After numerous revolutions, helical winding results in a balanced lamina with multiple crossovers as shown in the detail of *Figure 4-47*. When winding on a domed mandrel such as that shown in *Figure 4-51*, the layup angle for the cylindrical section of the mandrel is given by the fact that at the bosses $\alpha = 90°$. With equal-diameter bosses as shown in the figure, the layup angle for the cylindrical section of the mandrel is:

$$\alpha = \sin^{-1}\left(\frac{r_b}{r}\right) \tag{4-2}$$

where r_b is the boss diameter.

Hoop, or circumferential, winding (see *Figure 4-51*) is really a high-angle helix pattern where the angle is close to 90° to place bands side by side, thus covering the mandrel; for each mandrel revolution the guidance eye is indexed one band width. Hoop winding requires an essentially constant mandrel diameter to avoid slippage.

Polar winding is limited to mandrels with domed ends, such as pressure vessels, see *Figures 4-50* and *4-51*; the bands are placed side by side for each circuit. To avoid slippage, polar winding is normally limited to mandrel length-to-diameter ratios of less than 1.8.

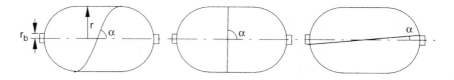

Figure 4-51 Helical (left), hoop (center), and polar (right) winding patterns

The full pattern for winding a component is traditionally determined through an iterative procedure involving trial-and-error experiments and compromises between desired and achievable angles. With the advent of fully computer-controlled winding machines and well-established computer-based design and analysis procedures, a possibility of connecting the two has arisen. It is thus possible to take a given component design, which has already been through iterative design and analysis steps, and numerically generate the machine code required by the winding machine. While such procedures are not yet commonplace, software packages are commercially available.

One of many attractive characteristics of filament winding is that winding angles and even volume fractions can be varied between laminae. It is, for example, common to start with a hoop layer to get a good inner surface, to continue with one or more $\pm\alpha$ laminae (which may have different angles), see *Figure 4-48*, and to finish with a hoop layer to get as good an outer surface as possible. To change from one winding pattern or angle to another obviously requires deviations from the geodesic path, which thus relies on friction. A means of artificially introducing friction is to use pins protruding out of the mandrel; this is common in for example winding of pipes (i.e. on mandrels without domed ends) to minimize the scrap at the ends caused by a gradual, friction-restricted turnaround (from $+\alpha$ to $-\alpha$). Without pins it is normally not possible to wind with angles smaller than 35°.

To wind rotationally symmetric structures with regular layup angles, it is sufficient for the machine to have the two degrees of freedom, often called axes, illustrated in *Figure 4-47*; mandrel rotation and transverse movement of the guidance eye. However, further axes, such as two more translational axes and one or two rotational axes for the guidance eye, are common on modern machines. Added axes allow winding on widely different mandrel diameters, with wide roving bands, and on more complex geometries. New machines tend to be computer controlled with fully independent servo motors for each axis. However, many older machines (which still are much used) have few axes and are fully mechanical without independent control of the axes, thus placing restrictions on the capabilities of the machine; through changes in driving gears, it is nevertheless possible to accommodate a range of mandrel sizes and layup patterns.

For large angles, large diameter mandrels, and simple geometries linear winding speeds may be in excess of 1 m/s and layup rates a few hundred kilograms of impregnated reinforcement per hour if many parallel rovings are used. However, lower winding speeds and layup rates are more common, especially with small-angle helix winding, small diameter mandrels, and complex geometries.

Although horizontally oriented mandrels are by far the most common, machines with vertical mandrels may be used to eliminate problems with mandrel deflection of long and heavy mandrels due to gravity. Another process variation allows winding of continuous pipes using specialized machinery. These machines use nonrotating and removable mandrels and winding heads that counter-rotate around the mandrel. The resin is heated to speed up crosslinking to allow mandrel removal.

Raw Materials and Mandrels

The entire process description above assumes wet winding, i.e. with on-line impregnation, since it is by far the most common impregnation method. However, in some applications, e.g. for high-performance aerospace components, it may be preferable to wind with prepreg tape. The use of prepregs brings on the expected basic advantages as discussed in *Section 2.3*, but also winding-specific advantages such as little or no slippage (i.e. higher friction due to the prepreg tack), higher winding speeds (which for high winding angles may be limited by resin spraying off the mandrel due to the centrifugal force), and no concerns of spreading of resin volatiles or carbon fiber particles since the fibers are encapsulated in a B-staged resin. Apart from the cost disadvantage, prepregs also may create problems in laminate compaction since the resin viscosity is so high. In high-performance applications some processors instead impregnate the reinforcement off-line, i.e. make their own prepreg, since it can be carried out faster and under better controlled conditions than with on-line impregnation. It is finally possible to wind without any resin whatsoever to manufacture preforms for use in some other manufacturing process, such as RTM, but this is rare.

The most common material combinations are glass-reinforced unsaturated polyester for relatively low-cost components and glass-, aramid-, and carbon-reinforced epoxy for more demanding applications. It is quite simple and not uncommon to combine reinforcement types to achieve optimum performance-to-cost ratio. The reinforcement form is continuous yarns or rovings in all but the cheapest components, where part of the reinforcement may be sprayed on (see *Section 4.2.1.2*) concurrently with the winding. It is rare, albeit technically possible, to wind with bands of fabrics or mats. Resin viscosities are normally in the range 1–10 Pa·s, but lower viscosities are also used. The resin pot life should be long—on the order of several hours—especially if bath impregnation is used.

The mandrels used in filament winding are unusual in that they are internal and that almost any material that can sustain the low loads—essentially mandrel and component weight and roving tension—may be used. For prototypes the mandrel may be of cardboard, wood, plastic, etc., but for series production other mandrel types are normally used.

However, the critical aspect is how to get the mandrel out of the finished component, if applicable. To facilitate mandrel removal following crosslinking, it is normally treated with a release agent or may be overwound with a hoop layer of release tape. If a metal mandrel is used it may be sufficient to carefully polish or chrome-plate it if some mechanism to pull or push the crosslinked composite off the mandrel is used. If the inner dimension is not critical, a slight mandrel taper greatly facilitates removal. If the mandrel has a higher CTE than the composite and crosslinking is carried out at elevated temperature, mandrel removal is also facilitated. It is thus normally easier to remove carbon- than glass-reinforced filament wound composites.

Water-soluble sand and salt mandrels as well as low melting-point metal alloy mandrels may be dissolved or melted following crosslinking of the wound component. The disadvantage is that each mandrel has to be cast in a separate mold, which adds to the final component cost. Cast plaster mandrels may also be used, but they must be manually chipped out, which may be quite difficult if the end openings are small; chipping may also inadvertently damage the component. For large-diameter structures it is possible to construct collapsible and reusable steel mandrels.

In many applications the mandrel is intended to become an integral part of the component; such applications include drive shafts and pressure vessels. In pressure vessels, for example, which normally have the general domed geometry illustrated in *Figures 4-50* and *4-51,* the metal or thermoplastic liner acts as a barrier to gas and moisture to allow the vessel to be leak-proof. The liner thus is a barrier on the atomic level while the wound structure carries the pressure load. The liner may be designed to adhere to the wound structure or to be partially independent of it. Liners are often capable of sustaining the loads of the winding process, but pressurization or inflation of flexible mandrels during winding may help matters.

As previously mentioned, pins protruding from the mandrel, usually at the ends, may be used to allow instantaneous changes of winding angle. It is also common practice to integrate fittings with the mandrel whenever possible; the fittings are then overwound to become integral parts of the final component (see *Figure 5-30*) thus eliminating many load-introduction problems. Where required, fittings may be treated with adhesive to improve adhesion to the composite.

Crosslinking

While some of the matrices used in filament winding may crosslink at room temperature, most require elevated temperature to crosslink as intended; this is normally achieved using an oven. Since many filament winding operations produce components continuously (albeit batch-wise), the mandrels are often placed in a rack that carries the mandrel from one end of the oven to the other at a rate that ensures that the wound component is fully crosslinked by the time it exits the oven. Whether crosslinked at elevated temperature or not, the mandrel is slowly rotated at least until the resin gels to cancel out gravity effects which otherwise might cause resin to bleed out of the laminate.

The vast majority of filament wound components are thus crosslinked without external pressure. However, properties will improve if external pressure is used. In the simplest case shrink tape may be hoop wound over the entire structure following completion of winding but prior to crosslinking. It is naturally also possible to crosslink the component in an autoclave using vacuum bag, bleeder plies, etc. as described in *Section 4.2.2*. A further way to apply consolidation pressure during elevated temperature crosslinking is to select a mandrel material with a higher CTE than the wound structure.

While lateral pressure, i.e. a convex surface, is required to compact the laminae during winding, it is possible to wind onto large-radius and flat surfaces as long as some means of pressurization is used during crosslinking, as described in the previous paragraph. If the mandrel, or at least part of it, is removed prior to crosslinking, molding procedures such as those described in *Sections 4.2.2* and *4.2.4* may be used to produce concave sections. However, these procedures are not common, probably since the cost-efficiency of filament winding is then rapidly lost, thus making some alternate manufacturing technique more attractive.

Technique Characteristics

Equipment cost:	Intermediate
Mold cost:	Low
Labor cost:	Low
Raw material cost:	Low
Feasible series length:	Short to intermediate
Cycle time:	Long
Crosslinking requirements:	Complex
Health concerns:	High

Since one of the most common material combinations in wet filament winding is carbon-reinforced epoxy, two special considerations arise. One is that carbon particles invariably will be produced from abrasion during guidance prior to impregnation, thus causing the problem of shortouts of electrical and electronic equipment. The other concern is that of worker health when dealing with liquid epoxy resin, see also *Section 8.4.1.3*.

Component Characteristics

Geometry:	Convex and closed
Size:	Any
Holes and inserts:	Inserts possible, holes difficult
Bosses and ribs:	Not possible
Undercuts:	Not possible

Surfaces:	Inner surface good, may be improved with gelcoat; outer surface poor
Fiber arrangement:	Planar arrangement; oriented and continuous (balanced laminae)
Typical fiber volume fractions:	0.5–0.6 with helical and polar winding, 0.6–0.7 with hoop winding
Void fractions:	Low
Mechanical properties:	Good
Quality consistency:	Good

Applications

Despite the limitations of the process, filament winding has been used in an impressive array of applications courtesy of repeatable and very high mechanical properties in combination with relatively low component cost. The more common applications include:

- Drive shafts for small trucks, helicopters, and general industry
- Rocket motor cases
- Launch tubes for anti-tank weapons, mortars, etc. (see *Figure 1-30*)
- Pressure vessels (see *Figures 1-28* and *4-50*), including auxiliary fuel tanks for military aircraft and air tubes for fire fighters (see *Figure 4-52*)
- Pipes and storage tanks
- Fishing rods

While the application examples above are quite common and usually produced in long series, there are some impressive examples that illustrate the state of the art. Although never used in production, filament winding was certified as one manufacturing technique to produce the entire fuselage of the Beech Starship business aircraft. A few prototypes, 85 percent of full scale, were wound from carbon-reinforced epoxy and one was flight tested [12]. Another example is farm storage tanks too large to transport, which are (vertically) wound on site using mobile machinery. Winding is performed in open air in parts of the United States where good weather can be relied on for the days it may take to complete winding and crosslinking. Although most wound components obviously are rotationally symmetric, very large and highly nonsymmetric windmill blades have also been wound.

4.2.6 Pultrusion

Being the only common continuous composite manufacturing technique, pultrusion is the most cost-effective of all methods for mass-producing composites. Despite essentially being limited to components with constant

Figure 4-52 Firefighters wearing compressed air breathing apparatuses with filament wound pressure cylinders. The cylinders, which have air-impermeable
thermoplastic liners, are wound from carbon-reinforced epoxy overwrapped with a
scratch-resistant glass-reinforced epoxy surface layer. Compared to conventional
steel cylinders, these composite cylinders allow increased time of operation, significant weight reduction, and extended service life. Photograph courtesy of
Interspiro, Lidingö, Sweden

cross-section, pultrusion is a fast-growing manufacturing technique and
pultruded composites are used in a wide variety of fields spanning every
conceivable application from electrical and civil engineering to sports and
medicine.

Process Description

In essence, pultrusion is a technique by which continuous fibrous reinforcement is impregnated with a resin and then is continuously consolidated into
a solid composite. While there are several different ways to achieve impregnation and consolidation, *Figure 4-53* illustrates the different steps in a basic
pultrusion process, while *Figure 4-54* shows a production facility.

The reinforcement is pulled from packages in creels (in the background of
Figure 4-54) and is then gradually brought together and pulled into an open
resin bath wherein the reinforcement is impregnated. Emerging from the
impregnation bath, the impregnated reinforcement usually requires some
additional guidance or shaping before entering the die, see *Figure 4-55*. The
die has a constant cross-section cavity throughout most of its length; the

exception is the tapered entrance, where excess resin is squeezed out of the reinforcement (see *Figure 4-53*). The die is heated and the heat that is transferred to the (liquid) reinforcement-resin mass initiates crosslinking. The composite emerges from the die as a hot solid (see *Figure 4-56*) and is allowed to cool off before being pulled by the pulling mechanism, which is followed by a saw to continuously cut the composite to desired lengths (in the fore-

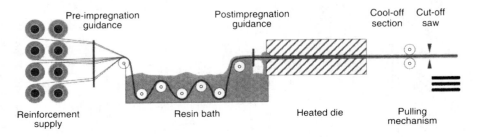

Figure 4-53 Schematic of pultrusion

Figure 4-54 Pultrusion facility. From background to foreground; creels (reinforcement supply), guidance devices, impregnation station, die, reciprocating pullers, and saw. Photograph courtesy of MMFG, Bristol, VA, USA

ground of *Figure 4-54*). The length of an entire pultrusion facility is in of order of 10 m, but may be considerably longer in pultrusion of large and complex cross-sections. The short process description above is based on the most basic but also most common pultrusion configuration. However, there are several variations on the theme and many details have been intentionally left out in the name of simplicity. In the following, the process is described in some detail and different processing and machinery options are briefly discussed. The variety in machinery solutions reflects that many pultrusion machines are designed and built in-house by the pultruder, although a few dedicated machinery manufacturers are responsible for the bulk of new machines.

The reinforcement creels are of the simplest possible rack construction since no significant loads, except the weight of the reinforcement, need to be supported. Creels often have wheels to allow off-line preparation and rapid reconfiguration of a pultrusion facility.

An overall consideration of pre-impregnation reinforcement guidance is that the reinforcement is fragile and in the case of glass also very abrasive. The dry rovings are often guided by ceramic eyelets to reduce friction and wear on both the fibers and guidance device, although steel rings suffice in many applications. Fabrics, mats, veils, and, close to the die, the rovings may also be guided by plastic sheets with machined slots or holes; these

Figure 4-55 Reinforcement guidance between impregnation station and die. Note that dry surface veils are added onto the impregnated reinforcement (from top) counting on there being sufficient resin to complete impregnation. Movement is from left to right. Photograph courtesy of MMFG, Bristol, VA, USA

Figure 4-56 Crosslinked components emerging from the dies of a multi-die facility. The components are rails for ladders such as that shown in *Figure 1-36*. Photograph courtesy of MMFG, Bristol, VA, USA

sheets, which are sometimes called carding plates, are often specific to a cross-section. Several sets of consecutive guides are employed to gradually shape and position the different reinforcement layers prior to impregnation.

There are four possible impregnation options, the first three of which are commonplace. The most common impregnation method is the one illustrated in *Figure 4-53*, i.e. the reinforcement is guided down into an open resin-filled bath. The impregnation is accomplished by capillary forces and the fact that the reinforcement is guided over and under rods located below the resin surface. The major advantage of this impregnation method is simplicity and good impregnation results, while the major disadvantage is significant volatile (usually styrene) emissions. The second impregnation option also employs an open resin-filled bath, but the reinforcement is not guided down into the bath, but rather horizontally enters and exits the bath through holes and/or slots in the walls of the resin bath and thus remains horizontal, see *Figure 4-57*. The major advantage of this method, which is often used in production of hollow composites, is that the reinforcement does not need to be bent, which otherwise may exclude the use of, or at least risk wrinkling or ripping, vertically oriented fabrics, mats, and veils. Also this impregnation method has the disadvantage of significant volatile emissions. Resin that inevitably leaks out with the reinforcement exiting the resin bath and that is squeezed out at the die entrance (see *Figure 4-55*) is normally collected in a drip pan and pumped back into the resin bath.

The third impregnation method, which is often referred to as injection or RIM pultrusion, employs a different kind of die. Unimpregnated reinforcement is guided into a narrow opening in the die, which further "downstream" widens into a cavity where the resin is injected under pressure, see *Figure 4-57*. The cavity, which is maintained at a temperature that ensures that the resin does not start to crosslink, is tapered towards the latter section of the die, which has a geometry similar to a traditional pultrusion die and is heated. The advantages of this impregnation method are low resin loss, no stagnant resin, that more reactive resin systems may be used (hence RIM pultrusion), and that the work environment is greatly improved since impregnation takes place in a closed mold resulting in very low volatile emissions. The major disadvantages are complex and expensive dies, that all guidance must be performed prior to impregnation, and potential problems in successfully impregnating large amounts of closely packed, longitudinally oriented rovings. A fourth but unusual impregnation alternative involves using prepregs. The reason for it ever having been considered is the improved composite properties which may be achieved.

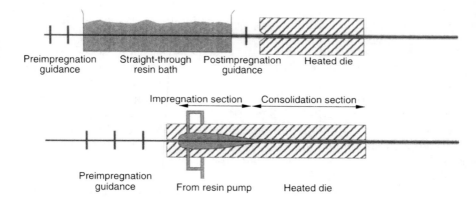

Figure 4-57 Straight-through (top) and injection (bottom) impregnation in pultrusion

The first two impregnation methods clearly dominate in North America and probably in most of the world, while injection pultrusion is favored in several European countries where permissible volatile levels in the workplace tend to be lower (cf. *Table 8-1*). Although injection pultrusion may be the simplest way of solving the emission problem, it is naturally possible to reduce this disadvantage of the first two impregnation methods through proper ventilation and air treatment, see also *Chapter 8*.

Once the reinforcement has been impregnated it is much less sensitive to friction than when it is dry and thus may be guided by for sheet-metal guides, for example. However, mats in particular tend to rip easily when wet (since the binder then is partially or completely dissolved) and therefore must be carefully guided. When manufacturing composites with complex geometries (e.g. multi-cavity, thin-walled sections) or with complex reinforcement orientations (e.g. large degree of off-axis reinforcement), the gradual forming of the impregnated reinforcement before it enters the die is one of the most critical and most difficult aspects of pultrusion, see *Figure 4-55*. In injection pultrusion all positioning of the reinforcement obviously must be accomplished before it enters the die, thus potentially making it a more complex issue.

The die is separated from the pulling mechanism by a long section to ensure that the consolidated composite cools off enough to be gripped by the time it reaches the pulling mechanism, see *Figure 4-54*. Caterpillar belt pullers, where successive rubber-coated pads are mounted on the belt, are common. However, probably the most common pulling concept involves reciprocating hydraulic clamp pullers with rubber-coated pads. Either one or two of these may be used; with one pulling mechanism the pulling is intermittent, with two coordinated reciprocating pullers the pulling is continuous, see *Figure 4-54*. Except for the simplest of composites, the rubber wheels or pads that grip the composite are tailored to fit the specific geometry of the composite, thus reducing the lateral pressure needed to grip the composite and reducing the risk of crushing a hollow component. A run-of-the-mill pulling mechanism may have a pulling capacity of 50–100 kN, although larger machines may have a capacity of several hundred kN.

Since pultrusion is a truly continuous manufacturing technique, composites may be produced in any length that can be handled; small, flexible cross-sections may even be wound onto drums "endlessly". Nevertheless, the composite normally is cut, in the simplest case with a hand-held hacksaw, but more commonly to predetermined lengths with an automatic saw that follows the moving composite during cutting, see *Figure 4-54*.

To satisfy particularly high torsional requirements, roving winding units that overwind the predominantly axial reinforcement with an angle pattern of choice may be used, see *Figure 4-58*. This technique is referred to as pulwinding (or pull-winding) due to the relation with filament winding. The figure does not show any impregnation unit, but possible solutions include a straight-through bath or even resin injection from within the mandrel and radially toward the exterior surface of the component.

Figure 4-59 illustrates typical pulling speeds as a function of wall thickness for unsaturated polyesters and vinylesters; the curve for an epoxy would show the same dependency but would lie below the vinylester curve. The figure illustrates how the slower crosslinking of vinylesters on the one hand and of thicker sections on the other hand lead to lower pulling speeds since longer die residence times are required. (These dependencies really apply to any manufacturing technique, but then it is the time to demolding that is affected by resin reactivity and component thickness.) Despite the very conservative data of *Figure 4-59* only showing speeds up to 10 mm/s, speeds up to 100 mm/s may be reached for simple geometries and reactive resins.

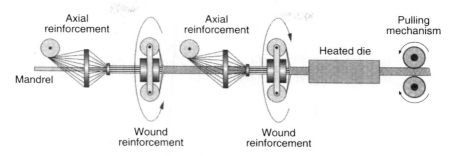

Figure 4-58 Schematic of pulwinding

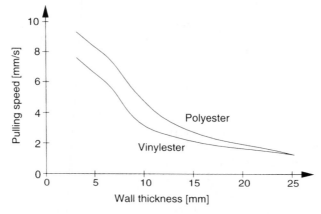

Figure 4-59 Typical pulling-speed dependencies for varying wall thicknesses for pultrusion with unsaturated polyesters and vinylesters

Raw Materials and Dies

Virtually any reinforcement–matrix combination feasible in any other composite manufacturing technique also may be used in pultrusion. However, glass-reinforced unsaturated polyesters heavily dominate (over 70 weight percent of total production).

Due to the nature of the process the reinforcement must be continuous, either in the form of rovings or rolls of fabrics, mats, and veils. Most of the reinforcement used is in roving form, although it is standard practice to interleave layers of roving with CSMs to improve transverse properties, and surfacing mats, veils, and non-wovens to create a resin-rich, and thus cosmetically appealing and environmentally resistant, surface. For applications requiring higher transverse or torsional properties, woven fabrics, noncrimp fabrics, and braids, which may all be stitched together into continuous pre-forms, are used. Pulwinding of rovings is used for the highest torsional requirements. While fabrics may be woven or braided to the desired widths, CSMs and veils are cut to desired widths by the user. At the end of a roving package a new roving is spliced to the end of the former one with an air-splicing device; in less critical applications a simple knot may be tied. Fabrics, CSMs, and veils are usually hand-sewn together. If required, the operator can easily track the composite section containing the splice through the process to ensure that it is discarded.

Unsaturated polyester is the dominating resin not only due to its attrac-tive performance-to-cost ratio, but also because it is the easiest to process. Improved properties are naturally achieved with vinylesters and epoxies, but processing becomes progressively more difficult and pulling speeds lower, since the resin reactivity determines the maximum speed all other things unchanged, see *Figure 4-59*. Pultrusion of epoxies is particularly prob-lematic since they do not shrink as much as unsaturated polyesters and vinylesters, meaning that the solid surface of the partially crosslinked com-posite is in full contact with the die for a greater portion of the die length; epoxies also tend to stick to the die to a greater degree than other resins, leading to inferior surface finish. Perhaps more in pultrusion than in other processes, it is up to the processor to formulate the resin. To achieve high processing speeds without significantly sacrificing product properties it may be necessary to use several different initiators as well as accelerators and inhibitors. It is also normal practice to use various additives, including internal mold releases. In addition to the previously mentioned resins, sev-eral others, including phenolics, acrylics, and PURs, are used in pultrusion.

For open-bath impregnation the resin must have a long pot life, prefer-ably a day or more, since parts of it will remain in the impregnation bath for a long time and cannot be allowed to gel. In contrast, with injection pultru-sion highly reactive resins with very short pot lives may be used. In either case the resin viscosity must be low, preferably less than 0.5 Pa·s, to enable satisfactory reinforcement impregnation. The key issue is to tailor the resin formulation to obtain an acceptable compromise between good properties, long pot life, and rapid crosslinking.

A pultrusion die is usually machined from tool steel and typically has a length of 600–1,500 mm. With the exception of the tapered or rounded entrance, the die has a constant cross-section cavity with highly polished surfaces that are chrome-plated to decrease wear, lower friction, and improve composite surface finish. Properly designed and maintained a die may last 15–30 km before requiring renewed chrome-plating to further extend its life. The die is usually sectioned to facilitate machining, plating, inspection, and cleaning. Normally the die has several independently con-trolled temperature zones. Most of the die is heated, typically employing two or more control zones, while the final section of the die may need active cooling. Since excess resin is squeezed out of the reinforcement at the tapered entrance to the die, this section may also require active cooling to prevent premature gelation of the resin (i.e. before it reaches the constant cross-section cavity). The die may be heated and cooled through clamping it between two temperature-controlled platens, through stripping of plate heaters onto its surface (see *Figure 4-56*), or through inclusion on heating and cooling provisions within the die steel. The first two solutions tend to be the most common in production, since they reduce the amount of machining that goes into each die. To increase throughput when pultruding reasonably simple components, it is common practice to mount several dies in parallel in the same pultrusion machine, see *Figures 4-54* and *4-56*.

For pultrusion of hollow cross-sections, cantilever mandrels that are mounted "upstream" of the die and that end close to the end of the die are used, see *Figure 4-58*. In the transverse direction, the cantilever mandrel (which may be heated) is oriented solely by the amount of reinforcement surrounding it. The fact that it is difficult to produce perfectly concentric pipes and hollow sections with well-defined wall thicknesses underscores the significance of reinforcement guidance and positioning.

As mentioned in the impregnation section above, a die for injection pultrusion is of a different design. One may look upon such a die as a conventional die with an added impregnation section, see *Figure 4-57*. Such a die therefore is longer, more complex, and more expensive than its conventional counterpart.

Crosslinking

Once the impregnated reinforcement reaches the heated portion of the die crosslinking is initiated and the resin gradually solidifies from the perimeter of the composite toward the center. Although the die initially heats the reinforcement–resin mass, the crosslinking exotherm causes the temperature of the solidifying composite to exceed that of the die toward its end and the die thus cools the composite. Correctly performed, the temperature peak caused by the exotherm will take place within the confines of the die and the contraction of the resin due to crosslinking will cause the composite to shrink away from the die to eliminate friction. Since crosslinking is not fully completed within the die it is separated from the pulling mechanism by a generous distance, generally 3 m or more, to allow completion of crosslinking and cooling before the composite is gripped by the pullers. The faster the resin crosslinks while still keeping the maximum exotherm temperature within reasonable limits, the faster and more economical the process will be.

To prevent cracking and excessive, exotherm-induced residual stresses in thick composites as well as to allow increased pulling speed, preheating may be used to heat either the reinforcement prior to impregnation or to heat the liquid reinforcement–resin mass through the thickness. Common heating principles are based on radio and induction frequency, depending on whether the reinforcement is conductive (i.e. carbon) or not. Alternatively, the resin may be heated prior to impregnation to lower its viscosity and thus facilitate reinforcement impregnation; this is most common with injection pultrusion where long pot life is not a necessity.

To ensure that the resin crosslinks with an exotherm temperature high enough to achieve complete crosslinking, but not so high as to cause excessive thermally induced stresses or even cracks in the composite, precise multi-zone temperature control is essential. While direct control of the temperature of the reinforcement–resin mass would be ultimately desirable, one must resort to controlling the die temperature at a few discrete points (where thermocouples are embedded). To get some feedback on the resulting composite temperature profile as a function of location in the die, it is common to embed bare thermocouples with long leads within the reinforcement–resin mass prior to entry into the die; the thermocouple thus monitors the temperature development within the composite throughout impregnation and consolidation. Die temperatures are in the range 100–160°C for unsaturated polyesters and vinylesters and around 300°C for epoxies. The consolidation pressure is not explicitly controlled and rarely known.

Technique Characteristics

Equipment cost:	Intermediate
Mold cost:	Intermediate
Labor cost:	Low
Raw material cost:	Low
Feasible series length:	Long
Cycle time:	Long
Crosslinking requirements:	Complex
Health concerns:	High with open bath impregnation; low with injection impregnation

For the same reasons as with filament winding, pultrusion using carbon and epoxies brings on special considerations. Pultrusion with these materials (not necessarily in combination) is therefore essentially only carried out by a small number of companies specializing in it.

The key technology issues of pultrusion are usually considered to be resin formulation, temperature control, material guidance, and die design (not in order of importance). It is typically the skills in these areas that distinguishes a successful pultruder from his hapless competitor.

Component Characteristics

Geometry:	Straight, constant cross-section
Size:	Any length; cross-section limited by machine size

Holes and inserts:	Longitudinal inserts and holes possible
Bosses and ribs:	Longitudinal ribs possible
Undercuts:	Longitudinal undercuts possible
Surfaces:	All surfaces good
Fiber arrangement:	Longitudinal, planar arrangement; random or oriented, continuous or discontinuous (bulk of fibers longitudinal)
Typical fiber volume fractions:	0.2–0.5 with roving-mat combination and fabrics, up to 0.65 with rovings only
Void fractions:	Intermediate
Mechanical properties:	Poor to good
Quality consistency:	Good

Applications

The major reason for the successes of pultrusion are low component cost, but when required it is possible to pultrude structurally very capable components. By composite standards pultruded components stand out since they are the only ones that to any significant degree are stock items. It is thus possible to buy numerous standard shapes (e.g. as shown in *Figure 4-60*) in much the same manner as one can find extruded aluminum profiles (with

Figure 4-60 Pultruded standard components. Photograph courtesy of Creative Pultrusions, Alum Bank, PA, USA

which pultruded components compete) at a hardware store. While there are many standard components, custom cross-sections are common but may require rather long series to write off the cost of the die. Common applications include:

- Truck and bus components, such as body panels and drive shafts
- Construction members, such as building panels, window and door lineals (frames), beams, gratings, pipes, walk ways, cable trays, etc. (see *Figures 1-31* through *1-34*)
- Electrical equipment, such as ladders (see *Figure 1-36*), booms for cherry picker trucks, tool handles, etc.
- Sporting goods, such as ski poles and fishing rods

About half of all pultruded products are used in non-structural areas, predominantly in electrical applications, such as equipment housings and spacers for transformer windings. Nevertheless, some recent outstanding examples of what can be achieved with structural components are several bridges (cf. *Figures 1-31* and *1-32*), a 48-m long airfoil-shaped windmill blade weighing 1.5 tons, and 450 km cable trays for the tunnel under the English Channel.

4.2.7 Other Techniques

The techniques for manufacturing thermoset composites that have been covered so far are the commercially most relevant ones, but a range of less common techniques are used as well. Further, many techniques logically belong in between the more conventional techniques, making classification difficult.

In recent years there has been much written about a family of conceptually similar manufacturing techniques often referred to as fiber placement. While there are endless variations to the processes envisioned, they all tend to involve the use of a multi-axis robot to deposit a continuous supply of wet-impregnated roving or prepreg onto a complex mold to create reinforcement orientations and geometries not achievable with filament winding, e.g. complete automobile space frames. A lot of effort has been spent on this concept, but it appears as if few, if any, commercial techniques have yet emerged. (It deserves to be pointed out that in trade and scientific literature some versions of filament winding and tape laying are also referred to as fiber placement—definitions obviously vary significantly between authors.)

A double-belt press (DBP) such as described in *Section 2.3.2* may be used in a number of ways to consolidate composites. Although quite rare in consolidation of thermoset composites, DBPs are used to produce sandwich components. In this case, face sheets are usually coiled up in very long lengths. Two rolls of face sheets on either side of a core and two adhesive films are uncoiled and guided in between the belts of the press. Face materials may be sheet metal, unreinforced polymers, and composite laminates or prepregs. To obtain a truly continuous core it may prove convenient to in-situ foam it between the faces through injection and subsequent expansion of PUR, for example, instead of using discrete blocks of expanded foam or wood. This method is used to produce flat building panels, for instance.

Another continuous manufacturing technique that is similar in concept to pultrusion uses an oven instead of a die. The impregnated reinforcement is sandwiched between two polymer films prior to entering the oven where forming shoes are used to shape articles that are essentially flat. Following crosslinking the two polymer films are removed. Products include corrugated paneling, road signs, etc.

In centrifugal casting, resin and reinforcement is deposited onto the inside of a rotating mold exploiting the centrifugal force. Either the reinforcement, usually in the forms of mats, is first placed in the mold and resin added later during rotation, or a spray gun may be used to deposit a mixture of resin and chopped reinforcement. Crosslinking is initiated through mold heating and/or injection of heated air into the mold cavity. Large-diameter pipes and tanks with well-controlled external surfaces may be produced; the external surface may be improved using a gelcoat.

4.3 Thermoplastic–Matrix Techniques

In the beginning of this chapter it was mentioned that thermosets clearly dominate in composite applications and that thermoplastics are relative newcomers to the field. In fact, it is only in injection and compression molding and to some degree in prepreg layup that thermoplastics-based composites are currently of any real commercial significance. However, since thermoplastic prepregs became readily available in the 1980s there has been an immense interest in this material family. The main overall reasons for the interest are related to potential improvements in composite properties, manufacturability, recyclability, and work environment, cf. *Table 2-1*. From a manufacturing viewpoint the main attraction of thermoplastics lies in the possibility of achieving very short demolding times since no chemical reaction needs to take place for the composite to solidify. Autohesive bonding (cf. *Section 2.1.3.5*) between two thermoplastic surfaces (or prepreg plies) may be achieved in a fraction of a second provided the surfaces are at sufficiently high temperature and are instantaneously brought into intimate contact. While autohesion has been achieved very rapidly in laboratory composite consolidation experiments, it is really only in filament winding and to some degree in tape laying that similarly short time frames may be relevant in real manufacturing situations. In most cases it is not the time required to achieve virgin material strength across an interface that limits the processing rate; the limits are usually set by the time required to heat and cool the material. The main disadvantages of thermoplastics are their high viscosity, which in most cases precludes in-manufacturing impregnation, their high requirements in terms of processing temperature, and last but not least the high raw material cost, which to a significant degree is a result of the general need to use prepregs or molding compounds. While it is possible to impregnate the reinforcement as part of the process using low molecular-weight prepolymer that following impregnation is fully polymerized (as briefly discussed in *Section 2.1.5*), this processing route has so far seen little commercial success.

Although the interest in thermoplastic composites has not resulted in any great number of commercially important manufacturing techniques, the large research and development efforts in industry and academia throughout the world have proven the technical—if not always the economical—feasibility of a number of more or less innovative processing routes. Many of these manufacturing techniques, and of course also the commercially successful ones, are the subject of the remainder of this section. The reason for discussing a range of so far commercially unsuccessful or uncommon techniques is that it is this author's belief that all the manufacturing-related research and development efforts, together with legal impediments in terms of requirements for good work environment and recyclability, will gradually lead to a greater acceptance of thermoplastic composites. Most manufacturing techniques that have been tried with thermoplastics are heavily inspired by the commercially successful techniques used with thermosets. Other techniques have been borrowed from sheet-metal forming; these include compression molding and rollforming (although the former obviously via thermoset composite manufacturing). The only major technique that so far has proven promising and that has been developed exclusively for thermoplastics is diaphragm forming, which therefore is treated a little more thoroughly than perhaps otherwise would have been warranted. The technique descriptions below essentially follow the same logical sequence as *Section 4.2* and just as in that section the order of the technique descriptions in no way indicates the degree of commercial success or how common they are.

4.3.1 Prepreg Layup

The most straightforward way of laminating a composite from thermoplastic prepreg no doubt is to cut the prepreg by hand and likewise to lay it up by hand onto the mold. Due to the research and development status of most techniques for consolidation of thermoplastic prepregs, hand layup is indeed the most common layup method. However, as with the corresponding thermoset technique there are also automated means to lay up thermoplastic prepreg tapes. For consolidation of a laid-up prepreg stack there is a range of feasible routes, some of which are addressed as independent manufacturing techniques in later parts of this chapter.

Process Description

In most respects layup of thermoplastic prepreg is quite similar to the thermoset counterpart. However, while cutting of the material may be carried out with the same or similar techniques and tools, one of the major differences between the material types is the complete lack of tack in thermoplastic prepregs; moreover, with melt- and solvent-impregnated prepregs the drape is poor. To allow conformation to the mold during layup, drape may be artificially achieved through localized heating, for example using a hot air gun (such as those used for paint stripping and welding of plastic bathroom flooring) to soften the matrix. Similarly, to ensure that prepreg plies remain where placed, tack may be artificially achieved through localized melting of adjacent plies using a soldering iron (or a hot air gun and some means of compressing plies during cooling). In contrast, fabrics made from powder-impregnated and commingled material forms possess drape, but tack must still be achieved through localized melting.

Equipment conceptually similar to that used for automated layup of thermoset prepreg tapes may also be used to lay up thermoplastic prepregs. In this case the mating surfaces are first heated and then joined under pressure, see *Figures 4-61* and *4-62*. If both the previously laid ply and the incoming tape have completely melted surfaces when they are joined, postprocessing may not be needed and the component thus is ready for demolding as soon as layup is completed. However, due to the localized and highly non-uniform heating most attempts at such a one-step procedure have resulted in considerable residual stresses causing component warpage. It is therefore more common to only aim for partial consolidation to ensure that the prepregs stay in place and then achieve full consolidation in a separate processing step. If only partial consolidation is the aim, layup can proceed much more rapidly with the same amount of heat input, thus increasing the processing rate.

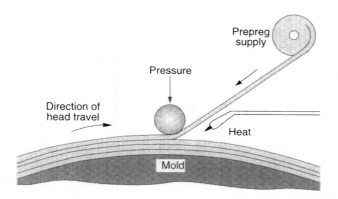

Figure 4-61 Schematic of automated thermoplastic tape laying

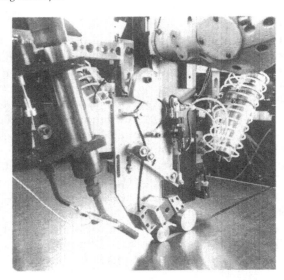

Figure 4-62 Experimental facility for thermoplastic tape laying. The black prepreg enters vertically in the center of the picture and goes through tensioning and cut-restart mechanisms before being pressed onto the horizontal mold by two rollers. In this case the sole heat source is a high-power nitrogen torch visible to the left. The cylinder enclosed in coiled plastic tubing to the right is a focused non-contact thermometer measuring the temperature of the newly laid material. The entire layup head is mounted on an industrial robot which positions the layup head over the stationary mold. Photograph reprinted from reference [13] with kind permission from VDI Verlag, Düsseldorf, Germany

Raw Materials and Molds

All kinds of prepregs may be used in hand layup, while tape laying machinery requires prepreg in tape form. Being a high-cost manufacturing technique, high-performance and thus high-cost carbon-reinforced materials are predominantly used, although glass-reinforced low-temperature polymers certainly may be processed in this fashion as well. The molds are likely to be identical to those used in the corresponding thermoset process with the exception that higher temperature tolerance will be required. Mold heating during layup may be used to decrease heating requirements (or increase layup rate) and to reduce processing-induced residual stresses.

Consolidation

Consolidation of a laid-up stack of prepregs may be achieved in a number of fashions. One of these techniques is the already mentioned tape laying incorporating complete consolidation concurrent with the layup, while other consolidation possibilities include compression molding, diaphragm forming, and rollforming, which will be covered in the appropriate contexts in later sections. Moreover, since much of the interest in layup of thermoplastic prepregs has its origin in the military aerospace industry, it should not come as a surprise that autoclave consolidation is another technically proven consolidation route. Vacuum bagging of a thermoplastic laminate is simpler than the thermoplastic counterpart in the sense that little, if any, resin should bleed out of the laminate (if processed correctly) so it may be possible to exclude the bleeder (cf. *Figure 4-26*). However, for many thermoplastics the processing temperatures are so high that conventional release films, vacuum bags, and sealing tapes cannot be used. Whereas PA films may be used as vacuum bags for thermosets and low-temperature thermoplastics, PI films may be required with high-temperature thermoplastics. A recommended—and surely very conservative—autoclave consolidation cycle for carbon-reinforced PEEK prepreg is illustrated in *Figure 4-63*.

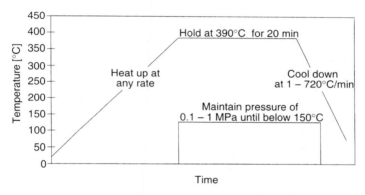

Figure 4-63 Suggested autoclave cycle for carbon-reinforced PEEK. Redrawn from reference [14]

Technique Characteristics

Comparing thermoplastic prepreg layup followed by autoclave consolidation to its thermoset relation one finds that the former offers few, if any, processing advantages; indeed, the disadvantages may even predominate. This processing route therefore may be considered an example of a less than ideal manufacturing technique. The only likely reason to select this route would be to utilize the inherent properties of the thermoplastic matrix, not any processing advantage.

Component Characteristics

Hand laid-up and autoclave consolidated composites essentially should have all traits that are not matrix-specific in common with their thermoset cousins. Due to the tack and drape problems it may however be more difficult to produce geometrically complex components with as good results as are common with thermosets.

Applications

Prepreg layup followed by autoclave consolidation has proven to be a feasible technique for the manufacture of range of aircraft components, but it is not likely in widespread commercial use. Specific application examples include wing skins (see *Figure 1-22*), access doors, and landing gear doors, where the impact properties of the composite rather than the manufacturing technique may have been the deciding factor.

4.3.2 Liquid Molding

As obvious from the process descriptions for thermoset liquid molding of structural composites (i.e. RTM, vacuum injection molding, and SRIM) in *Section 4.2.3*, the processes require resins with very low viscosities to enable reinforcement impregnation during molding. Since molten thermoplastics have very high viscosities (cf. *Table 3-3*), direct adaptation of the conventional thermoset liquid molding techniques would appear impossible. Indeed, with a couple of exceptions of sorts this is correct.

The first of these exceptions employs the aforementioned low-viscosity prepolymers that are not fully polymerized until the reinforcement has been fully impregnated, in which case the technique should be quite similar to RTM or possibly SRIM. Although technically possible this processing route is not yet of commercial significance.

The second exception is injection molding, although it is highly questionable whether injection molded composites can be called structural. While this treatment is essentially limited to structural composites, thermoplastic injection molding is of interest to this treatment at least as an excellent way to recycle production scrap from other thermoplastic composite manufacturing operations (see also *Chapter 7*) and will therefore be discussed in some detail. As was alluded to in *Section 4.2.3.1*, thermoplastic injection molding is significantly more common than its thermoset counterpart and is widely used to manufacture all kinds of plastic articles.

Process Description

Thermoplastic injection molding is quite similar to its thermoset relation and the following sections will therefore concentrate on the differences with respect to *Section 4.2.3.1*. With reference to *Figure 4-32* the machinery is similar, with the most notable differences being that barrel and manifold are heated to maintain the resin in the liquid state and that the mold is cooled to solidify the molded component. Screws used for thermoplastics are generally longer than those used for thermosets and the compression ratio tends to be higher. The stages illustrated in *Figure 4-33* are common to both thermoset and thermoplastic injection molding.

Raw Materials and Molds

Essentially all thermoplastics may be injection molded, but the emphasis is on commodity and engineering polymers such as PE, PP, and PAs, which are often reinforced. Commodity-grade resins are generally glass-reinforced, while high-performance thermoplastics such as PPS and PEEK sometimes are carbon-reinforced. The raw material form is normally pellets that have been continuously produced with continuous reinforcement (e.g. through miniature pultrusion) and then chopped so as to contain collimated fibers of the same length as the pellet. Sprue and runners are normally chopped and recycled together with virgin raw material without any property degradation being noticeable (see also *Chapter 7*). Just as in thermoset injection molding, molds are machined from tool steel and include cooling channels.

Consolidation

Since the mold is cooled the material essentially solidifies instantaneously as it touches the mold surface, while solidification of the interior of the component is dependent on conduction through the thickness. Cycle times range from a few seconds to a minute or two. Due to the higher viscosity of thermoplastics, injection pressures tend to be higher than with thermosets.

Technique Characteristics

The predominant characteristic of thermoplastic injection molding is its capability of economically producing components in long series. Since no chemical reaction needs to take place, cycle times are shorter and the process therefore even better suited for long series than its thermoset relation. However, machinery and molds remain expensive.

Component Characteristics

Injection molded components may be extremely geometrically complex. Though most injection molded components are unreinforced, inclusion of reinforcement significantly enhances properties to allow components to support modest structural loads. However, the work of the screw tends to significantly damage fibers, but it is possible to obtain fiber lengths of 0.5–5 mm in the molded part. The fibers are aligned by the flow to create a "sandwich" with fibers in the surface layer parallel to the surface and a gradual change to an essentially perpendicular or random fiber arrangement in the interior.

Applications

While almost every plastic household article is manufactured through thermoplastic injection molding, they are often unreinforced and rarely intended to be load-carrying. Semistructural short-fiber reinforced articles include components for light machinery, including for example equipment housings and sprockets, and a range of automobile components, such as air intake manifolds. The latter are interesting in that they are becoming extremely complex from a geometrical point of view. A commercial example is where a conventional manifold consisting of over 80 components (for a 32-valve V8 engine) has been replaced by a single-piece composite consisting of discontinuous glass fibers in a PA 6,6 matrix. Such manifolds are injection molded using a so-called lost-core technique, where a metal alloy with low melting point is cast to the shape of the interior surface of the component. This core is then placed in the mold of the injection molding machine and the glass-reinforced thermoplastic is injected into the space between mold and core. Following solidification the core is carefully melted to produce a hollow structure (and the metal alloy cast into a new core).

4.3.3 Compression Molding

The range of techniques that may be used to compression mold thermoplastic composites are conveniently categorized by whether they involve a significant amount of material flow to fill the mold (as with molding of SMC and BMC) or whether the material predominantly is deformed to conform to the mold. This difference in processing and naturally also raw material form warrants separate treatments.

4.3.3.1 Compression Molding of GMT

Compression molding using GMT as raw material is currently the only technique for manufacturing of structural thermoplastic-based components in widespread commercial use. The technique is mainly used in the automobile industry, where the similarities with sheet metal stamping and molding of SMC and BMC have facilitated its acceptance. In analogy with the ever more common practice of referring to the techniques of compression molding of SMC and BMC merely as SMC and BMC, respectively, compression molding of GMT has unfortunately become known as GMT.

Process Description

Compression molding of GMT employs presses and matching molds nearly identical to its thermoset counterparts, but there are notable differences in the thermal history experienced by the material; *Figures 4-42* and *4-46* could thus equally well apply to compression molding of GMT. For large-scale manufacturing operations the raw material is delivered already cut to size and stacked on pallets. A robot (or a human) picks the raw material off the pallet and places each sheet individually on a conveyor belt that brings the material through an oven, normally IR or hot air, see *Figure 4-64*. By the time the material exits the oven at the other end the sheets are thoroughly melted (and have expanded in the thickness direction due to fiber relaxation) and are stacked to form the charge, which is rapidly placed in the mold. The press then very quickly closes and the material flows to fill the cooled mold before it solidifies and the component is ejected. Since the charge only covers approximately half the mold surface, material flow is notable, as is consequently reinforcement orientation. Properties therefore may vary significantly within a component. Contrary to the scale of *Figure 4-64*, the preheating oven in a large-scale production facility easily takes up more room than all the other stages together, since it may have to be over 10 m long to produce a continuous supply of molten GMT to the molding stage at a rate to match a cycle time of less than a minute.

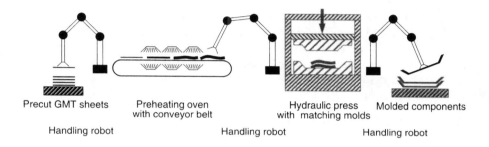

Precut GMT sheets   Preheating oven   Hydraulic press   Molded components
with conveyor belt   with matching molds

Handling robot   Handling robot   Handling robot

Figure 4-64 Schematic of automated thermoplastic compression molding line

Raw Materials and Molds

By far the most common resin in GMT is PP, although PBT, PAs, and other resins are available as well; they are all normally reinforced with continuous but randomly oriented glass fibers. Melt-impregnated GMT is the most common material form but powder-impregnated discontinuous-fiber reinforced GMT is also used. In applications where increased stiffness is required, GMT with varying degrees of oriented and continuous reinforcement or interleaved UD prepreg layers may be used.

While it is the norm that GMT sheets are bought from a dedicated raw material supplier, at least a couple of solutions for in-house manufacture of thermoplastic molding compounds have been tried in order to reduce cost. One of these methods employs a proprietary extrusion-type compounder to melt-impregnate discontinuous reinforcement which is then extruded and stored in a metered amount right by the press. Apart from the potential cost reduction in raw material, the compounded material does not require (renewed) melting since it exits the compounder molten and is kept molten until molding. The process is said to be capable of supplying fiber lengths up to 50 mm. Another method involves powder impregnation of chopped fibers onto a preforming screen employing machinery not unlike that used in preform manufacture (cf. *Figure 2-47*).

Matched steel molds are almost identical to those used in molding of SMC and BMC. To maintain a constant mold temperature requires heating during startup and cooling during the remainder of the shift. Since material that comes into contact with the cooled mold rapidly solidifies—sometimes before having had time to flow at all—it may be possible to see the charge footprint (the shape of the bottom layer of the charge) on the surface of the molded component.

Consolidation

Consolidation takes place within the closed mold under pressures in the range 10–20 MPa, although higher pressures are used in some cases. For PP-based GMT the surface temperature of the material when it exits the preheater should be in the range 200–230°C (some 15°C less in the interior), while the mold temperature should be in the range 30–70°C. Although material flow should be accomplished within less than 3 seconds after it has commenced, cycle times are usually between 20 and 60 seconds, most of which is taken up by cooling.

Technique Characteristics

Equipment cost:	High
Mold cost:	High
Labor cost:	Low to intermediate
Raw material cost:	Intermediate
Feasible series length:	Long
Cycle time:	Short
Consolidation requirements:	Simple
Health concerns:	Low

Component Characteristics

Geometry:	Any
Size:	Limited by press size
Holes and inserts:	Inserts and vertical holes possible
Bosses and ribs:	Possible
Undercuts:	Difficult
Surfaces:	All surfaces controlled, but appearance poor
Fiber arrangement:	Planar arrangement; random and continuous or discontinuous (continuous and oriented fibers possible)
Typical fiber volume fractions:	0.1–0.3, slightly higher with oriented reinforcement
Void fractions:	Intermediate

Mechanical properties: Poor

Quality consistency: Good

Applications

Most of the reasons for the popularity of SMC and BMC molding apply to molding of GMT as well. However, since cycle times are even shorter with GMT and no concerns of chemistry are present, GMT is becoming increasingly popular in the automobile industry. Courtesy of the poor surface finish of GMT-based composites, they are often hidden from direct view. Common applications of compression molded GMT composites include:

- Automobile, truck, and bus components, such as battery trays, bumper beams, noise shields, front ends, dashboards, etc. (see *Figures 1-13* and *1-16*)
- Containers and housings
- Electrical and machinery components
- Helmets

The front end in *Figure 1-16*, which originally was compression molded from SMC, is a good example of how the shorter cycle time in compression molding of GMT is a considerable advantage. When production rates were to be increased, two parallel production lines for molding of the SMC component would have been required. An attractive alternative was then offered by the GMT solution, which in this case required a cycle time of about a third of the SMC operation; a single production line was thus sufficient and substantial cost savings were consequently achieved.

4.3.3.2 Compression Molding of Prepregs and Flat Laminates

When compression molding material containing aligned and continuous fibers the material cannot flow adequately to fill the mold and the deformation modes discussed in *Section 4.1.3* must be responsible for material conformation to the mold lest the fibers be fractured or buckled. This significant difference brings on a host of new processing issues and extensive research and development work has resulted in a range of different technique variations.

Process Description

Compression molding of material with continuous and aligned reinforcement, also known as hot stamping due to the similarity with sheet metal stamping, employs the same kind of molds as used for GMT molding, but the deformation modes available to the raw material impose severe restrictions on what geometries may be molded. The general processing steps illustrated in *Figure 4-64* also apply to compression molding of stacked prepregs or already consolidated laminates, or blanks, with the difference that the charge essentially already covers the entire mold surface and that the final component thickness is a direct function of the thickness of the raw material at any given location, see *Figure 4-65*. For stacks of unconsolidated prepregs, particularly powder-impregnated and commingled prepregs where the reinforcement is not yet melt-impregnated, two presses—one heated and one cooled—may be required to successfully impregnate the reinforcement and then consolidate the component.

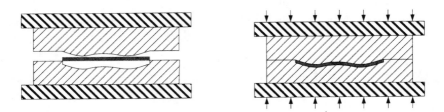

Figure 4-65 Schematic of compression molding of material with continuous and aligned reinforcement

As an aside it may be interesting to learn that compression molding using matching molds is also a common technique for consolidating flat test laminates in research and development work. The materials consolidated in this fashion tend to be high-performance prepregs of all types and sometimes layers of unimpregnated fibers interleaved with resin films; with the latter forms of raw material the technique is called film stacking. In consolidation of flat laminates the raw material is wrapped in a suitable release film and then placed in a picture-frame mold. The mold is then closed and placed in a heated press under slight pressure until the desired equilibrium temperature is reached within the material (alternatively, the mold itself may of course be heated). When the molding temperature has been reached, molding pressure is applied for a given length of time before the mold is rapidly moved to a cold press (or the mold itself cooled) and the part is demolded. This is clearly not a viable technique for anything but test specimen manufacture.

Returning to techniques potentially realistic for commercial use, rubber-block molding and hydroforming reduce the risk of wrinkles in the part and employ cheaper molds; *Figures 4-66* and *4-67* illustrate the two concepts. To enable molding of complex geometries in rubber-block molding the block may be contoured to match the rigid lower mold half. In hydroforming a pressurized liquid is contained behind a flexible membrane that is capable of conforming to the shape of the lower mold half. Pressures may be significantly higher in hydroforming than in rubber-block molding.

Either of the compression molding techniques described above may be modified to produce a continuous component of essentially constant cross-section through an incremental molding procedure incorporating both

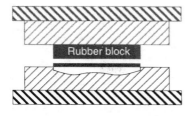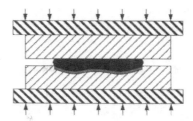

Figure 4-66 Schematic of rubber-block molding

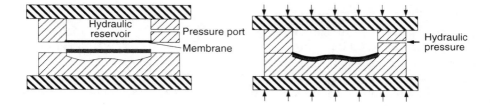

Figure 4-67 Schematic of hydroforming

heated and cooled molds (or a longer mold having both heated and cooled zones). This concept has been used to produce curved I beams, for example.

In deep drawing the material to be formed is mounted in a frame that keeps it under slight tension until forming is completed. As the molten material leaves the preheater it is placed over a female "mold" essentially consisting of a frame the shape of the projected final product, see *Figure 4-68*. A male mold then rapidly punches the molten material through the frame. This crude forming technique produces components with rather poor surface finish, especially the outer surface, and allows limited control of material movement during forming. Nevertheless, it is a low-cost method that has proven its commercial feasibility in for example manufacturing of hinge covers for cargo bays of civilian aircraft.

Work is also underway to develop compression molding techniques capable of mass-producing all-thermoplastic sandwich components in which the foam or honeycomb core is based on the same materials as used in the faces. To manufacture such components through compression molding, the sandwich may either be created in the same step as the forming operation or preconsolidated sandwich blanks may be formed in a second processing step. The main drivers behind such developments are postforming and recycling possibilities.

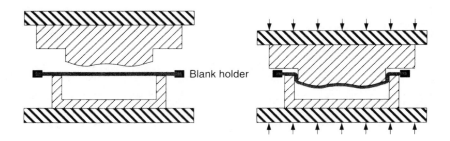

Figure 4-68 Schematic of deep drawing

Raw Materials and Molds

Compression molding using matching molds has been successfully employed to consolidate prepreg stacks into flat blanks as well as to mold both prepreg stacks and blanks into moderately curved sheet-like components. All types of prepregs, as well as film-stacked materials, have been used and all material combinations have proven viable. Rubber-block molding, hydroforming, and deep drawing all are likely to require preconsolidated blanks to produce well-consolidated components.

Matching metal molds are machined from steel just as in molding of GMT, whereas rubber-block molding and hydroforming employ one rigid mold half and one flexible. While the rigid mold certainly needs to be metal in a production environment, lower strength materials, including wood, have proven satisfactory for prototype runs. Rubber blocks used tend to be silicone or silicone-coated PUR, while the membrane containing the hydraulic reservoir in hydroforming is of high-strength rubber.

Consolidation

If the raw material is in the form of powder-impregnated and commingled prepregs (in which the reinforcement is not melt-impregnated), the aforementioned procedure with a heated and a cooled die (or a die that is first heated and then cooled) is required to satisfactorily impregnate the reinforcement and consolidate the laminate. In this case the time in the heated mold easily ranges from fractions of an hour and upwards to an hour, while pressures are in the range 1–10 MPa. The cooling time is likely to range from a few seconds with a separate cooled mold to fractions of an hour if only one mold is used (limited by cooling capacity).

If the raw material is unconsolidated melt-impregnated prepregs the procedure is slightly different. Assuming that the prepreg stack has a uniform temperature through the thickness when placed in a cooled mold, consolidation times should be measured in seconds for all the aforementioned techniques. However, due to the poor heat transfer capabilities of rubber, cycle times will be longer with rubber-block molding and hydroforming than with matching metal molds. The actual forming time, on the other hand, is likely to be on the order of seconds. In consolidation of stacks of melt-impregnated prepregs, low pressure is sufficient; pressures in excess of 1 MPa appear not to improve consolidation.

Improved component properties are achieved if preconsolidated blanks are used as raw material, since the degree of consolidation clearly will be better. Use of preconsolidated blanks also allow more rapid preheating and is likely to be a prerequisite in the molding of anything other than flat and moderately curved components.

Technique Characteristics

Despite rubber-block molding and hydroforming offering certain advantages over compression molding with matching metal molds, the main industrial interest is in the latter due to its close resemblance with established molding techniques. The technique shares most characteristics with molding of SMC and GMT, but the raw material cost is notably higher. Compared to SMC molding there are advantages such as even shorter cycle times and no health concerns brought on by the resin.

Component Characteristics

Thermoplastic composites compression molded from prepregs are by necessity moderately curved shell-like structures with small differences in thickness. Courtesy of the aligned and oriented reinforcement and the high fiber content, properties may be very good. The component characteristics are otherwise similar to compression molded SMC and GMT components, although inserts, holes, bosses, ribs, and undercuts would be difficult to incorporate.

Applications

The main interest in compression molding of thermoplastic composites with aligned and oriented reinforcement comes from the automobile industry. Components under consideration include ones similar to those currently molded from GMT, but for which redesign could facilitate use of prepreg-based raw material to significantly increase the component's structural capability while reducing weight.

4.3.4 Filament Winding

Filament winding is arguably the manufacturing technique for thermoplastic composites about which there is the most to be found in the scientific literature. Countless studies in academia and industry have addressed filament winding and almost every conceivable processing option appears to have been tried.

Process Description

Most filament winding facilities have the elements of *Figure 4-69* in common. The prepreg tape is unwound from its spool and is preheated on its way to the mandrel. At the contact point the previously wound layer is heated and the incoming prepreg often given a further thermal boost. As the two molten surfaces merge they are compacted through back tension on the prepreg or through the use of an optional pressure roller or sliding shoe.

The most basic filament winding facility for low-temperature thermoplastics may employ a single hot air gun aimed at the contact point. However, to reach higher winding speeds a combination of heat sources is commonly used to melt the two surfaces to be mated. In most cases the preheater is a tunnel oven employing IR elements or a forced flow of heated gas. To compensate for the slow heat transfer of thermoplastics the tunnel oven may be in excess of a meter long. To achieve good consolidation requires that also the surface onto which the prepreg tape is to be placed is molten (cf. *Section 2.1.3.5*) and some degree of such preheating is common; IR elements or heated gas have proved feasible. It is further common to use a contact-point heater to finally ensure that the desired temperatures are reached at the contact point. Hot gas and focused IR heaters work well, but more refined techniques involve lasers and open flames. To reduce the amount of heat that needs to be introduced into the already wound material, the mandrel may be heated.

As long as only convex surfaces are wound it may be sufficient to use back tension on the prepreg as the sole means of compaction. However, since the prepreg tape may be continuously consolidated in fractions of a second, flat and concave surfaces may be wound without tape bridging if an external pressure source is used, see *Figures 4-70* and *4-71*. This continuous consolidation is often referred to as on-line, or in-situ, consolidation and apart from concave winding also virtually eliminates concerns of a geodesic path, thus giving the component designer a much greater freedom in geometries and reinforcement orientations. In helix winding of open-ended components such as pipes, on-line consolidation allows much faster turnarounds at the mandrel ends without the need for protruding pins. On-line consolidation

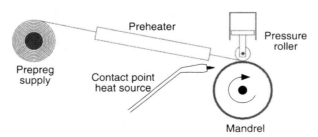

Figure 4-69 Schematic of thermoplastic filament winding

also may result in considerably lower residual stresses of thermal nature than if the entire component went concurrently through the thermal cycle from processing temperature to ambient conditions. It is thus at least theoretically possible to build up any wall thickness without having to worry about thermal residual stresses deteriorating component properties.

While on-line consolidation clearly offers many advantages it has proved technically feasible to wind thin-walled components at room temperature and consolidate them in a secondary operation, as well as to place both raw material supply and mandrel in an oven maintained at processing temperature. Although technically feasible under certain conditions, these techniques have serious limitations and are likely to be not viable processing routes for general filament winding operations.

While many experimental filament winding facilities are custom-built (such as the one in *Figure 4-70*), the more successful ones usually involve modifications of existing machinery built for the related thermoset process, so that the research and development work may concentrate on consolidation aspects rather than on machine kinematics. A completely different

Figure 4-70 Experimental facility for thermoplastic filament winding of convex-concave components. The prepreg tape is unwound from its spool and is guided vertically through IR preheaters and is placed on the convex-concave mandrel by a pressure roller. The mandrel rotation is controlled independently of the filament winding head which is positioned by an industrial robot. Photograph reprinted from reference [15] with kind permission from SAMPE, Covina, CA, USA

machine concept has also been developed where a gantry-mounted robot positions a winding head that continuously rotates around a stationary mandrel. This concept permits winding of otherwise prohibitively undulating snake-like components.

Figure 4-71 Convex-concave mandrel and glass-reinforced PP component wound with the facility in *Figure 4-70.* Photograph reprinted from reference [16] with kind permission from SAMPE, Covina, CA, USA

Raw Materials and Mandrels

For filament winding the raw material must naturally be in continuous tape or yarn form. Since the resin is molten for such a short time, fully melt-impregnated prepregs often result in better component properties than when commingled or powder-impregnated yarns are used. All material combinations available in prepreg tape or yarn form have been used with some degree of success.

More or less successful attempts at on-line reinforcement impregnation have been carried out. One such concept impregnates a continuous roving with resin in powder form, whereupon the reinforcement is melt-impregnated in a miniature pultrusion die right before being on-line consolidated onto the mandrel. Similarly, on-line impregnation using conventional thermoset impregnation baths and low-molecular weight prepolymer has proved technically possible, but has not yet seen a commercial breakthrough.

Mandrels used in thermoplastic filament winding are no different than their thermoset counterparts, with the exceptions that they must tolerate higher temperatures and possibly may include provisions for heating.

Consolidation

When winding using on-line consolidation, the two molten surfaces are brought into contact and consolidated in a fraction of a second (assuming winding speeds similar to thermoset filament winding). Although the cooling rates following mating of the two surfaces clearly are high, the thermal history is made very complex by repeated heating of the material when the next layer is added to the already wound structure.

Winding speeds in pure hoop winding range from 1 mm/s for simple experimental facilities to in excess of 1 m/s for sophisticated facilities employing laser or open flame heating of the contact point. Helix winding generally takes place at considerably lower speeds. In most cases increased winding speed results in less good consolidation and consequently less good component properties. With helix winding there is also a considerable problem of voids at the points where one prepreg crosses over another. In thermoset winding this is not much of a concern since the resin is liquid and thus may fill in these voids, but with thermoplastics the previously wound material is essentially solid and must be heated to a certain depth to make it sufficiently compliant to allow elimination of crossover voids. Although clearly undesirable from a process economy point of view, the void content of a wound structure may be reduced through reconsolidation employing vacuum bagging and an autoclave.

Technique Characteristics

The technique characteristics are for the most part shared with the thermoset counterpart. However, there are notable exceptions such as greatly reduced cycle time due to elimination of a secondary processing step (crosslinking in oven), elimination of health concerns, and significantly costlier raw material.

Component Characteristics

The component characteristics are also quite similar to the thermoset relations. Distinguishing differences are that flat and concave surfaces may be wound, that elimination of geodesic considerations give significantly greater freedom in reinforcement orientation, that void contents normally are higher, and that fiber volume fractions in the components essentially are the same as in the prepreg (i.e. normally lower than achievable with thermosets).

Applications

The technical feasibility of thermoplastic filament winding has been illustrated beyond all reasonable doubt by numerous independent organizations. Despite this overwhelming body of evidence there are few commercial applications. Potential applications naturally are as replacements for filament wound thermoset components where some property inherent to the matrix, such as toughness, is desired. Most components wound to date are rotationally symmetric cylinders wound at large angles. Examples of filament wound prototypes are pressure vessels (such as those shown in *Figure 4-52*) and a bicycle frame (see *Figure 1-40*).

A potential application where thermoplastic filament winding has been investigated for its ability to produce very thick-walled sections without introducing prohibitively large residual stresses is a submarine hull. A study was undertaken for the United States Navy and thick-walled cylindrical components with approximately 2-m diameters and integrated internal stiffening ribs were successfully wound.

4.3.5 Pultrusion

For obscure reasons thermoplastic pultrusion has not yet been the subject of anywhere near as much research and development work as thermoplastic filament winding. The technical feasibility of thermoplastic pultrusion has nevertheless been established by a number of independent organizations.

Process Description

The configuration of a basic facility for thermoplastic pultrusion is illustrated in *Figures 4-72* and *4-73*. The raw material in prepreg form is unwound from a creel and pulled into a preheater which heats the feedstock to a temperature near or in excess of the softening point of the matrix. The material then enters a heated die, which has a taper where the material is gradually shaped to the final cross-section before it is consolidated in a cooled die.

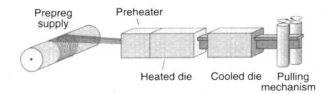

Prepreg supply   Preheater

Heated die   Cooled die   Pulling mechanism

Figure 4-72 Schematic of thermoplastic pultrusion

The machinery used in thermoplastic pultrusion is the same as that used with thermosets with the exceptions that preheating is nearly always used and that the already impregnated feedstock is consolidated in different type of dies. As with thermoset pultrusion, the preheater is used to increase pulling speed, but the temperature required to melt the thermoplastic matrix is obviously considerably higher than temperatures common in the thermoset technique; IR or hot air heating are common.

In order to improve process efficiency, the reinforcement of course ideally should be impregnated as part of the pultrusion operation, as with the thermoset technique. Although such on-line impregnation units have been developed, the concept still must be considered to be still in the early stages of development due to the difficulties encountered in impregnating with

Figure 4-73 Experimental facility for thermoplastic pultrusion. The prepreg tapes enter from the left and are guided into the preheater which is covered by insulation to minimize heat loss. Following the preheater are dies, cool-off section, and pulling mechanism. Photograph reprinted from reference [17] with kind permission from SAMPE, Covina, CA, USA

highly viscous thermoplastic melts. However, as a process to manufacture melt-impregnated prepregs it is rather common. Such miniature pultrusion-like processes may employ direct melt-impregnation or powder-impregnation followed by melt-impregnation. It is also common to chop such prepreg or rod as part of the process so as to produce long-fiber reinforced pellets for use in injection molding. Various attempts at scaling up such processes to produce large-scale components have so far yielded meager results. However, use of low-molecular-weight prepolymer in injection pultrusion is a technically proven alternative with a great chance of future success.

Raw Materials and Dies

All prepreg types have been tried, but there is little doubt that melt-impregnated prepregs result in the best component properties and the highest pulling speeds as long as the reinforcement essentially is longitudinal. To include off-axis reinforcement it may prove necessary to employ woven and braided material forms and these are more readily available based on powder-impregnated and commingled yarns. With raw material that needs to be melt-impregnated as part of the pultrusion operation the pulling speeds are likely to be lower and impregnation results less good. While much of the early pultrusion work concentrated on high-performance materials such as carbon-reinforced PEEK and glass- or carbon-reinforced PPS, later studies appear to aim for commodity-type applications and materials such as glass-reinforced PP and PAs.

Although one-piece dies have been tried in thermoplastic pultrusion, it is more common that at least two separate dies are used, the last of which is cooled, see *Figure 4-74*. Heated dies have significant tapers and cooled dies have constant cross-section cavities. While dies used in thermoset pultrusion are between 600 and 1,500 mm long, the length of an individual die used in thermoplastic pultrusion is on the order of 100 mm. However, since many different sources have addressed the technique, huge differences in die design are to be expected.

Figure 4-74 Close-up of preheater and dies in the facility of *Figure 4-73*. The heated prepreg tapes (black) enter from the left and continue into a heated die, which has a significant entrance taper, followed by a constant cross-section cooled die; the dies are separated by a narrow air gap. For visibility reasons the insulation has been removed from one side of preheater and heated die. Photograph reprinted from reference [17] with kind permission from SAMPE, Covina, CA, USA

Consolidation

Consolidation in pultrusion is a gradual process which starts in the taper of the heated die and continues until the cooled die solidifies the matrix. The consolidation time therefore is a function of die lengths, die temperatures, and pulling speed. While speeds near 100 mm/s have been reported for pultrusion of simple components using melt-impregnated prepregs, speeds an order of magnitude lower are more common. Ultrasonically vibrating dies are said to allow higher pulling speeds when melt-impregnated prepregs are used. Not surprisingly, composite properties deteriorate with increasing pulling speed due to lower component temperature (despite constant die temperatures) and shorter consolidation times, see *Figure 4-75*.

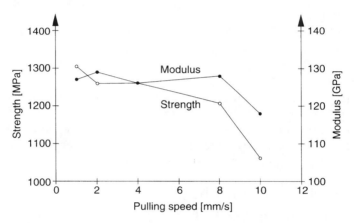

Figure 4-75 Flexural strength and flexural modulus of unidirectionally carbon-reinforced PEEK composites pultruded with the facility of *Figures 4-73* and *4-74* as functions of pulling speed while all other independent process variables were held constant. Data from reference [18]

Technique Characteristics

Ever since the early days of thermoplastic pultrusion (mid-1980s) its proponents have harbored the belief that a significantly higher processing rate than is common in thermoset pultrusion would compensate for the higher raw material cost. The argument in favor of a higher processing rate has always relied on the fact that with thermoplastics no chemical reaction needs to take place and only physical phenomena need to be considered. Despite the early optimism and some rare but reasonably successful exceptions, the high pulling speeds basically remain as elusive as ever and in addition to this several other impeding factors remain. It is likely that the only way in which thermoplastic pultrusion can ever become truly competitive is by not being compared to the notoriously economical technique of thermoset pultrusion, but rather by finding or creating niche markets unavailable to thermoset pultrusion, e.g. where the inherent properties of a thermoplastic matrix are sought.

The characteristics of thermoplastic pultrusion are essentially the same as for its thermoset parent, except that health concerns have been eliminated and that, unfortunately, pulling speeds are rarely as high. Furthermore, the raw material cost is significantly higher.

Component Characteristics

The current state of the art of thermoplastic pultrusion lags far behind its thermoset origin. With some rare exceptions only relatively simple and small components with most or all of the reinforcement longitudinally oriented have been pultruded. While fiber volume fractions tend to be very high by pultrusion standards (the same as in the prepreg or slightly higher), void fractions tend to be higher and surface finish inferior.

Applications

The commercial availability of pultruded thermoplastic composites appears to be limited to small rods and square profiles. Considerably more complex components, including a multi-cell missile body made from commingled and stitched preforms, have been pultruded in short lengths, but the commercial breakthrough of the process remains elusive.

4.3.6 Diaphragm Forming

The only technique that has been developed exclusively for thermoplastic composite manufacturing is diaphragm forming, which may be seen as a refined form of vacuum bag consolidation in an autoclave. The main characteristic of the technique is that deeply drawn and geometrically complex components may be formed. The process has been investigated by several independent organizations and with encouraging results.

Process Description

Since diaphragm forming has been investigated by many, technical solutions vary; two of the more common versions are described in the following. The (flat) composite material to be formed is placed between two flexible diaphragms. The diaphragms, but not the composite material which remains free-floating, are clamped around the entire perimeter using a clamping frame and the air is evacuated between the diaphragms. Thus far most versions of diaphragm forming remain the same; the main differences are in the heating and forming techniques.

In the first diaphragm forming version described, the diaphragm–material stack is placed in an oven and heated to a temperature in excess of the softening point of the matrix. The stack is then rapidly placed onto a one-sided female mold and vacuum is drawn in the space between the lower diaphragm and the mold and pressure applied above the upper diaphragm to force the blank to conform to the mold, see *Figure 4-76*. Either vacuum and pressure or just one of them is used in forming. Since the mold is normally unheated the component solidifies as it comes in contact with the mold.

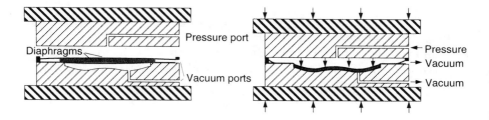

Figure 4-76 Schematic of diaphragm forming

In the second diaphragm forming version described, the diaphragm–material stack is placed on top of the mold (see *Figure 4-77*) and then this entire assembly is placed in an autoclave, which is likely to be purpose-built. Following evacuation of the air between the diaphragms, the internal atmosphere of the autoclave is heated to melt the matrix, whereupon the combined forces of vacuum below the lower diaphragm and pressure above the upper diaphragm make the material conform to the mold, conceptually as illustrated in *Figure 4-76*. When forming is completed the heated gas surrounding the mold and the diaphragms is replaced with cool (ambient) air to solidify the component.

Raw Materials and Molds

All types of prepregs have been used in diaphragm forming, but it appears as if investigations using melt-impregnated prepregs and fabrics woven from commingled and powder-impregnated yarns have been particularly successful. To date most of the work has concentrated on high-performance material combinations such as carbon-reinforced PEEK, although the interest in commodity-type materials such as glass-reinforced PP appears to be growing. Results are generally enhanced if consolidated blanks are used as raw material.

The first diaphragm material used was superplastic aluminum, but most later investigations have employed sheet rubber and polymer films, particularly PI. Rubber diaphragms have an advantage in that they may be reused a few times, whereas aluminum and polymer films may not be reused since they become permanently deformed. When using polymer diaphragms it is important that the processing temperature of the prepreg matrix is such that the diaphragms allow large plastic deformations, i.e. between T_g and T_m of the diaphragms. PEEK-based prepregs are thus very much compatible with PI diaphragms, whereas they do not allow sufficient deformation at the processing temperatures commonly used with PP. With rubber diaphragms this is not a concern since they are highly elastic at all realistic forming temperatures, but instead the upper use temperature of the rubber must be considered.

Since low pressures are involved, simple tooling may be used. Materials include sheet metal, which allows rapid temperature changes and, for experimental work with the first diaphragm forming version, plaster and wood molds, which have their predominant advantage in low cost.

Figure 4-77 Part of experimental diaphragm forming equipment employed in autoclave forming. The prepreg plies are sandwiched between two transparent diaphragms which are clamped around the perimeter. The tubing visible at the bottom of the photograph is used to draw vacuum between the lower diaphragm and the mold (which is hidden from view). Photograph courtesy of C. Ó Brádaigh, Department of Mechanical Engineering, University College Galway, Ireland, and Center for Composite Materials, University of Delaware, Newark, DE, USA

Consolidation

When forming outside of an oven and against an unheated mold, forming times are measured in seconds and cycle times in minutes (excluding pre-heating). When forming a component in an autoclave, the forming part of the processing cycle is only a few minutes, but the full cycle time may approach an hour, primarily depending on heat transfer limitations of the equipment. More rapid processing may be achieved if the mold incorporates heating and cooling provisions.

The forming step is naturally crucial to the success of the technique. In order to prevent buckling and wrinkling, which may occur in areas that go through large deformations (cf. *Sections 4.1.2* and *4.1.3*), it is important to use low forming rate, low pressure (typically less than 1 MPa), and raw material that is near net shape. Too high pressures and too stiff diaphragms may cause excessive transverse flow resulting in uneven part thickness; see *Figure 4-10* which was diaphragm formed. The buckling and wrinkling issues are nevertheless less critical in diaphragm forming than in compression molding, for example, partly since the diaphragms keep the material both under slight tension and slight lateral compression and partly since the material remains pliable for a comparatively long time (until it comes into contact with the mold in the first version and for any desired time in the latter).

Technique Characteristics

Equipment cost:	Low to intermediate
Mold cost:	Low
Labor cost:	High
Raw material cost:	High
Feasible series length:	Short
Cycle time:	Short to long
Consolidation requirements:	Simple
Health concerns:	Low

The spread in equipment cost and cycle time depends on the processing route. Equipment cost is low and cycle times short when forming outside of an oven and against an unheated mold, while they are intermediate and long, respectively, when forming in an autoclave.

Component Characteristics

Geometry:	Any
Size:	Limited by oven or autoclave size
Holes and inserts:	Not possible
Bosses and ribs:	Not possible
Undercuts:	Difficult
Surfaces:	One good surface

Fiber arrangement:	Planar arrangement; oriented and continuous
Typical fiber volume fractions:	0.35–0.6
Void fractions:	Low
Mechanical properties:	Good
Quality consistency:	Intermediate

Applications

The scope for applications of diaphragm formed components should be great, but the commercial acceptance of the technique remains to be seen. *Figure 4-78* shows some diaphragm formed components, while the model trailer of *Figure 4-79* illustrates the complexity of components that may be formed. Components that appear to be nearing commercial reality on a small scale are glass-reinforced PP blister fairings and leading edges for jet engines of civilian aircraft to replace hand laidup and autoclave consolidated epoxy-based components.

Figure 4-78 Components diaphragm formed in autoclave. Photograph courtesy of B. P. Van West, Boeing Defense and Space Group, Seattle, WA, USA, and Center for Composite Materials, University of Delaware, Newark, DE, USA

4.3.7 Other Techniques

Since manufacturing of thermoplastic composites is a field enjoying considerable research and development efforts there are naturally a range of other, but less common, manufacturing techniques than the ones discussed above. Some of these techniques are briefly discussed in the following.

Rollforming is a relation of pultrusion that just like compression molding has its origin in sheet metal forming. In rollforming several consecutive pairs of contoured rollers, normally four or more, gradually deform a molten blank to the desired shape, see *Figure 4-80*. The rollers, which are driven, are normally unheated and consequently they gradually cool the

Figure 4-79 Model trailer diaphragm formed in autoclave. Photograph courtesy of B. P. Van West, Boeing Defense and Space Group, Seattle, WA, USA, and Center for Composite Materials, University of Delaware, Newark, DE, USA

component. Advantages of rollforming over pultrusion include reduced shear and solid-to-solid friction experienced by the material since the rollers rotate with the material, potentially higher forming rates, and that curved components can be manufactured. It appears as if all work to date has utilized preconsolidated blanks as raw material, despite there being little reason why consolidation of prepregs should not be continuously carried out prior to the actual forming operation. Rollforming is potentially capable of manufacturing any constant cross-section geometry. Nevertheless, so far thermoplastic rollforming has been used to manufacture hat and Z shapes only. Compared to most of the previously mentioned manufacturing techniques, little work on rollforming of thermoplastic composites is ongoing despite the technique's feasibility being proven.

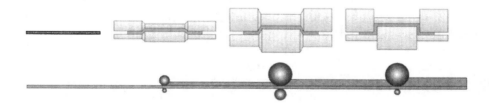

Figure 4-80 Schematic of thermoplastic rollforming. At the top, frontal views of preheated blank and consecutive roller pairs are shown, while the bottom shows side views of the gradually formed laminate and roller pairs (movement is from left to right)

Trapped elastomer molding is an innovative but unusual method to consolidate a stack of laidup prepregs. The entire mold-prepreg-vacuum bag assembly is placed in a rigid vessel that is completely filled with a pourable elastomer powder. As temperature is raised to the processing temperature of the prepreg, the elastomer powder wants to expand but is largely prevented from doing so by the constraining vessel, thus pressurizing the consolidating laminate.

Localized postforming offers a simple yet effective means of making limited changes to already consolidated components. For example, sheet and bar stock may be folded, flattened, etc. through localized heating and forming, see *Figure 4-81*. Despite the fact that if continuous reinforcement fibers are involved they are likely to buckle or fracture, line heating and subsequent folding of sheets—even honeycomb sandwich panels, where the core ends up severely damaged—has proved commercially viable in manufacturing of, for example, aircraft interiors. Localized postforming, or rather reconsolidation, may also be thought of as including repair of unsatisfactorily molded or consolidated components as well as ones damaged in service (see also *Section 5.4*).

Material fixture

Local heating

Figure 4-81 Schematic of localized postforming of consolidated laminate (or sandwich panel)

The technique of fiber placement was briefly introduced in *Section 4.2.7* as a method of manufacturing geometrically complex geometries largely ignoring concerns of geodesic paths. Since thermoplastics can be on-line consolidated they are potentially even more appropriate for use in fiber placement, but just as with the thermoset sibling, few, if any, commercial techniques have yet resulted despite considerable research and development work. Also, as with the thermoset technique there is a lack of consensus on what versions of thermoplastic filament winding and tape laying (if any) that should be referred to as fiber placement.

DBPs (cf. *Figures 2-51* and *2-52*), which are much used in manufacturing of thermoplastic prepregs and molding compounds, are also used in consolidation of laminates. While flat laminates have limited uses, they are important as raw material (blanks) in forming techniques such as compression molding, diaphragm forming, roll forming, etc. A DBP may also be used to manufacture flat thermoplastic sandwich panels through sandwiching of a polymer film containing a foaming agent between face laminates; as the temperature is increased the polymer film expands into a cellular structure.

In rotational molding, resin in powder form is first introduced into a hollow, thin-walled metal mold. The closed mold is then placed in an oven to heat it to a temperature in excess of the melting point of the resin. While in the oven the mold slowly rotates around two perpendicular axes to coat the entire insides of the mold with resin. In rotational molding the component consequently is not formed by centrifugal forces (as in centrifugal casting, cf. *Section 4.2.7*); instead the powder is gradually picked up from the bottom of the slowly rotating mold as it melts and sticks to its heated inside surface. Following complete melting of the resin and coating of the mold, the mold is removed from the oven and allowed to cool at room temperature under continued rotation until the component can be demolded. Almost all rotational molded components contain no reinforcement whatsoever, but introduction of short fibers is possible. Common unreinforced applications include storage pallets, fish tubs used on fishing boats and in food-processing plants, buoys, and balls. Both storage pallets and fish tubs are sandwich structures, often with PE faces and PUR or PE foam core that is in-situ foamed in a secondary operation, although one-step manufacturing of a sandwich with PE faces and foamed PE core is technically possible.

Another process that is predominantly known as a thermoplastics manufacturing technique is extrusion, wherein machinery very much like an injection molding machine is used (cf. *Sections 4.2.3.1* and *4.3.2*). The significant difference is that in an extrusion machine the screw only has a rotational degree of freedom to continuously push (i.e. extrude) material at a constant rate through a constant cross-section die and into a water bath where it solidifies. Although reinforcement is uncommon in extrusion, fillers and short fibers may be introduced. Extruded unreinforced thermoplastics, predominantly PVCs, are extensively available; indeed, practically every unreinforced constant cross-section plastic profile is extruded. In contrast, thermoset extrusion is quite rare. (It is likely from extrusion that the name pultrusion has been derived; pull-extrusion becoming pultrusion.)

While thermoset composite manufacturing techniques to a significant degree are mature, most techniques used to manufacture thermoplastic composites are considerably less developed and the scope for technique improvements and innovation of completely new techniques is therefore much greater.

4.4 Manufacturing-induced Component Defects

During manufacturing there is a significant risk of somehow failing to fulfill the requirements of the component design. While this is true for manufacturing with any kind of material, the fact that with composites the raw material and the component are often made in the same manufacturing step means that the opportunity for simultaneous introduction of defects is significantly greater. In broad terms manufacturing-induced defects may be categorized into structural, geometrical, and cosmetic defects. Unfortunately, several of these defect types are difficult to foresee and the only way to avoid them is through regular quality control procedures to ensure consistent raw material quality, close control of the manufacturing operation, and, last but not least, experience. Since there is an immense array of processing difficulties that potentially may lead to some kind of defect and many of these are relevant to a specific technique or material only, this section merely points to some of the more common and largely generic defect types and discusses ways to avoid or minimize them.

4.4.1 Structural Defects

For the purpose of this treatment the term structural defect refers to a flaw that is detrimental to the load-carrying capability of a component. For composites intended to carry loads this type of defect is generally the most undesirable.

4.1.1.1 Deviations in Reinforcement Orientation and Content

Perhaps most detrimental to the structural capability of a composite is when the actual fiber orientation does not correspond to the intended orientation, since even a seemingly small misalignment may result in drastically reduced properties (cf. *Figure 3-10*). Similarly, if the design counts on straight and aligned fibers, fiber waviness may significantly decrease properties. Both these defect types may be introduced during manufacturing of preform and prepreg or during conformation of the material to the mold, but may also occur due to fiber washout during impregnation (RTM and SRIM). One way or another these defect types essentially are relevant to all manufacturing techniques.

Another reinforcement-related defect is variation in fiber content (both within a component and between components), which naturally also impairs composite properties (cf. for example *Figures 3-6* and *3-14*). This defect type may have its origin in numerous sources, including but not limited to:

- Manufacturing of preform, molding compound, and prepreg
- Conformation of reinforcement or prepreg to the mold
- Use of incorrect amount of reinforcement
- Worker inconsistencies in layup or rolling, or excessive resin bleeding in vertical areas due to too low viscosity (wet layup)
- Excessive or insufficient resin bleeding (vacuum-bagging)
- Fiber washout and mold deflection (RTM and SRIM)
- Fiber tension (filament winding)

While defects due to variations in fiber content are more common with thermoset matrices due to in-process impregnation and low resin viscosity, they certainly cannot be ignored in thermoplastic techniques. In any technique where there is a potential for an uneven pressure distribution, there is also a potential for this type of defect.

When the raw material is a molding compound—be it thermoset or thermoplastic—there is also a possibility of the aforementioned problems. The potential for precise control of fiber orientation is limited with randomly reinforced molding compounds, and tolerance to variations in orientation therefore should be considered already in the design phase. However, charge pattern, size, and placement strongly influence mold filling and therefore fiber orientation. Through intelligent charge design, fibers may largely be aligned in directions where loads are greater and knitlines may be eliminated. Since molding compounds are used with matching molds, there should be little concern about global variations in fiber content, but locally, for example in ribs and bosses, resin can bleed out of the reinforcement.

4.1.1.2 Voids

All composites, regardless of constituents, raw material form, and manufacturing technique, contain microscopic voids, which may be due to air entrapped in tightly compressed fiber yarns, volatiles from the resin, air in the resin (e.g. from mixing), and incomplete consolidation. Acceptable amounts of voids are typically in the range 1–5 volume percent, where the lower figure tends to be required in aerospace applications and the higher figure is tolerated in lightly stressed (or unstressed) applications where component failure does not have catastrophic consequences. Voids are highly unwanted since they cause stress concentrations and act as crack initiators.

In techniques where the reinforcement is impregnated as part of the process, void contents tend to be the highest; the main reasons are air entrapped in the reinforcement and air accidentally whipped into the resin during mixing and/or impregnation. In wet layup vigorous rolling is crucial to work out voids. In RTM and SRIM it is easy to entrap voids in tightly compressed fiber yarns if mold filling is too rapid. In filament winding and pultrusion where the reinforcement is continuously impregnated there is also a significant risk of entrapping air in the reinforcement.

Even prepregs contain some voids which to a degree may be eliminated during manufacturing. Especially with vacuum bagging it is possible to reduce the void content, although too vigorous use of high vacuum may bleed out too much resin. At the other extreme, too low or unevenly applied vacuum and pressure, as well as too low temperature, may lead to insufficient consolidation between prepreg plies. Too high void content is a major reason for the high rejection rates of composites for aerospace applications. Consolidation between prepregs plies is particularly difficult in thermoplastic filament winding due to the crossovers in helix winding and the fact that the high resin viscosity makes it difficult to completely fill these gaps. With components compression molded from molding compounds, the void content is likely to be the same as in the raw material.

4.1.1.3 Incomplete Impregnation and Mold Filling

Although a certain degree of voids is tolerated (since voids are inevitable), (macroscopic) dry spots and unfilled sections of the mold are usually not acceptable at all. Dry spots, meaning large sections of unimpregnated reinforcement detectable with the naked eye, in RTM and SRIM may result from too high injection rate, merging flow fronts entrapping air, too high (initial) resin viscosity, premature resin gelation, etc. Although such problems can be caused by improper reinforcement placement or preform design, they may also be caused by more fundamental problems in terms of location and number of injection and ventilation points. In both thermoset and thermoplastic compression molding prematurely gelating or solidifying resin may result in incomplete mold filling; such problems are generally due to incorrect charge design or too low processing temperatures and pressures. Since the viscosity of thermosets is dependent on both temperature and time (cf. *Section 2.1.3.4*), it is important to complete impregnation and/or mold filling when the viscosity is low. A too high viscosity tends to lead to incomplete impregnation or mold filling, while a too low viscosity may cause resin bleeding. The risk of a too high resin viscosity is especially relevant to techniques where fully formulated resin is kept in impregnation baths for extended periods of time (filament winding and pultrusion).

4.1.1.4 Variations in Constituent Properties

As discussed in *Chapter 2*, resin properties are highly dependent on processing conditions and in particular the temperature history. Thermosets that have been crosslinked at too low a temperature (or for too short a time at the specified temperature) end up having a lower crosslink density than intended and will therefore possess inferior stiffness, strength, temperature tolerance, etc. Conversely, it may also be possible to achieve a too high crosslink density leading to an overly brittle matrix. Excessively high temperatures, which for example may be caused by the exotherm in thick composites, can lead to chemical degradation of the matrix and to matrix cracking due to differential thermal shrinkage, but also due to matrix shrinkage from crosslinking.

With semicrystalline thermoplastics the cooling rate determines degree of crystallinity, where a lower cooling rate leads to a higher degree of crystallinity (cf. *Section 2.1.2*). A higher degree of crystallinity leads to a stiffer matrix with a higher temperature tolerance but also to a lower strain to failure (cf. *Section 2.1.3.2*). The properties of amorphous thermoplastics, on the other hand, are essentially insensitive to the temperature history. However, all thermoplastics have viscosities that are highly dependent on temperature and shear rate (cf. *Figure 2-17*) and will start to degrade if molten for too long, particularly in the presence of oxygen (cf. *Figure 2-19*). It is consequently of paramount importance that reinforcement impregnation, conformation to the mold, and mold filling, respectively, take place at the right viscosity, i.e. at recommended temperature and, to a lesser degree, shear rate.

Section 2.2.6 covered the importance of interface and interphase between fiber and matrix. Poor fiber–matrix interaction may from a processing point of view be influenced by incomplete fiber wetting, contaminants, moisture, and excessively high temperatures, all of which may lead to poor interfacial bonding and microcracking. The issue of contaminants and moisture is relevant to all constituents in that it points to the importance of proper storage. Resins, reinforcements (particularly organic and glass fibers), additives, and the interface/interphase are all hygroscopic to some degree and the composite properties suffer accordingly unless absorbed moisture is removed from the constituents prior to processing. While thermoplastics have an infinite shelf life, thermosets, including prepregs and molding compounds, age. Thermosets must therefore be stored at the specified temperature and be used within a given time frame lest their processability deteriorate due to increased viscosity and reduced tack.

In techniques where the reinforcement is impregnated as part of the process there is a potential for damaging the fibers during handling, conformation (e.g. wet layup, RTM, and SRIM), and guidance (filament winding and pultrusion). Fractured fibers in critical locations, such as sharp corners, may be detrimental to component properties.

4.4.2 Geometrical Defects

Geometrical defects include component warpage and angles or thicknesses deviating from specifications. Components may be warped due to the layup accidentally being unbalanced and due to residual stresses arising from differential shrinkage caused by spatial differences in temperature history. Geometrical deviations may result in a component not fitting into a larger assembly. If it is possible to force a warped component to fit into such an assembly or if the residual stresses for any other reason cannot be relieved through warpage or angle changes, these stresses may be high enough to notably reduce the structural capability of the component or may lead to failure.

The warpage problem is particularly obvious in flat laminates that have been on-line consolidated in thermoplastic tape laying and to some degree also with wet hand layup, since both these techniques introduce spatial differences in temperature history during consolidation and crosslinking, respectively. Due to the difficulty in producing perfectly flat composites and the human eye's capability of detecting minute deviations from flatness, it is recommended that components that fulfill some aesthetic function, such as automobile body panels, always have some curvature. The warpage problem is also obvious in the spring-forward phenomenon, where internal angles end up smaller than the angle of the mold. Warpage can be reduced through a general reduction in the maximum processing temperature and through reduction of spatial temperature differences.

A deviation in component thickness is probably the most critical when the thickness is less than specified, since even if the required amount of reinforcement is present, the flexural stiffness (which is proportional to the thickness cubed) may suffer significantly. Nevertheless, an excessive component thickness, which is often synonymous with a too low fiber content and possibly a high void content, is also generally undesired.

Assuming that the correct amount of reinforcement has been used, too low a pressure may lead to too much resin remaining in the laminate and vice versa. Critical issues are therefore moderation in rolling in wet layup, degree of vacuum and pressure in vacuum bagging, and resin viscosity and yarn tension in filament winding. In pultrusion thickness variations are an issue only when hollow components are produced and the need for cantilevered mandrels arise. The position of the freely floating mandrel is entirely determined by the amount of material around it. If part of the reinforcement is out of place, a pipe will therefore not be concentric.

Correctly performed, RTM and SRIM offer the best thickness control of all techniques, but molds that are not sufficiently stiff or sufficiently supported may deflect from the injection pressure, thus increasing both wall thickness and matrix content. When sandwich structures are molded in RTM and SRIM, cores may shift during injection giving rise to changes in face thicknesses. In compression molding using molding compounds, component thickness is a direct function of the volume of the charge, so any deviation in charge volume directly affects component thickness.

4.4.3 Cosmetic Defects

The most common cosmetic defects, which are often symptoms of structural defects, may be categorized into poor surface finish and discoloration. While discoloration generally is a result of incompletely dispersed pigments, contaminants, or too high a processing temperature, poor surface finish may arise from numerous sources.

A source of poor surface finish that all techniques have in common is poor mold finish; the component surface will never be better than that of the mold. It is therefore critical to keep the mold in good general condition, to keep its surface spotlessly clean, and to apply mold release evenly. If the component sticks to the mold, the surface of the component (and sometimes also the mold) may be damaged upon demolding. This is a particular problem in pultrusion with epoxies, since they do not shrink as much as the more easily processed unsaturated polyesters and vinylesters and also tend to adhere well not only to the reinforcement but also to the die; the same problem can easily arise in thermoplastic pultrusion. In thermoplastic tape laying, and sometimes also in thermoplastic filament winding, a pressure roller is used and resin easily sticks to the roller if it is allowed to heat up; such sticking invariably leads to a very poor surface. In compression molding of GMT the charge footprint problem accentuates the already poor surface of most commonly used material systems.

When gelcoats are used poor surface is often associated with a too thin or too thick layer, under- or overcuring, moisture, and contaminants. The reinforcement structure may be visible on the part surface if the outermost reinforcement layer is too coarse and if the gelcoat or the surface film is too thin or undercured. Dry spots on the surface may be a problem in wet layup and in RTM and SRIM. Wrinkles in vacuum bags and diaphragms are also detrimental to good surface finish.

4.5 Summary

The manufacturing technique selected to produce a polymer composite strongly influences component properties and cost. Thus, for a composite designer to be able to do a good job requires that he has a relatively good knowledge of the limitations and opportunities offered by candidate manufacturing techniques.

For all techniques, manufacturing is a matter of maintaining the liquid reinforcement–matrix mass in the desired shape—usually using a rigid mold—at a specified temperature for the time required to allow it to become dimensionally stable. This requirement may be rephrased as controlling temperature and pressure throughout the process and throughout the component. When manufacturing thermoset-based composites, time must be provided to allow the crosslinking reaction to progress far enough to give dimensional stability to the component before it is demolded. With thermosets, this time frame ranges from a couple of minutes to several days. In the manufacturing of thermoplastic-based composites, only physical phenomena need to be considered; the matrix is first melted, then formed, and finally cooled to provide dimensional stability. Demolding times for thermoplastics range from fractions of a minute to a few minutes.

Although this chapter concentrates on a few commercially common and emerging manufacturing techniques, any means of making the pliable reinforcement–matrix mass conform to a given shape while it solidifies may be employed to manufacture a composite. In order to aid in understanding the essence of any given manufacturing technique, it should be helpful to consider answers to the following five basic questions:

- What is used as the mold to give the composite its shape?
- How is the raw material made to conform to the mold?
- How is the impregnation pressure gradient applied (if applicable)?
- How is the consolidation pressure applied?
- How is the heat applied (and removed)?

From a composite property point of view there are numerous good reasons for selecting a thermoset matrix, which partially explains why this matrix family heavily dominates in composite applications. In manufacturing, thermosets offer further advantages such as low viscosity and reasonably moderate temperature and pressure requirements, which translate into modest requirements on the mold used. The predominant disadvantage is the active chemistry, which brings on another variable in manufacturing, results in relatively long demolding times, and potentially brings on health concerns.

Thermoplastic matrices offer certain advantages over thermosets from a composite property perspective. In manufacturing, thermoplastics offer further advantages such as shorter demolding times and elimination of health concerns, while the high processing temperatures required and the high viscosity, which for most practical purposes results in a need to use costly preimpregnated reinforcement, are major disadvantages. With the exception of the commercially accepted compression molding of GMT, all techniques for manufacturing of thermoplastic composites are in their infancy or still in the laboratory.

The commercially relevant techniques for the manufacture of structural composites are briefly summarized below.

Wet Layup

In this treatment, wet layup has been separated into three related but discrete techniques: hand layup, spray-up, and hand layup of sandwich components. In all three techniques, reinforcement impregnation is an intimate part of manufacturing. All techniques are characterized by low mold costs, low capital costs, and high labor costs, where the latter result from the significant degree of craftsmanship involved. The techniques are best used in manufacturing of components in short series where the cycle time is not critical. Components are characterized by geometrical complexity and reasonably good mechanical properties (except for spray-up). The most common applications are marine vessels.

Prepreg Layup

In the aerospace industry virtually all composite components are hand laid up using prepregs, which are consolidated in autoclaves resulting in long cycle times. The technique is characterized by high mold costs, high capital costs, and high labor costs. Also in this case the high labor costs result from the significant degree of craftsmanship involved. The technique is normally used to manufacture components in the relatively short series typical of the aerospace industry, but is also used to manufacture sporting goods in long series. Components are characterized by geometrical complexity and very good mechanical properties.

Liquid Molding

While there are many liquid molding techniques, it is mainly RTM, vacuum injection molding, and SRIM that are of relevance to manufacturing of structural composites. All three techniques have in common that dry reinforcement is placed in the mold, whereupon the resin is gradually introduced into the mold to complete impregnation. While all techniques require a moderate degree of manual labor, RTM is characterized by low to intermediate mold costs and low capital costs, vacuum injection molding by low mold costs and low capital costs, and SRIM by high mold costs and high capital costs. While vacuum injection molding tends to be used in short series, RTM is used in series of intermediate length, and SRIM requires long series to write off the high initial costs. All techniques are used to manufacture geometrically complex vehicle components and RTM and SRIM particularly have found acceptance in the automobile industry.

Compression Molding

Compression molding of preimpregnated molding compounds (such as SMC, BMC, and GMT) has been widely accepted by the automobile industry for manufacturing of sheet-like components in long series since cycle times are relatively short. Characteristics of the technique include high mold costs, high capital costs, and low to intermediate labor costs. Since the randomly reinforced molding compound flows to fill the mold, limited control of fiber orientation is possible, resulting in relatively poor mechanical properties but on the other hand also permitting molding of rather complex geometries. While thermoset compression molding (i.e. with SMC and BMC) yields components with good surface and cycle times on the order of a few minutes, the thermoplastic relation (with GMT) exhibits cycle times less than half of those common with thermosets, but yields components with inferior surface finish. Common applications of compression molded thermoset composites are vehicle body panels, while their thermoplastic siblings are used in automobile components hidden from view, such as under-hood components and seat back structures.

Filament Winding

Filament winding is essentially limited to manufacturing of rotationally symmetric components, such as pipes, tanks, and pressure vessels, which are often intended for military use. Characteristics include low mold costs, intermediate capital costs, and low labor costs. While limited in achievable geometries, precise placement of continuous reinforcement and high reinforcement contents translate into very good mechanical properties.

Pultrusion

Due to its continuous and automated nature, pultrusion is the most economical of all composite manufacturing techniques, but it is limited to producing components of constant cross-section, which nevertheless may have high reinforcement contents. Technique characteristics include intermediate mold costs, intermediate capital costs, and low labor costs. Pultruded components for structural applications, such as rods, bars, I beams, box sections, etc., are the only composites that to any significant degree are standard off-the-shelf stock items.

While the techniques summarized above are the major commercially accepted ones, extensive research and development work has shown that numerous other techniques are technically—if not always economically—feasible. Composite manufacturing is a very dynamic field that constantly undergoes changes and there is no doubt that new techniques for manufacturing of both thermoset and thermoplastic composites will more or less continuously emerge and reach commercial maturity. The scope for inventing new manufacturing techniques reinforces the thesis that with composites the only limitation is the imagination.

Suggested Further Reading

For more detailed information on mature manufacturing techniques with emphasis on thermoset composites, there are several sources worth mentioning. The first three sources below are encyclopedias containing information on composites in general, including manufacturing-related issues. While these encyclopedias mainly cater for high-performance applications, the following two references in contrast lean more towards mass-produced low-cost components.

International Encyclopedia of Composites, Ed. S. M. Lee, VCH Publishers, New York, NY, USA, 1990–1991 (6 volumes).

Engineered Materials Handbook, Volume 1, Composites, ASM International, Metals Park, OH, USA, 1987.

Engineered Materials Handbook, Volume 2, Engineering Plastics, ASM International, Metals Park, OH, USA, 1988.

Introduction to Composites, SPI Composites Institute, Washington, DC, USA, 1992.

Plastics Engineering Handbook of the Society of the Plastics Industry, Van Nostrand Reinhold, New York, NY, USA, 1988.

Since manufacturing of thermoplastic composites is largely far from maturity and a field of active research and development, information on manufacturing-related issues is scattered throughout the scientific and trade literature. The two texts below review the field up to 1991 and 1990, respectively:

A. B. Strong, High *Performance and Engineering Thermoplastic Composites*, Technomic, Lancaster, PA, USA, 1993.

F. N. Cogswell, *Thermoplastic Aromatic Polymer Composites*, Butterworth-Heinemann, Oxford, UK, 1992.

Mold manufacturing, particularly for high-performance applications, is treated in:

J. J. Morena, *Advanced Composite Mold Making*, Van Nostrand Reinhold, New York, NY, USA, 1988.

For most manufacturing techniques that are reasonably mature, quite detailed manufacturing and design guides are published by the larger raw material suppliers, by trade organizations, and in some cases also by composite manufacturers and independent authors. The following list of sources for further information on composite manufacturing and design is far from comprehensive.

Liquid Molding

C. D. Rudd, A. C. Long, K. N. Kendall, and C. Mangin, *Liquid Moulding Technologies, Resin Transfer Moulding, Structural Reaction Molding and Related Techniques*, Woodhead Publishing, Cambridge, UK, 1997.

C. W. Macosko, *RIM Fundamentals of Reaction Injection Molding*, Hanser, Munich, Germany, 1989.

Compression Molding of SMC

SMC Design Manual Exterior Body Panels, SMC Automotive Alliance, AF-180, SPI Composites Institute, Washington, DC, USA, 1991.

Compression Molding of GMT

Technopolymer Structures Design and Processing Guide, GE Plastics, General Electric Company, Pittsfield, MA, USA.

Filament Winding

S. T. Peters, W. D. Humphrey, and R. F. Foral, *Filament Winding, Composite Structure Fabrication*, Society for the Advancement of Material and Process Engineering (SAMPE), Covina, CA, USA, 1991.

Pultrusion

Pultrusion Today, Ed. T. F. Starr, Chapman & Hall, London, UK, 1998.

R. W. Meyer, *Handbook of Pultrusion Technology*, Chapman & Hall, New York, NY, USA, 1985.

There are several sources dealing with modeling of phenomena relevant to composite manufacturing. The first book below deals with modeling of manufacturing-related phenomena in general, and modeling of autoclave processing, liquid molding, filament winding, and pultrusion specifically. The second book is a compilation of thermoplastic sheet forming techniques, while the third book concentrates on modelling of material flow in general, and specifically as applied to injection molding, RTM, compression molding, filament winding, and sheet forming.

Processing of Continuous Fiber Reinforced Composites, Eds. R. S. Davé and A. C. Loos, Hanser, Munich, Germany, 1997.

Composite Sheet Forming, Ed. D. Bhattacharyya, Elsevier, Amsterdam, The Netherlands, 1997.

Flow and Rheology in Polymer Composites Manufacturing, Ed. S. G. Advani, Elsevier, Amsterdam, The Netherlands, 1994.

Manufacturing-related research and development work is often published in national and international trade journals, as well as in refereed scientific journals such as:

Composites, Butterworth-Heinemann, Oxford, UK, to 1995.

Composites Manufacturing, Butterworth-Heinemann, Oxford, UK, to 1995.

Composites Part A: Applied Science and Manufacturing, Elsevier, Oxford, UK, 1996 onwards

Composites Part B: Engineering, Elsevier, Oxford, UK, 1996 onwards

Journal of Composite Materials, Technomic, Lancaster, PA, USA.

Journal of Reinforced Plastics and Composites, Technomic, Lancaster, PA, USA.

Journal of Thermoplastic Composite Materials, Technomic, Lancaster, PA, USA.

Polymer Composites, Society of Plastics Engineers (SPE), Brookfield, CT, USA.

SAMPE Quarterly, SAMPE, Covina, CA, USA.

and in the proceedings from recurring international conferences such as:

International SAMPE Symposium, USA.

International SAMPE Technical Conference, USA.

Society of the Plastics Institute (SPI) Composites Institute Annual Conference, USA.

American Society of Composites (ASC) Annual Technical Conference, USA.

SPE Annual Technical Conference, North America.

International Conference on Composite Materials, worldwide.

International Conference on Sandwich Constructions, worldwide.

Flow Processes in Composite Materials, Europe.

International Conference on Automated Composites, Europe.

References

1. J. J. Morena, "Mold Engineering and Materials—Part I," *SAMPE Journal,* **31**(2), 35-40, 1995.

2. J. J. Morena, "Mold Fabrications," in *International Encyclopedia of Composites,* Ed. S. M. Lee, VCH Publishers, New York, NY, USA, **3**, 394-420, 1990.

3. J. Z. Yu, Z. Cai, and F. K. Ko, "Formability of Textile Preforms for Composite Applications. Part 1: Characterization Experiments," *Composites Manufacturing,* **5**, 113-122, 1994.

4. B. P. Van West, "A Simulation of the Draping and a Model of the Consolidation of Commingled Fabrics," Report 90-7, Center for Composite Materials, University of Delaware, Newark, DE, USA, 1990.

5. F. N. Cogswell, *Thermoplastic Aromatic Polymer Composites,* Butterworth-Heinemann, Oxford, UK, 1992.

6. M. R. Monaghan, C. M. O'Bradaigh, P. J. Mallon, and R. B. Pipes, "The Effect of Diaphragm Stiffness on the Quality of Diaphragm Formed Thermoplastic Composite Components," International SAMPE Symposium, **35**, 810-824, 1990.

7. U. W. Gedde, *Polymer Physics,* Chapman & Hall, London, UK, 1995.

8. T. W. McGann and E. R. Crilly, "Preparation for Cure," in *Engineered Materials Handbook, Volume 1, Composites,* ASM International, Metals Park, OH, USA, 642-644, 1987.

9. J. Corden, "Honeycomb Structure," in *Engineered Materials Handbook, Volume 1, Composites,* ASM International, Metals Park, OH, USA, 721-728, 1987.

10. A. D. Murray, "Injection Molding," in *Engineered Materials Handbook, Volume 1, Composites,* ASM International, Metals Park, OH, USA, 555-558, 1987.

11. Anon., "Over the Long Haul," *DuPont Magazine,* Jan/Feb., 1996.

12. S. T. Peters, W. D. Humphrey, and R. F. Foral, *Filament Winding, Composite Structure Fabrication,* SAMPE, Covina, CA, USA, 1991.

13. K. V. Steiner, "Einsatz einer robotergestützten Anlage zum Bandablegen von thermoplastischen Verbundwerkstoffen," Fortschritt-Berichte VDI, Reihe 2, Nr. 369, VDI Verlag, Düsseldorf, Germany, 1996.

14. *Thermoplastic Composites Materials Handbook,* ICI Composites, 1991.

15. K. D. Felderhoff and K. V. Steiner, "Development of a Compact Robotic Thermoplastic Fiber Placement Head," International SAMPE Symposium, **38**, 138-151, 1993.

16. J. Hümmler, S. K. Lee, and K. V. Steiner, "Recent Advances in

Thermoplastic Robotic Filament Winding," International SAMPE Symposium, **36**, 2142-2156, 1991.

17. B. T. Åström, P. H. Larsson, and R. B. Pipes, "Experimental Investigation of a Thermoplastic Pultrusion Process," International SAMPE Symposium, **36**, 1319-1330, 1991.

18. B. T. Åström, P. H. Larsson, P. J. Hepola and R. B. Pipes, "Flexural Properties of Pultruded Carbon/PEEK Composites as a Function of Processing History," *Composites*, **25**, 814-821, 1994.

CHAPTER 5

SECONDARY PROCESSING

For the purposes of this chapter the term secondary processing is taken to include tasks following the (primary) processing operations covered in *Chapter 4*. This chapter reflects the fact that only under very rare circumstances is a composite component ready for use directly following demolding. Secondary processing operations normally required include machining, joining, surface treatment, and sometimes repair. In many cases these operations bear significant resemblance to those used with traditional construction materials, but for the most part the differences are great and secondary processing tends to be considerably more difficult with composites than with metals, wood, plastics, etc.

5.1 Machining

Since composites may (and should) be manufactured to net shape or at least near net shape, ideally there should not be any need for machining of the demolded component. Reality is different, however, and composites manufactured by almost every technique require some degree of post-molding machining. Often the required machining is limited to edge trimming of flash (material that has penetrated in between mold halves), but in many other instances extensive machining cannot be avoided. As an example of the former situation, compression molded parts for automobiles and other vehicles manufactured in short to intermediate series may be hand-trimmed along edges using a hand-held knife or coarse file while the component is still warm and the matrix has not yet reached full strength. The other extreme may be illustrated by the carbon-reinforced epoxy main wings of the fighter aircraft in *Figure 1-23* that require approximately 8,000 holes to be drilled for mechanical fastening.

The term machining includes any process by which material is gradually removed from a workpiece through any of a number of techniques and mechanisms. The material may be dislodged through localized shear, erosion, melting, chemical degradation, and vaporization. Due to the material removal a health concern arises; removed airborne particles, vapors, and gases are generally considered potential health hazards and good ventilation consequently is essential (see further *Chapter 8*).

Since it is inevitable that a discussion of machining of composites compares these techniques to the well-known techniques used for metals, an appropriate starting point is to assume that from essentially every point of view machining of composites is more difficult. Most of the difficulties arise due to composites being anisotropic, nonhomogenous, and containing very abrasive or tough fibers. These difficulties lead to a range of machining-induced defects in the workpiece that are unique to composites, including delamination, cracking, fiber pullout, fiber reorientation, fiber fuzzing, matrix chipping, and thermally induced matrix degradation. Needless to say all these defects are highly unwanted—not least if there are 8,000 potential sources of defects in the wings of a very expensive aircraft—so special procedures and tools* must be used to reduce their occurrence. Defects are created when the composite is subjected to forces and increased temperature and the general approach in reducing machining-induced flaws therefore is to minimize forces and heat generation.

The main difficulties in composite machining have their origin in the great differences in properties between reinforcement and matrix, see *Table 5-1*. Since all machining operations (except water jet cutting which is treated in *Section 5.1.2*) heat the material, the thermal properties are of paramount importance. If the heat generated cannot effectively be conducted away from the point of machining, the temperature rapidly increases which may lead to thermal degradation; on the positive side the damage becomes localized—albeit more serious. This is the situation with composites reinforced with organic fibers and to a lesser degree with glass fibers, since the heat-transfer capability of these fiber types (as well as the matrix they are embedded in) is so poor, see *Table 5-1*. With carbon reinforcement the situation is different since the high thermal conductivity of the fibers distributes the generated heat over an extended area thus simultaneously reducing peak temperatures. However, long before significant degradation occurs the matrix softens (i.e. around T_g; cf. *Table 3-3*), which effectively places an upper limit on the speed of the machining operation. In addition to the aforementioned thermal aspects, the heat generated may also cause matrix cracking and delamination due to the large differences in thermal expansion (CTE) between con-

---

* In composites manufacturing-related activities the words tool and tooling are used in both the context of this chapter, i.e. as a device used to machine a workpiece, and in the context of *Chapter 4*, i.e. as a mold.

Table 5-1 Thermal and mechanical properties of selected constituents. Data, which have been averaged and rounded off to two significant digits for clarity, from *Chapter 3*

	ρ kg/m^3	k_l $W/m°C$	C_p $kJ/kg°C$	α_l 10^{-6} $°C^{-1}$	E_l GPa	σ^*_l MPa	ε^*_l %
Matrices							
Polyester	1200	0.20	1.8	78	3.8	62	3.8
Epoxy	1200	0.18	1.0	55	3.2	72	4.8
PP	900	0.14	2.1	90	1.4	36	350
PEEK	1300	0.25	1.3	44	3.2	93	50
Reinforcements							
E glass	2600	12	0.45	5.4	70	3600	4.7
Kevlar 49 (aramid)	1400	0.72	0.77	-4.3	130	3800	2.8
Carbon (PAN)	1800	38	0.8	-0.6	240	4200	1.5
Carbon (pitch)	1800	310	0.8	-1.2	580	2700	1.0

stituents. Furthermore the great differences in mechanical properties between fibers and matrix (cf. *Table 5-1*), i.e. strong and stiff fibers in a weak and ductile matrix, create difficulties in most machining operations.

Most work on machining of composites has been performed on thermoset-based composites, and differences between various thermosets in terms of machining behavior is normally small or nonexistent, whereas comparatively little work has been devoted to thermoplastic-based composites. With high-performance thermoplastics having high temperature tolerance (i.e. high T_g; cf. *Table 3-3*), machining is similar to the thermoset situation, although the much higher strain to failure may create some difficulties. With engineering thermoplastics the low temperature tolerance makes machining an even more delicate operation from a thermal point of view. As an example, PP is difficult to machine even at room temperature and as temperature increases during machining the matrix may quickly clog the tool. A redeeming feature of thermoplastics is that their greater ductility and damage tolerance make them less susceptible to machining-induced delamination, cracking, and matrix chipping.

5.1.1 Conventional Cutting Techniques

By far the most common machining techniques for composites are modified metal-working operations where the workpiece is cut with a tool that moves relative to the workpiece, see *Figure 5-1*. Although most traditional techniques work fairly well with many composites, different tools must be used and special procedures followed. Compared to metal machining, small chips should be machined off at high speed, low feed rate, low but positive working normal rake, and low depth of cut.

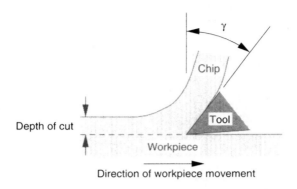

Figure 5-1 Cutting definitions. The angle γ, which still is still often called rake angle, is according to recent standards to be named working normal rake

5.1.1.1 Drilling

By far the most common machining operation is drilling, which therefore is treated more thoroughly than other cutting operations. From a strength point of view a composite should not contain any holes at all since they are seriously detrimental to component strength. Unfortunately holes are often a necessity and the (unintended) damage created by drilling consequently must be minimized.

When the drill first enters the laminate it may peel up part of the uppermost laminae, see *Figure 5-2a*, and when the drill exits the laminate it may act as a punch, thus causing delamination also on the other side of the laminate as illustrated in *Figure 5-2b*. Delaminations on the exit side of the laminate may be reduced by lowering feed rate near the exit and by supporting the back side of the laminate during drilling. *Figure 5-3* illustrates that the extent of delaminations on the exit side is reduced when using a drill with a more pointed tip which makes penetration more gradual. In addition to machining-induced surface damage, there is likely also to be internal damage to the laminate, as shown in *Figure 5-4*. Due to laminate anisotropy damage is also a function of fiber orientation. Where the local motion between tool and workpiece is parallel to the fibers (0°) they tend to be pulled out of the matrix; the best surface is obtained where fibers are sheared off at a right angle (90°), while the worst surface is obtained where fibers are compressed and bent, which occurs at intermediate angles (20–45°) [1]. This directional dependency is mainly confined to unidirectionally reinforced composites; with angle-ply laminates it is easier to achieve a good machined surface.

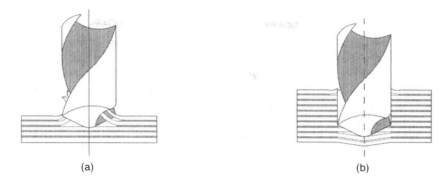

(a) (b)

Figure 5-2 Schematics of delaminations caused by drilling; (a) Upon entry (b) Upon exit. Drawings reprinted from reference [2] with kind permission from Elsevier Science Ltd, The Boulevard, Langford Lane, Kidlington OX5 1GB, UK

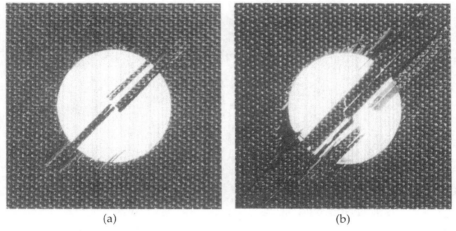

(a) (b)

Figure 5-3 Photographs of delaminations on the exit side of a carbon-reinforced epoxy laminate caused by drilling with (a) pointed and (b) blunt drills. Photographs reprinted from reference [2] with kind permission from Elsevier Science Ltd, The Boulevard, Langford Lane, Kidlington OX5 1GB, UK

In addition to the mechanically induced defects so far discussed, the possibility of thermally induced defects must be considered as well. At high temperature delaminations can arise from the difference in CTE between constituents and between plies, and in the vicinity of the hole the heat generated may soften (or melt) the matrix and may very well also chemically degrade it. More localized and more severe thermal damage is to be expected with less conductive reinforcement. With aramid and carbon reinforcement the negative longitudinal CTE of the fibers may actually cause the laminate to squeeze the drill as temperature increases. Due to these thermal effects the permissible drilling speed is largely limited by the temperature induced in the workpiece. Less heat is generated by reducing turning speed and feed rate. Of the total heat generated in drilling of metals, the chips absorb 75 percent, the tool 18 percent, and the workpiece 7 percent. With composites the situation is quite different; in drilling carbon-reinforced epoxies the tool absorbs half of the heat while the remainder is equally absorbed by chips and workpiece [1]. This points to a significantly greater need for the tool to be able to conduct heat away from the cutting zone in composite machining.

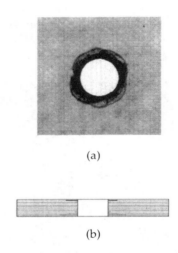

(a)

(b)

Figure 5-4 Drilling-induced delaminations in carbon-reinforced epoxy laminate; (a) X-ray radiograph showing the superimposed damages throughout the thickness (in black) (b) Schematic representation of the damage distribution through the thickness. Radiograph reprinted from reference [2] with kind permission from Elsevier Science Ltd, The Boulevard, Langford Lane, Kidlington OX5 1GB, UK; drawing reprinted from reference [3] with kind permission from the author

One of the main issues in composite machining is the rapid tool wear that is caused by the abrasive reinforcement; carbon is the most damaging to machine tools and organic reinforcement the least abrasive. Tool wear increases with increased thrust and torque, and as a drill wears increased thrust is required, thus accelerating the wear process and increasing the risk of delaminations on the exit side of the laminate. Although high-speed steel (HSS) drills initially perform well, they are already worn out after a few holes, whereas cemented carbide drills fare somewhat better. Polycrystalline diamond (PCD) drills (see *Figure 5-5a*) have relatively large diamond crystals (on the order of a few millimeters) fastened to the drill. Although considerably more expensive than carbide drills, the life expectancy of a PCD drill is two orders of magnitude greater than for carbide drills. A number of drill types have been developed exclusively for drilling in composites, see *Figure 5-5*.

While glass- and carbon-reinforced composites can be drilled with relatively good results using several types of drills, aramid fibers create problems in that their low compressive strength allows them to recede into the matrix rather than being cleanly cut, thus creating fuzz. To successfully drill in aramid-reinforced composites the fibers first have to be stretched and kept under tension as they are sheared off; *Figure 5-5d* shows a drill

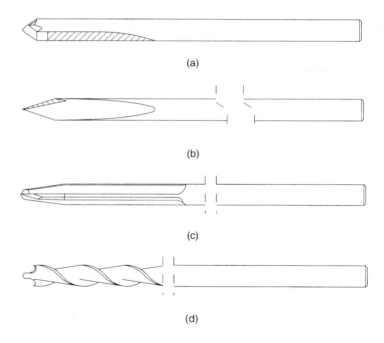

(a)

(b)

(c)

(d)

Figure 5-5 Drill types; (a) Eight-faceted PCD drill (b) Dagger drill providing improved hole quality (c) Four-flute tapered drill for hand-held equipment (d) Modified twist drill for drilling in aramid-reinforced composites. *Figures 5-5a* and *b* reprinted from reference [2] with kind permission from Elsevier Science Ltd, The Boulevard, Langford Lane, Kidlington OX5 1GB, UK7; *Figures 5-5c* and *d* reprinted from reference [3] with kind permission from the author

designed for this task. Specific drilling conditions for a given combination of composite material and drill, including turning speed and feed rate, may normally be obtained from the drill supplier. While cooling of both tool and workpiece using air, water, or some specialty liquid is normally required under most conditions, it has further been shown that machining under cryogenic conditions (very low temperature normally locally induced with liquid nitrogen) to embrittle aramid fibers results in significantly enhanced surface finish with a minimum of fuzz even when conventional tools are used [4].

Despite the damages incurred to the composite, drilling using traditional twist (helix) drills are the norm in most applications, particularly in components that are not highly stressed, see *Figure 5-6*. In critical applications, such as aerospace, more refined drills are used. While PCD drills (*Figure 5-5a*) offer greatly extended life, dagger drills (*Figure 5-5b*) provide better hole quality. Unconventional hole-generation techniques, such as the one schematically illustrated in *Figure 5-7*, are also being evaluated. This "twin-spin" technique employs a small grinding cylinder coated with microscopic diamond particles which both axially and radially grinds a hole through simultaneous rotation about its own axis and excentrically about a principal axis [5]. The main advantage of this technique is greatly improved hole quality compared to traditional drilling techniques.

To illustrate the importance of minimizing machining-induced flaws, it is instructive to compare the damage propagation in fatigue-loaded laminates with varying degrees of initial flaws around a machined hole. To this end, conventional aerospace techniques (cf. *Section 4.2.2*) were used to hand layup and autoclave-consolidate quasi-isotropic 24-ply carbon-reinforced epoxy laminates. Holes were then machined in these laminates with three different techniques using no laminate back support; drilling using a conventional cemented carbide twist drill, drilling using a PCD drill (cf. *Figure 5-5a*), and grinding using the technique illustrated in *Figure 5-7*. The machining-induced delaminations, apparent as black regions, are shown in the first line of *Figure 5-8* ("0 cycles"); the twist drill clearly induced the gravest damage and the twin-spin technique hardly any at all. The specimens were then subjected to a fatigue load introduced through a steel pin inserted in the machined hole. (The sinusoidal fatigue load had constant amplitude and zero ratio between minimum and maximum loads, i.e. the load was purely tensile.) After 250,000 cycles delaminations had clearly propagated upwards in the figures (where the pin loaded the hole edge). The final line in the figure shows the samples right after failure, which was defined as a maximum deformation of the hole edge occurring just prior to catastrophic failure. The relation between the relative extent of damage is maintained throughout loading, i.e. the sample with the greatest extent of initial delaminations saw the fastest propagation of these delaminations and exhibited the shortest life [6].

Figure 5-6 Drilling with traditional twist drill in resin transfer molded glass-reinforced polyester component with foam core inserts containing sheet metal at load-introduction points. Photograph courtesy of Lear Corporation Sweden, Ljungby, Sweden

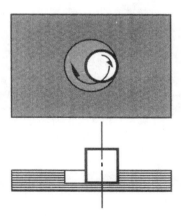

Figure 5-7 Schematic of twin-spin cutting technique to machine holes. Drawing reprinted from reference [2] with kind permission from Elsevier Science Ltd, The Boulevard, Langford Lane, Kidlington OX5 1GB, UK

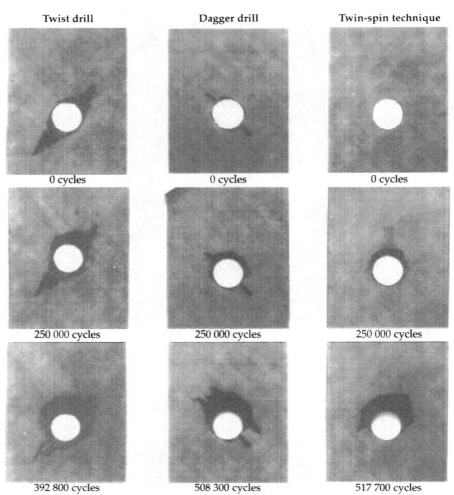

Twist drill	Dagger drill	Twin-spin technique
0 cycles	0 cycles	0 cycles
250 000 cycles	250 000 cycles	250 000 cycles
392 800 cycles	508 300 cycles	517 700 cycles

Figure 5-8 X-ray radiographs showing the superimposed through-the-thickness delaminations (black regions) around holes machined in carbon-reinforced epoxy laminates subjected to fatigue loading introduced with pins inserted into the holes. "Twist drill" is a conventional helix drill, "Dagger drill" the one shown in *Figure 5-5b*, and "Twin-spin technique" the technique illustrated in *Figure 5-7*. Radiographs reprinted from reference [6] with kind permission from Technomic Publishing Co, Lancaster, PA, USA

5.1.1.2 Sawing

In addition to the previously discussed generic machining defects, sawing can introduce fiber reorientation near the cut. Common band saws as well as hand-held hack saws, circular saws, and cutting disks (see *Figure 5-9*) are used for general sawing tasks, but blades and disks should be diamond-coated for long life. Hand-held hack saws or automated circular saws or cutting disks are used in pultrusion.

It has been found that in sawing the cutting speed should be as high as the matrix can tolerate, since a higher speed leads to a better cut with less severe and less widespread damage to the interior of the laminate. The reason for this is that a higher cutting speed requires lower forces applied both perpendicular to the workpiece and in the feed direction, which consequently reduces the risk of manufacturing-induced damages [1]. Sawing of aramid-reinforced composites creates problems for the same reason as for drilling. Running the blade backwards, i.e. reversing the relative motion of the teeth to the workpiece, has proved to be a simple yet effective solution, since run backwards the teeth of the blade act as knives that effectively shear the fibers. However, before first use the blade should be honed to eliminate burrs that otherwise could catch and pull out fibers to create fuzz.

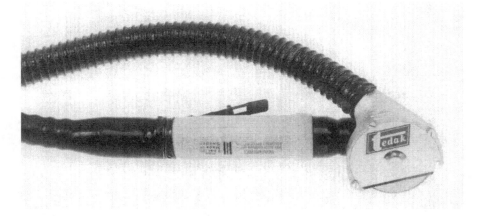

Figure 5-9 Diamond-coated cutting disk with on-tool extraction. Photograph courtesy of Nederman High Vacuum Systems, Sweden

5.1.1.3 Edge Trimming

Almost all composite components must be edge trimmed following demolding, but the way in which edge trimming is carried out varies greatly. Partially crosslinked wet laid-up glass-reinforced polyester laminates may conveniently be cut with a knife along a steel edge in the mold to remove overspray, for example. After the component is fully crosslinked, hand-held files and hacksaws as well as grinding disks (see *Figure 5-10*) may be used. Edge trimming of flash in glass-reinforced polyester components manufactured through compression and resin transfer molding may also be carried out with such techniques (see *Figure 5-11*), but in long series hand-held or robotized routers are used (see *Figure 5-12*). Routing may introduce the same defect types as drilling and also with routing the quality of the cut

Figure 5-10 Grinding of glass-reinforced polyester pipe. Photograph courtesy of Nederman High Vacuum Systems, Sweden

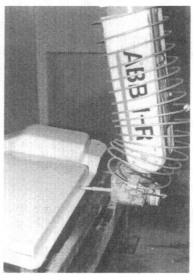

Figure 5-11 Sanding to remove flash of compression molded glass-reinforced polyester component (shown demolded in *Figure 4-45*). Photograph courtesy of Lear Corporation Sweden, Ljungby, Sweden

Figure 5-12 Routing of resin transfer molded glass-reinforced polyester component. Photograph courtesy of Lear Corporation Sweden, Ljungby, Sweden

depends on fiber orientation; the worst results are achieved at a 45° angle between fibers and edge [1]. As with other cutting techniques, tool wear is high and cemented carbide and PCD tools should be used to enhance tool life and cut quality. For aramid reinforcement special tools that achieve the required tensile-shearing sequence must be used to achieve good results.

5.1.1.4 Other Cutting Operations

Most conventional cutting techniques may be used with composites with some degree of success. However, since net-shape manufacturing is such an important advantage of composites, extensive machining is normally avoided. Composites may be milled (see *Figure 5-13*) and turned, but in unidirectionally carbon-reinforced epoxy workpieces the material fractures ahead of the tool, resulting in compression-induced cracks in the machined surface. Little plastic deformation is noted and chips are formed through a series of fractures. In turning glass-reinforced epoxy composites the cutting force is low when tool motion is parallel to the unidirectional reinforcement,

Figure 5-13 Milling of autoclave consolidated carbon-reinforced epoxy component. Photograph courtesy of Saab Military Aircraft, Linköping, Sweden

to reach a minimum at 30°, and a maximum at 90°. Cutting forces and depth of machining-induced damages are drastically reduced when ultrasonically vibrating the tool. As with other cutting methods, the cutting speed is limited by temperature, both in terms of workpiece damage and tool wear [1].

5.1.2 Waterjet Machining

A waterjet system consists of a filtration device, a high-pressure pump, an orifice that forms the waterjet, and a catching device. The water pressure generated by the pump may be up to 400 MPa, but the flow rates are low, in the range 4–8 liters per minute. The orifice has a diameter in the range 0.8–8 mm and the resulting jet may reach velocities up to 850 m/s [7]. A waterjet removes material through localized shear and is capable of cutting rather thick glass- and aramid-reinforced composites, but is much less effective with carbon reinforcement. In most composite applications abrasive particles are added to the waterjet to achieve higher cutting speeds and to enable cutting of thicker laminates through a combination of localized shear and erosion. The principle of an abrasive waterjet (AWJ) is illustrated in *Figure 5-14*. After forming the primary waterjet, abrasive particles, normally garnet, are added. The particles are accelerated by the waterjet and emerge from the mixing nozzle as a coherent particle-containing waterjet. On the other side of the workpiece the catching device catches the jet. The catcher may either be a device whose movements are synchronized with the delivery system, be an integral part of it, or be a (stationary) water tank located beneath the workpiece.

Hydraulic operating parameters with an AWJ include pressure and nozzle diameter, where a higher pressure results in a smoother cut and higher cutting speed, while at the same time exposing the workpiece to higher

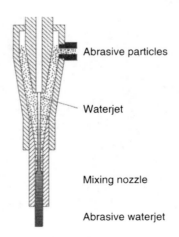

Figure 5-14 Schematic of AWJ. Redrawn from reference [8]

forces with the associated risk of increased damage. Abrasive parameters include flow rate and concentration, size, shape, and hardness of the particles. Harder particles certainly are more effective in cutting but also wear out the mixing nozzle more rapidly.

A complete AWJ head may weigh as little as 5 kg and may be hand-held (see *Figure 5-15*), mounted above an x-y table that positions the workpiece, or robot-mounted for positioning relative to a stationary workpiece. The width of a machined cut is 0.5–2.5 mm with a tolerance of ±0.4 mm, although the actual accuracy is often limited by the relative positioning of nozzle and workpiece rather than by the accuracy of the jet. The advantages of AWJ machining are high cutting speeds, good surface finish, no heat-affected zone (HAZ), and that the removed material is contained by the water instead of becoming airborne to pose a health hazard. Disadvantages include that cuts are tapered, that the laminate may absorb water, and that aramid reinforcement sometimes create fuzz. Further, while the mechanical forces and the hydrostatic pressure that the jet exposes the material to are

Figure 5-15 Hand-held AWJ with integrated catcher (to the left of the workpiece). Photograph courtesy of Saab Military Aircraft, Linköping, Sweden

modest, they may be sufficient to cause delaminations on the exit side of the component due to jet deflection if feed rates are too high, see *Figure 5-16*. There are also disadvantages associated with the equipment, including high cost, dangers due to the high water pressure, and noisy pump and jet.

To generate holes with an AWJ, called piercing, the jet is suddenly turned on (and off) and this impact load may cause delaminations on the exit side for the same reasons as in drilling. The delamination risk is reduced by lowering pressure and jet diameter, by increasing standoff distance (cf. *Figure 5-16*), and through use of laminate back support [9].

Although the use of AWJs in composite machining is a fairly recent occurrence, they have been widely accepted in both aerospace and automobile industries. In composite applications AWJs are generally used for machining of holes and edge trimming of flash, and consequently compete with the conventional techniques of drilling, routing, and sawing. In all cases AWJ cutting is faster and the quality of the cut generally better, especially bearing in mind that there are no thermally induced damages.

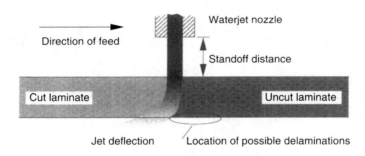

Figure 5-16 Jet deflection, or trailback, at too high feed rates. Redrawn from reference [7]

5.1.3 Laser Machining

Laser machining is rather well established for a range of materials, including composites. The concentrated monochromatic raw light beam, with composites usually supplied by a carbon dioxide (CO_2) laser, is focused into a spot size of 0.1–1 mm, see *Figure 5-17*. The intense heat generated by the beam removes material through melting, chemical degradation, and vaporization and the waste products are removed by a gas jet that is coaxial with the laser beam (cf. *Figure 5-17*); with organic workpieces the gas is normally air. Laser machining is very effective with (unreinforced) thermoplastics, since the HAZ becomes very limited due to the poor thermal conductivity of the polymer leading to localized melting. Thermoset polymers require higher energies since they do not melt and therefore must be chemically degraded and vaporized at considerably higher temperatures, see *Table 5-2*.

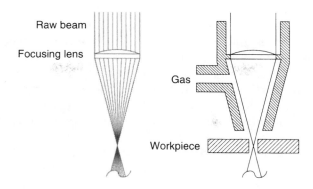

Figure 5-17 Schematic of laser machining. Reprinted from reference [3] with kind permission from the author

Table 5-2 Vaporization properties of selected constituents [9]

	Vaporization temperature °C	Heat of vaporization kJ/kg
Polyester	350 – 500	1000
E glass	2300	31000
Aramid	950	
Carbon	3300	43000

Once fibrous reinforcement is introduced into a polymer, laser machining becomes much more difficult due to the vast differences in thermal (*Table 5-1*), vaporization (*Table 5-2*), and light absorption properties between constituents. *Table 5-2* shows that very high temperatures are required to vaporize inorganic reinforcement and due to the higher thermal conductivity of inorganic reinforcement the HAZ increases in size, which further increases the power required. Due to a smaller difference in thermal conductivity and vaporization temperature, it is less difficult to laser machine aramid- than carbon-reinforced composites, whereas glass-reinforced composites represent an in-between situation.

The heat generated potentially leads to the same kind of thermally induced damage as discussed for conventional machining, but also to some additional concerns. *Figure 5-18* illustrates the damage incurred in an aramid-reinforced epoxy laminate. On the top layer there is an extended HAZ with partial matrix degradation, while closer to the cut the matrix has been removed and bare fibers protrude. On the cut surface a charred layer representing the remainder of degraded matrix and fibers is observed. Although the figure illustrates the damage incurred in an aramid-reinforced epoxy, the damage pattern should be similar for all polymer composites, while the extent to which bare fibers protrude from the remaining matrix will be a function of the difference in vaporization temperature and thermal conductivity between fibers and matrix [9].

Advantages of laser machining include that no mechanical forces are introduced meaning that delaminations usually can be avoided, that cuts in otherwise inaccessible locations are possible, and that cuts may be as narrow as 0.8 mm. The cut tolerance is ±0.5 mm, but as with AWJ machining the limitation usually lies in positioning. While small spot sizes and therefore cuts are within reach, a smaller spot size also yields less depth of field, thus limiting workpiece thickness. The major disadvantages of laser machining no doubt are the thermal damage incurred and that also laser-machined holes are tapered (as suggested by *Figure 5-18*); the higher the power, the greater the taper, since higher power leads to a larger HAZ and rounded edges on the beam entry side. Further disadvantages are high equipment cost, dangers due to the high-intensity light, and that removed particles and fumes are potential health hazards.

Laser machining is used in composite machining operations similar to where AWJs are employed, i.e. in competition with conventional drilling, routing, and sawing—and of course AWJ machining. Laser and AWJ machining may be carried out at similar speeds, which are normally much higher than with traditional techniques. Milling and turning using lasers has been suggested, but laser machining works best in through-the-thickness machining since waste products are then more easily removed.

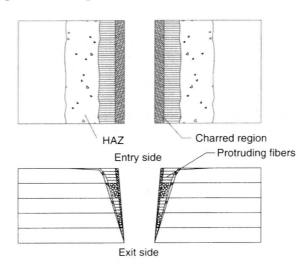

Figure 5-18 Schematic of damage incurred in aramid-reinforced epoxy laminate by laser machining. Redrawn from reference [9]. Figure courtesy of E Persson, Department of Aeronautics, Royal Institute of Technology, Stockholm, Sweden

5.1.4 Other Machining Techniques

Other machining techniques that have proven technically feasible with composites but are not in widespread use include electro-discharge machining (EDM), electro-chemical machining (ECM), and ultrasonic machining. The principle of EDM is illustrated in *Figure 5-19*. Both the workpiece and the male tool are immersed in a dielectric liquid and connected to a power source. A pulsed current causes electrical discharge between the two electrodes (workpiece and tool) and the very high temperatures generated by the spark gradually erode the workpiece through a combination of melting, degradation, and vaporization. The workpiece must be conductive and glass and aramid reinforcement are therefore effectively excluded. Instead of a male tool, the machining may be a metal wire that gradually works its way through the workpiece (in much the same way that expanded thermoplastics may be cut with a heated metal wire). Compared to the previously treated machining techniques EDM is slow.

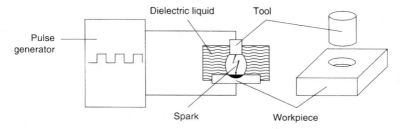

Figure 5-19 Schematic of EDM. Reprinted from reference [3] with kind permission from the author

ECM is similar in concept to EDM but also works with nonconductive workpieces. The workpiece is immersed in an electrolyte and placed between an anode and a cathode, the latter being the tool. When a DC current is applied hydrogen gas bubbles form at the cathode surface and sparking occurs across the bubbles. As with EDM the sparks generated at the cathode, which also in ECM may be a male tool or a wire that is placed near the workpiece, produce heat that gradually erodes the workpiece. With carbon reinforcement a separate anode may be omitted since the workpiece then can be used as anode. Also like EDM, ECM is slow in comparison to the aforementioned techniques.

In this context ultrasonic machining does not refer to ultrasonic vibration of a conventional machine tool, but rather to ultrasonic vibration of a male tool that together with the workpiece is immersed in a slurry containing abrasive particles. Since the tool is located close to the workpiece its vibrating motion grinds the abrasive particles against the workpiece to remove material through erosion.

5.2 Joining

One of the major advantages of composites over other material concepts is the possibility of designing and manufacturing large, geometrically complex, and highly integrated components to reduce the overall number of parts and thus joints. However, only in rare instances can the joining issue be entirely eliminated and in most cases the number of joints is merely reduced. Such a reduction is nevertheless highly desirable, since each eliminated joint translates into lower cost, lower weight, and exclusion of the inherent structural weakness that most joints constitute. The possibility of reducing the number of costly joining operations is indeed often the reason for composites being able to replace metal structures on strictly economical grounds. As an example, the automobile front end in *Figure 1-16* would in sheet steel have consisted of eleven spot-welded pieces, while the composite solution has integrated the entire front end into a single component. In this case the composite component not only saved money, but also reduced weight by 40 percent. Unfortunately the joining issue obviously was not eliminated; the composite component must still be connected to the steel chassis of the car. Traditional riveted aluminum airframe components have similarly been replaced with composite designs leading to two- to five-fold reductions in the number of parts and fasteners, thus saving precious weight while at the same time reducing overall manufacturing cost (in applications such as those shown in *Figures 1-22* to *1-26*).

The joints that cannot be avoided may be accomplished in several different ways. The two most common techniques are mechanical and adhesive joining (roughly corresponding to mechanical joining and welding with metals), but cocuring, lamination, and adhesively bonded or molded-in mechanical fasteners are other common joining techniques. A general comparison of mechanical and adhesive joints is given in *Table 5-3*.

5.2.1 Mechanical Joining

Mechanical joints rely on physical forces to hold components together through macro- and microscopic mechanical interference. In this case, macroscopic interference means shear and tensile forces in fasteners and contact forces between components, while microscopic interference is synonymous with friction. In most cases a mechanical joint requires an overlap of the components to be joined and a hole through the overlap so that a metal bolt or rivet can be inserted. In mechanical joining of metal components localized plastic deformation around the hole provides stress relief, but since a composite essentially is elastic to failure (assuming multidirectional reinforcement) no such stress relief is possible and problems should be expected due to poor interlaminar shear strength as well as poor transverse tensile and compressive strengths. These weaknesses are compounded by the damages likely to be incurred during hole generation (cf. *Section 5.1.1.1*).

Table 5-3 Qualitative comparison of mechanical and adhesive joining. "+" denotes an advantage

Property	Mechanical joints	Adhesive joints
Potential for damage from hole generation		+
Required shape match of components	+	
Required surface preparation	+	
Potential for stress concentrations		+
Peel strength	+	
Potential for creep	+	
Potential for galvanic corrosion		+
Sensitivity to environmental exposure	+	
Sensitivity to differences in CTE	+	
Possibility for disassembly	+	
Possibility of joining dissimilar materials	+	
Possibility of joining thin components		+
Smooth external joint surface		+
Weight		+
Cost		+

For each joint type there are several important parameters affecting strength, including laminate layup, hole pattern, component thickness, fastener-hole tolerance, overlap configuration, and clamping force. The key to mechanical joining of composites is to achieve load distribution over a large area. To this end, large washers that fit tightly onto the fastener and relatively high clamping forces should be used, but it is important that clamping forces are not excessive lest the components be damaged. Stress concentrations may further be lessened through a reduction in component anisotropy and use of doublers (local increases in component thickness). Use of interference fit (i.e. when the hole is slightly smaller than the fastener), which is common with metals, is generally not recommended with composites since it may cause delaminations during installation. A slight interference fit and very careful installation of fasteners has nevertheless proved to be beneficial to fatigue life in aerospace applications.

The degree of sophistication of mechanical joints varies tremendously with application; the highest requirements are found in aerospace applications. *Figure 5-20* illustrates some types of mechanical fasteners that are used in applications ranging from construction to aerospace; in components exposed to modest stresses it is possible to use self-tapping screws and inserts, see *Figure 5-21*.

Although commonly used, the straight single-lap joint shown in *Figure 5-20* may give rise to local compression damage when loaded in tension, as illustrated in *Figure 5-22*. It is therefore common to use slightly more sophisticated joint configurations, such as those shown in *Figure 5-23*. Particularly in aerospace applications it is normal to have several rows of bolts, screws, or rivets to distribute loads over as large an area as possible, see *Figure 5-24*. Despite sophisticated designs, joints normally remain the weakest point of an assembled structure. It has proved to be difficult to achieve a joint strength that is even half the laminate strength.

When dealing with carbon-reinforced composites there is always a potential problem in that the reinforcement conductivity may give rise to galvanic

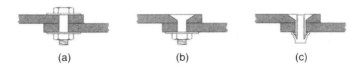

(a)   (b)   (c)

Figure 5-20 Examples of mechanical fasteners (a) bolt and nut with large washers; (b) countersunk screw and nut; (c) countersunk blind bolt (for use where the back side of the joint cannot be accessed)

Figure 5-21 Insertion of insert with interior thread in compression molded glass-reinforced component shown demolded in *Figure 4-45*. The insert is both self-tapping and adhesively bonded in place and is normally inserted using automated procedures. Photograph courtesy of Lear Corporation Sweden, Ljungby, Sweden

corrosion of metal fasteners and joining elements. The worst corrosion problems may occur with aluminum and cadmium-plated fasteners, while titanium fasteners fare much better. A solution to this problem is to coat the fastener with a non-conductive material (usually a polymer) or to avoid metal fasteners altogether; composite fasteners, although not yet very common, offer an obvious solution.

Figure 5-22 Single-lap joint loaded in tension illustrating where damage is likely to occur

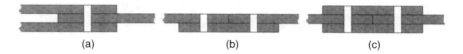

Figure 5-23 Examples of mechanical joint configurations (a) double lap; (b) butt lap; (c) double butt lap

Figure 5-24 Mechanical joining of carbon-reinforced epoxy canard wing and metal component in fighter aircraft shown in *Figure 1-23*. Photograph courtesy of Saab Military Aircraft, Linköping, Sweden

5.2.2 Adhesive Joining

Just like a mechanical joint, an adhesive joint normally requires an overlap of the components to be joined, the adherends. In adhesive joints the load transfer is chemical in nature and for chemical bonds to form the adherend surfaces must fit well to result in a uniform adhesive thickness and the surfaces should be free of contaminants. The preferred mode of load transfer through an adhesive joint is shear as shown in *Figure 5-25*.

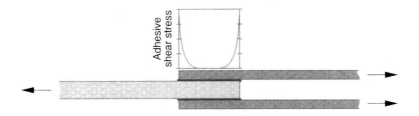

Figure 5-25 Shear stress distribution in adhesive of double-lap joint. Redrawn from reference [10]

Failure of an adhesive joint is normally initiated at the ends of the joint where shear stresses are the highest (cf. *Figure 5-25*). Failure may occur through shear within the adhesive, shear between adhesive and adherend, transverse tension within the adhesive, and transverse tension or compression within the surface plies of the adherend. The risk of the latter failure mode may be reduced by aligning the reinforcement of the adherend surface plies with the predominant load direction. Transverse tensile failure within the adhesive may occur when the joint is loaded transverse to the intended load direction, see *Figure 5-26a*. This peel load case also arises in tensile loading of a single-lap joint due to its deformation, see *Figure 5-26b* (cf. *Figure 5-22*), and may even cause delamination of the adherends, as illustrated in the figure. The poor peel properties of adhesive joints is a major disadvantage, but the risk of peel failure may be reduced through careful joint design.

Figure 5-26 Peeling of adhesively bonded single-lap joint (a) due to loading transverse to the joint; (b) due to joint deformation in tensile loading. The figure to the right also illustrates how a peeling load may lead to adherend delamination

The properties of adhesive joints are naturally highly dependent on the adhesive used. The first approach in selecting an adhesive may be to use the matrix resin in the adherends, but since many resins used as matrices are rather brittle with low strain to failure and therefore have poor peel properties, they are unlikely to perform well as adhesives. Most adhesives used to join composites are epoxy-based (cf. *Section 2.5*) and to improve fracture toughness, fatigue life, and peel properties they are often elastomer-modified, or toughened, which unfortunately also results in lower modulus, higher creep, and greater sensitivity to elevated temperature and moisture.

To achieve the previously mentioned intimate contact and contaminant-free surfaces that are prerequisites for a successful adhesive bond, careful preparation of the adherend surfaces is of paramount importance. Although there appears to be no complete consensus on what procedure to follow, the following steps represent one possible sequence of treatments that might be used in high-performance applications [11]:

1. Wipe with solvent
2. Rinse with water or alkaline solution
3. Dry in oven
4. Lightly abrade to roughen matrix surface without exposing reinforcement
5. Remove abraded particles
6. Wipe with solvent
7. Rinse with water
8. Dry in oven
9. Bond as soon as possible

Steps 1 and 2 remove surface contaminants, such as grease and mold release, while step 4 increases surface energy and bonding area. An alternative to this time-consuming series of steps is to use peel plies as discussed in *Section 4.2.2*. The peel ply is included in the vacuum bag layup directly on top of the outermost prepreg ply and consequently becomes part of the molded composite. Immediately prior to bonding the peel ply is peeled off to expose a clean and slightly rugged surface consisting of fractured matrix well suited for bonding. Less rigorous bonding preparations than those discussed above may be used in less critical applications; with polyester-based components, for example, surface treatments such as sanding, sandblasting, or flame treating are normally considered sufficient.

Adhesive composite-to-metal bonding is more difficult than composite-to-composite bonding for a couple of reasons. First, surface preparation of a metal adherend is even more critical than with composites; common treatments include etching and/or anodizing to produce an oxide layer on the surface followed by priming. A second difficulty arises from the fact that most effective adhesives must be crosslinked at an elevated temperature and the CTE difference between adherends (cf. *Tables 3-4* and *3-13*) therefore induces residual stresses in the adhesive.

Common adhesive joint configurations are illustrated in *Figure 5-27*. *Figures 5-27a* and *5-27b* illustrate that adherends sometimes are bevelled or tapered to reduce the stress concentrations at the ends of the joint and thus the potential for the failure mode illustrated in *Figure 5-26b*. It was previously mentioned that the reinforcement at the adherend surfaces should be aligned with the load to reduce the risk of failure within the surface ply; with the joints illustrated in *Figures 5-27c* through *5-27f* this is usually impossible to achieve throughout the joint.

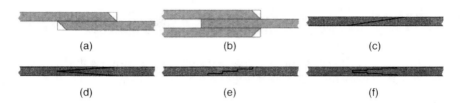

(a)   (b)   (c)

(d)   (e)   (f)

Figure 5-27 Examples of adhesive joint configurations; (a) Single lap, with optional beveling of adherends (b) Double lap, with optional beveling of adherends (c) Single scarf (d) Double scarf (e) Stepped lap (f) Double stepped lap. Redrawn from reference [10]

While the adhesive crosslinks (or solidifies) the adherends must be supported, usually by a dedicated bonding fixture, to ensure geometrical accuracy of the joint, see *Figure 5-28*. In most high-performance applications crosslinking takes place in an autoclave using sophisticated fixtures or molds. In less critical areas the adhesive may crosslink at room temperature, but the need for a more or less elaborate bonding fixture remains. It is further possible to use both mechanical and adhesive techniques in the same joint, thus eliminating or at least reducing the need for component fixation during crosslinking.

In applications where components are not intended to see excessively high structural loads it is often cost-effective to mold mechanical joining elements into the component, see *Figure 5-29*. With the exception of pultrusion, all major manufacturing techniques treated in *Chapter 4* allow such joining elements to be molded into the component. In particular filament winding conveniently allows end fittings to become integral parts of the wound component, see *Figure 5-30*.

Another joining alternative is to adhesively bond mechanical joining elements to the finished component surface, see *Figure 5-31*. It would be reasonable to assume that such adhesively bonded joining elements are capable of supporting lower loads than other adhesive (and mechanical) joining alternatives, but this solution is nevertheless extensively used where only low loads need to be supported.

Figure 5-28 Bonding fixture for the bus in *Figure 1-14*. (a) Body components in open fixture; (b) body components bonded in closed fixture. Photographs courtesy of Gottlob Auwärter, Stuttgart, Germany

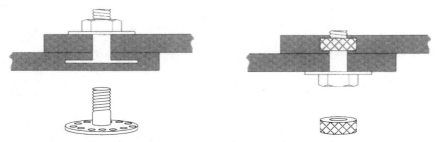

Figure 5-29 Examples of molded-in joining elements. Left figure redrawn from reference [12]

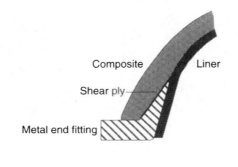

Figure 5-30 Example of wound-in metal end fitting in filament wound pressure vessel. The shear ply shown is likely an adhesive film applied prior to winding. Redrawn from reference [13]

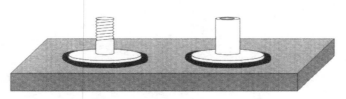

Figure 5-31 Examples of adhesively bonded joining elements

5.2.3 Cocuring and Lamination

In the aerospace industry it is becoming increasingly common to cocure, i.e. simultaneously crosslink and bond, for example wingskins and stiffeners in one and the same operation, as discussed in *Section 4.2.2*. In this manner a separate joining step may be eliminated, but there is a penalty in that molds become significantly more complex and layup more time-consuming. To ensure adequate bonding it is normally necessary to include adhesive films between components to be cocured (cf. *Figure 4-27*).

In lamination two fully crosslinked components are joined using wet hand layup techniques, see *Figure 5-32*. This is a common technique in, for example, joining of prefabricated bulkheads to a boat hull, see *Figure 4-22*. Similar in concept is the technique of placing a fully crosslinked composite component, such as a stiffener, onto a laidup but not yet crosslinked laminate. Just as with conventional adhesive bonding, lamination requires careful surface preparation of components to be joined. Both these lamination techniques are most likely in wet layup techniques.

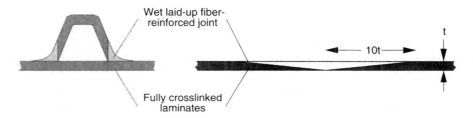

Figure 5-32 Examples of laminated joints. With the joint design to the right the laminates should be tapered over a distance equal to at least ten times the laminate thickness. Redrawn from reference [12]

5.2.4 Fusion Bonding

In fusion bonding molecular interdiffusion across the interface between the components to be joined is responsible for bond formation (cf. *Section 2.1.3.5*). Exploiting autohesion it is possible to locally melt the surfaces of two thermoplastic-based components and bring them together to form a joint without introduction of any foreign material (such as adhesive film or fastener). Fusion bonding, or fusion welding, has proven feasible with all kinds of thermoplastics, but fusion bonding of composites has mainly been tried with high-performance material systems. Numerous heating methods have successfully been used in fusion bonding of composites; the most promising being resistance, ultrasonic, and induction heating.

In resistance heating the most commonly used resistive heat source is carbon fibers. To this end, the resin at both ends of a unidirectionally carbon-reinforced prepreg ply is removed through pyrolysis or with a solvent and then electrical connectors are soldered onto the bare fiber ends. The two components to be joined, which are often manufactured from the same prepreg as used for bonding, are then placed on either side of the prepared prepreg ply, whereupon a large current is run through its fibers to generate the heat required to melt the matrix in both prepreg and component surfaces. If the components are also carbon-reinforced the reinforcement tends to conduct heat away from the bonding zone and it may be advantageous to place neat resin films on both sides of the heated prepreg to reduce heat loss and thus create the desired localized heating. Another reason for such extra resin films is that if fiber content in the components is high there may not be enough resin present along the bond line to form a good bond. With resistance heating it is difficult to obtain uniform heating throughout the bonding zone, which may lead to non-uniform bond strength.

With ultrasonic heating, energy is introduced into the joint through a vibrating tool, or horn. To obtain localized heating the mating surfaces of the components to be joined should be grooved to focus the energy to a smaller volume that therefore experiences a more rapid temperature rise and earlier melting. A problem with ultrasonic welding is that only relatively small sections can be welded at a time and that an incremental procedure therefore is necessary for large joints.

Carbon fiber-reinforced composites may be induction heated without special preparations, but heating is more efficient and the technique may also be used with any kind of reinforcement if a metal mesh impregnated with resin is placed between the components to be joined to concentrate heating to the bond line. Disadvantages include that heating is not as localized as with the aforementioned techniques and that a metal mesh remains embedded in the joint.

An innovative special case of fusion bonding is interlayer bonding where the mating surfaces of the components to be joined are coated with a thin thermoplastic resin layer that is chemically compatible with the component matrix but has a softening temperature below that of the matrix. It is then possible to use any of the aforementioned heating techniques (or some other) to fusion bond the components without much risk of deformation, which is a distinct possibility in other fusion bonding techniques. Interlayer bonding may be used to join both thermoset- and thermoplastic-based components; in the latter case the interlayer should ideally be amorphous with a T_g between T_g and T_m of the semicrystalline component matrix.

The characteristics of fusion bonding are largely shared with those of adhesive bonding. However, while fusion bonding is quite common with unreinforced polymers it is not yet of commercial significance with composites despite its technical feasibility having been proven.

5.2.5 Solvent Bonding

In solvent bonding the components to be joined are wiped with a solvent capable of dissolving the matrix. The intention is to soften the surfaces sufficiently to promote autohesion as the components are brought into contact. Following bonding the residual solvent diffuses out of the components over time. Solvent bonding is another technique that only works with thermoplastics and since most semicrystalline thermoplastics are difficult to dissolve the method is more likely to be successful with amorphous thermoplastics. Although used with unreinforced commodity and engineering thermoplastics and proven feasible also with high-performance thermoplastics, solvent bonding has not been much applied to composites.

5.2.6 Joining of Sandwich Components

While most of the aforementioned joining techniques may be used one way or another with sandwich components as well, there are many differences. In addition to joining sandwich components to each other, to laminates, and to non-composite materials, there is the issue of the closeouts along the edges of the component to protect the core and the face–core joint. In many manufacturing operations it is possible to include such closeouts in the original panel manufacture (see *Figures 5-33a* through *5-33d*), whereas in cases where this is impossible or impractical, closeouts may be added in a secondary operation (see *Figures 5-33e* through *5-33h*). Once closeouts have been incorporated, most of the previously discussed joint configurations for composite components may be used, but mechanical joining would be more common with sandwich components than adhesive joining. As an alternative to closeouts and subsequent joining, it is possible in lightly loaded applications, such as building panels, to join sandwich panels using custom-made profiles as illustrated in *Figure 5-34*. Such profiles may be pultruded composites or extruded aluminum and the panels may be adhesively or mechanically joined to the profiles. The same concept may naturally also be used with single-skin laminates (in lightly stressed applications).

For joining at points away from the edges of a sandwich panel it is common to use mechanical fasteners, which in most cases requires some kind of modification to faces or core to obtain a strong joint that does not cause

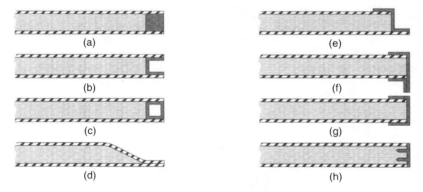

Figure 5-33 Examples of sandwich closeouts; (a) – (d) would be molded into the panel during original manufacture, while (e) – (h) would be adhesively bonded or possibly bolted on in a secondary operation. In (b), (c), and (e) – (h) the dark profile would be metal, plastic, or composite, whereas it in (a) likely would be a high-density block of expanded polymer, wood, etc. Redrawn from reference [14]

face–core disbonding. Some possible ways to include mechanical joints while still retaining most of the structural capabilities of the sandwich panel are shown in *Figure 5-35*. In all cases the sandwich panel has already been modified to accommodate load introduction when first manufactured. A less strong alternative (which is not shown in the figure) is offered by drilling a hole in a panel that has not been prepared for joining and then inserting a metal sleeve through the hole and using large washers for both nut and bolt.

Figure 5-34 Examples of joining of sandwich panels using custom-made profiles. Redrawn from reference [14]

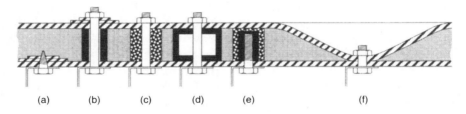

(a)   (b)   (c)   (d)   (e)   (f)

Figure 5-35 Examples of mechanical joints in a sandwich panel; (a) Self-tapping screw threaded into molded-in doubler; (b) Molded-in metal sleeve in core and external doubler; (c) Molded-in high-density core or wood insert; (d) Molded-in metal, plastic, or composite profile; (e) Molded-in polymer insert containing threaded metal insert. Doublers reduce stress concentrations, core inserts in (b) – (d) help support compressive loads from bolt tightening, and insert in (e) gradually distributes stresses into core through shear. Redrawn from reference [12]

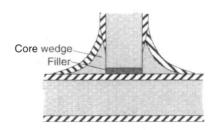

Figure 5-36 Schematic of common solution for T joints between sandwich components

Particularly in shipbuilding it is common to laminate already fully crosslinked sandwich panels together. Such a need arises where bulkheads are connected to the hull, where the deck joins the hull, and where possible deck structures are mounted onto the deck. *Figure 5-36* illustrates a common bulkhead–hull joint configuration.

5.3 Surface Finishing

The requirements for the surface finish of a composite may be both functional and cosmetic in nature. On the functional side, a surface finish may yield planeness, hardness, reflectivity, conductivity, or be a barrier to environmental degradation from sunlight, oxygen, water, cosmic radiation, etc. Following the normally inevitable edge trimming after demolding, few components have the required surface finish.

Where a gelcoat has been used, polishing of the surface using a polishing compound normally suffices to produce a high-gloss finish, although subsequent painting on top of the gelcoat for decorative purposes is possible as well. Where the surface finish is not considered critical, such as with most pultruded products, components are normally used without secondary surface finishing; however, pultruded components normally contain pigmented matrices. The outside surface of filament wound composites is unusually poor by composite standards. In applications where the distinct roving pattern is unacceptable, the component may first be carefully turned or sanded to smooth the surface before it is covered with an epoxy or a PUR, which may be pigmented.

Components manufactured through most other techniques are painted in a separate procedure. Proper adhesion between paint layer and component demands surface preparations similar to those required for successful adhesive bonding (cf. *Section 5.2.2*). The component is first degreased with a sprayed-on solution and then washed off with water to remove dirt, grease, and possible remaining mold release. Following surface cleaning, the component is dried in an oven, spray painted, and then brought into another oven to crosslink the paint. To achieve good results, it is critical that heating is slow and uniform so that the paint layer crosslinks evenly. With joined structures (such as automobile body panels which are often bonded together from two compression molded shells) and components of varying thickness, including sandwich panels, great care must be taken to achieve such uniform heating. Paint systems used for components compression molded from SMC crosslinks at temperatures up to 160°C and sometimes even 200°C, while components compression molded from GMT are painted with systems requiring maximum temperatures of 120°C. At these temperatures components and possible adhesive joints have little remaining stiffness and therefore require support during crosslinking, for example using the loading points and inserts intended for final assembly. Most components require a primer before application of the final paint layer, or top-coat. If in-mold coating was used as part of the molding process, this coat is usually used as primer. In mass production, for example in the automobile industry, components are automatically moved from one station to the next suspended from a continuously moving overhead rail; the entire painting cycle may well take a couple of hours. Although spray painting is the most common technique with composites, any conventional paint application technique, including dipping and brushing, is of course possible.

A thin metal surface layer may be required for both functional and cosmetic reasons. Several techniques may be used to deposit such a thin metal layer onto a composite surface, including electroless plating, electrolytic plating, chemical vapor deposition, and sputter deposition.

5.4 Repair

In contrast to conventional materials, for which repair techniques and procedures are well established, composites are perceived as being difficult to repair. While it may not be easy to repair a composite, the problems are exaggerated and the true obstacle lies in lack of knowledge, experience, and standardized repair procedures.

In general terms, a need for repair may arise from a manufacturing anomaly or from in-service damage. Of the manufacturing-induced defects discussed in *Section 4.4*, it is normally only cosmetic defects that are readily remedied, although a thermoplastic-based composite through reheating and reconsolidation may also be rectified in terms of geometrical and structural defects. For it to be economically defensible to spend time repairing a newly manufactured component it must either involve a quick repair only or the component must be expensive, since repair is very labor-intensive and time-consuming. Very expensive components thus may see extensive repairs of manufacturing-induced defects using procedures similar to those discussed below for in-service repair. Surface flaws, such as scratches and areas of shallow porosities, can be repaired by filling them with putty, often a filled polyester or epoxy formulated for rapid crosslinking, and then polishing the surface. Repair of manufacturing-induced defects is of course carried out prior to any final surface finishing. While polishing of a component normally is considered an expected finishing operation as discussed in *Section 5.3*, extensive polishing may border on repair; wonders can be achieved with vigorous polishing.

Repair due to in-service damage is a more common cause for repair than manufacturing-induced defects and is generally caused by impact rather than fatigue (which is more common with metals). Once a defect has been located in a component (see further *Section 6.4*) a decision must be made as to whether the part should be replaced or repaired. Since spare parts tend to be very expensive or infeasible (for example for composite boat hulls) and may not be available quickly enough or at all, repair often ends up being an attractive alternative. Most published work on repair deals with thermoset-based composites since they are significantly more common.

Repair techniques vary greatly depending on type of structure, materials, and applications. Another deciding factor is the type of damage; if the component is penetrated the damaged material must be removed and replaced with new material, whereas it sometimes may be salvaged if the damage is limited to for example delaminations. An example of the latter is an aramid-fabric reinforced polyester kayak dropped onto sharp rocks to cause extensive matrix fracture and significant loss of stiffness. Courtesy of the inherent toughness of aramid, fiber damage was assumed limited. Rather than cut out and replace the damaged regions, it was decided to locally laminate doublers reinforced with aramid fabric and glass CSM onto the inside of the hull to restore stiffness using the original laminating polyester. As in all repair, joining, and painting operations, the surface was carefully cleaned to remove dirt, grease, and possible remaining mold release (cf. *Section 5.2.2*). On the outside of the hull the gelcoat on and around the damaged regions was removed and a new gelcoat with matching color applied. The repair was successfully completed following crosslinking and polishing. Leaving the damaged material in place saved machining time, left the original reinforcement largely intact, thus enhancing load transfer around the damage, reduced the risk of leakage; and, due to the localized nature of the damages, eliminated the need for a mold to restore the original hull shape. For an application where weight is a major concern, it may be unacceptable to use doublers.

Another alternative to try to remedy delaminations in laminates, as well as face–core disbonds and limited core damage in foam–core sandwich components, is to drill holes in the laminate and inject resin to penetrate and fill the cavities, while applying vacuum at other holes. This RTM-like repair technique is in limited use, but it is very difficult to completely fill all cavities and the results are rarely completely satisfactory. The technique of bolting together a delaminated composite to try to prevent buckling must be considered an emergency operation and cannot be recommended, since new damages are induced in drilling.

Figure 5-37 illustrates a generic repair procedure for a situation in which the damaged regions must be replaced (for whatever reason). First all damaged material is removed using a knife, file, router, grinding disk, etc. Then the gelcoat (if any) is removed and the laminate is tapered. New material, which has been carefully cut to size and shape as shown in the figure, is then laminated in place using a back support as mold. When the resin has crosslinked completely, the new material is machined down to the original laminate thickness before gelcoat is applied. If there is no access to the back side of the laminate (which implicitly is assumed in the figure) a back support, often preconsolidated laminates, is bonded to the inside of the laminate; in most cases at least two laminates must be used side by side so that they can be inserted through the machined hole.

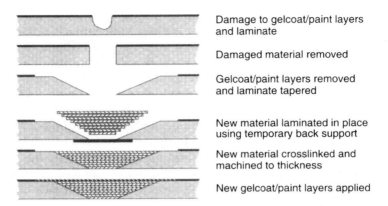

Damage to gelcoat/paint layers and laminate

Damaged material removed

Gelcoat/paint layers removed and laminate tapered

New material laminated in place using temporary back support

New material crosslinked and machined to thickness

New gelcoat/paint layers applied

Figure 5-37 Generic procedure for repair of damaged laminate. Not to scale

In general the laminate edges are tapered at least 1:10 (i.e. the length of the taper should exceed 10 laminate thicknesses) to provide an efficient single-scarf joint (cf. *Figure 5-27c*), but in aerospace applications tapers of 1:50, or 13 mm (0.5 in) per reinforcement ply, are specified. Stepped-lap repairs (cf. *Figure 5-27e*), typically with the same overall tapering as for single-scarf repairs, and preconsolidated patches with matching steps are sometimes used in aerospace applications.

A foam–core sandwich structure, e.g. a boat hull, is rarely fully penetrated following impact, thus often allowing return to repair facilities at reduced speed. If the core is undamaged it is used as back support for face lamination, but since the core surface often is damaged during removal of the damaged face, it is usually necessary to use a filled putty to build up an even surface for face lamination. If the core is significantly damaged it is removed all the way to the back face. A core plug is then fitted to the original core and is bonded both to the back face and laterally to the remaining core using a filled adhesive. Repair of a fully penetrating damage to a foam–core sandwich structure would (following removal of all damaged material) start with reconstruction of the core followed by subsequent lamination on both sides of the core using previously discussed techniques. The procedure for face repair in a foam–core sandwich structure is quite similar to that outlined in *Figure 5-37*, but there is an important difference in that, since damages to ship hulls are generally caused by groundings, lamination often must be carried out on vertical surfaces or even upside down. A filled resin with thixotropic additives may then be used to limit the effects of gravity. In such situations it may—provided that the outside face is largely undamaged—be worthwhile to perform the repair from the inside of the hull to avoid adverse effects of gravity even if more material then has to be replaced before reconstruction can commence.

Most of the treatment so far applies to non-flying applications. In the aerospace industry, where composites have been used for longer and where component failure is more likely to have catastrophic consequences, repair procedures and techniques are considerably better-developed than in other fields. However, in aerospace applications repair is made difficult by the abundance of material types and forms that are or have been in use and by their storage and crosslinking requirements. Since matrices and adhesives used in aerospace applications require storage in freezers and crosslinking under well-controlled temperature, vacuum, and pressure conditions, candidate repair techniques depend upon whether repair must be carried out in the field or if it can be completed at a dedicated composite repair facility. In field repair, material storage facilities are usually limited and crosslinking often has to be carried out under crude conditions (as illustrated by *Figure 5-38*), often with a mere resistance-heated blanket and a vacuum pump. Field repair thus requires simplified procedures and less demanding matrices and the repair is likely to result in poorer results than could have been achieved under more benign conditions, albeit often good enough. An intermediate alternative is to make a temporary repair in the field and then fly on to an appropriately equipped facility. If spare parts are available, the

Figure 5-38 Repair exercise on vertical stabilizer of fighter aircraft under conditions realistic to field repair. Photograph courtesy of Team KREP39, Sweden

damaged component can of course be removed and repaired in a dedicated composite repair facility. If a spare part is not available within the organization, parts for commercially common aircraft can often be leased until the damaged component has been repaired.

Small non-penetrating damages, such as scratches, can be satisfactorily repaired with a filled room-temperature adhesive used as putty. For larger non-penetrating or penetrating damages, preconsolidated quasi-isotropic carbon-reinforced laminates or metal plates may be bonded in place with a filled room-temperature adhesive either to constitute a permanent repair or merely to allow flying on to a better equipped facility; *Figure 5-39* illustrates the concept, which may be complemented with blind rivets. Repair procedure and allowable flying time before and after repair of a given defect type of given size and in a given location are generally documented in detail in repair manuals issued by the airframe manufacturer.

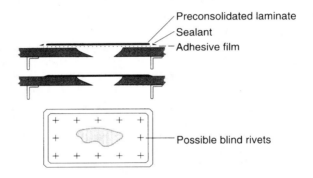

Figure 5-39 Schematic of repair with preconsolidated laminate or metal plate. Not to scale. Redrawn from reference [15]

When flat laminates or metal plates cannot be bonded or riveted on to cover a damage, e.g. for curved surfaces, and when a protuberance on the surface is not acceptable (for aerodynamic or weight reasons), repair may utilize prepregs and the procedure ends up being conceptually similar to that illustrated in *Figure 5-37*. If an autoclave is available, crosslinking requirements are easily met, but in field repair vacuum bagging, including a resistively heated blanket to achieve sufficiently high temperatures, must suffice.

Sandwich components with honeycomb cores can be repaired in a fashion similar to that described above for foam–core sandwich components. If one face has been penetrated or pushed into the core, the entire core beneath the damage is often removed. A core plug of the same honeycomb type as originally used is fitted to the remaining core. The plug is then bonded to the remaining face with an adhesive film and to the original core using a foaming adhesive to fill the open honeycomb cells on both sides of the joint.

The face is then built up on top of the core using gradually larger prepreg plies, as illustrated in *Figure 5-37*. If damage to the core is limited, it is instead possible to replace only part of it (in the thickness direction). The core–core bond may then be accomplished using two adhesive films sandwiching a glass-reinforced prepreg, while the sideways joining is still accomplished using a foaming adhesive. A simpler alternative that increases the total thickness of the structure instead uses a core plug that is flush with the original face (i.e. it has a thickness equivalent to the original core plus one face). The new face is then laminated on top of the original face, thus leaving a bulge, but also limiting the number of prepreg plies required, reducing the demands for accuracy in shape and size of the plies used, and reducing work involved.

The sequence of photographs in *Figure 5-40* illustrates how field repair of a Nomex honeycomb structure with carbon-reinforced epoxy faces may be carried out. *Figure 5-40a* reveals the damage incurred in the elevator of a Boeing 767 by a rock kicked up from the runway by one of its own tires. In *Figure 5-40b* the damaged laminate and part of the core is removed and the laminate edges tapered, a process taking a couple of hours, while *Figure 5-40c* shows the general working conditions of this field repair. Following removal of damaged material and laminate tapering, a new core section is bonded in place and gradually larger prepreg plies—one more than in the original laminate—are used to build up the new laminate. The vacuum bag assembly, including peel ply, separator, bleeder, barrier, aluminum caul plate, resistively heated blanket, breather, and vacuum bag (cf. *Figures 4-5* and *4-26*), is then placed over the repair, see *Figure 5-40d*. The complete crosslinking cycle, including 1–1.5 hours at 120°C, takes 3–4 hours. *Figure 5-40e* shows the repair following crosslinking and removal of peel ply, where the gradually larger prepreg plies of the laminate can be seen. Immediately after the peel ply is removed, a PUR primer is sprayed on and allowed to crosslink in about half an hour, whereupon a PUR topcoat is sprayed on and likewise crosslinked; *Figure 5-40f* shows the repair after painting. Under ideal conditions a complete repair such as the one described above can be completed in eight hours, whereas the one shown in *Figure 5-40* took twelve hours due to problems with the unreliable power supply shown in *Figure 5-40c* and strong winds.

Repair of thermoplastic-based composites is not very common, largely due to the fact that there are so few high-cost thermoplastic components (that are worthwhile repairing) in use. However, most of the repair techniques discussed above, including use of thermoset adhesives and prepregs as well as bolted-on patches, are clearly applicable to thermoplastic-based composites as well. Although thermoplastic adhesives and prepregs may be used for repair using the fusion bonding technique, the temperatures required normally vastly exceed those available in field repair, thus necessitating access to a dedicated composite repair facility with high-temperature capabilities.

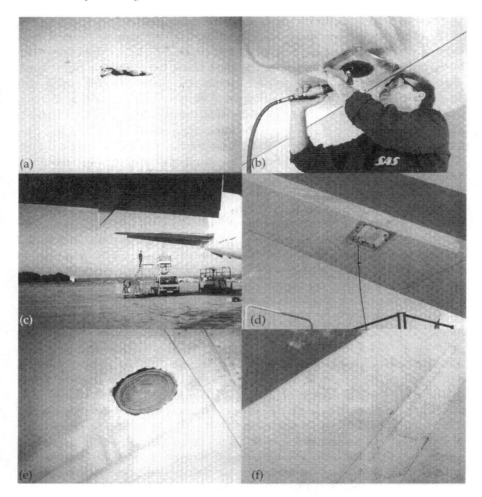

Figure 5-40 Repair of sandwich elevator of a Boeing 767 (a) original damage; (b) removal of damaged material and laminate tapering; (c) overview of repair area; (d) vacuum bagging; (e) repair following crosslinking; (f) repair after painting. Photographs courtesy of SAS Component, Stockholm, Sweden

The repair techniques outlined above are by no means the only ones possible or in widespread use; techniques vary widely between industries, applications, and materials. The intention of this section is only to point to some reasonably straightforward repair alternatives to illustrate that composites indeed can be repaired. Initiatives are ongoing in many fields to harmonize and standardize repair techniques, procedures, and materials to take some of the magic out of composites repair and make it into the standard procedure it ought to be.

5.5 Summary

When a composite component is demolded it is only in very rare instances ready for use immediately. In almost all cases the component must have its edges trimmed and holes machined, whereupon it normally needs to be joined to a larger structure and have its surface painted. A composite may also require repair due to damage incurred during manufacturing or due to in-service impact.

Composites present a true challenge in machining due to the large difference in properties between matrix and reinforcement. Traditional cutting techniques used for metals nevertheless often work reasonably well, but tools should be of different design. Particularly in machining of aramid-reinforced composites, where the reinforcement toughness makes machining even more difficult, special tools are required. Since glass and carbon fibers are very abrasive, harder and more durable tools are required for long life and better cuts; cemented carbide and PCD tools are normally used. In machining, special care must be taken to avoid machining-induced delaminations and thermal damage. To this end tool speed should be high, feed rate low, and during drilling a back support should be used. While conventional cutting techniques dominate in composite machining, AWJs and lasers offer several advantages, particularly if series are long and requirements on cut accuracy great.

Joining is usually accomplished through mechanical, i.e. bolting, screwing, and riveting, or adhesive means. The main concern in mechanical joining is to reduce stress concentrations as much as possible, since composites (unlike metals) cannot relieve stresses through local plastic deformation and instead easily delaminate. Stress concentrations are usually reduced through use of washers, doublers, and a multiplicity of joining elements. Adhesive joining requires exact shape fit between adherends, which must have well-prepared surfaces, and the joint must be exactly supported during crosslinking of the adhesive. For thin components, an adhesive joint is often preferred over a mechanical joint since it offers less stress concentrations and lower weight, but requirements on the adherends are stricter, there is potential for creep, and the joint cannot easily be disassembled without damaging the adherends. In contrast, mechanical joints are often preferred for thick components. Sandwich components are rarely joined through adhesive means, while mechanical joining tends to require procedures different from those used with single-skin composites. Cocuring (in aerospace-type applications) and lamination (in wet layup) are common alternatives to conventional adhesive joining of both single-skin composites and sandwich components.

Most composites need to have their surfaces finished, usually through sanding, polishing, and painting. Special paint systems and techniques have been developed for composites.

While composites are thought of as being difficult to repair, this is rarely true. With a nonpenetrating damage, it may be sufficient to laminate a doubler onto the inside of the component to strengthen the damaged region and then to refinish the exterior surface. In more severe cases, the damaged material is machined completely off and the laminate edges carefully tapered. A new laminate is laid up ply by ply and then crosslinked. In most cases the result is fully acceptable. Also sandwich structures may be repaired through removal and replacement of the damaged core; faces are repaired in the same fashion as with single-skin composites. Composite repair is particularly well established in aerospace applications where damaged components are generally far too expensive to not repair.

Suggested Further Reading

One of the few sources treating both machining and joining of composites is:

M. M. Schwartz, *Post Processing Treatment of Composites*, SAMPE, Covina, CA, USA, 1996.

Joining of sandwich components is treated in:

The Handbook of Sandwich Construction, Ed. D. Zenkert, EMAS, Warley, UK, 1997.

References

1. S. Abrate and D. A. Walton, "Machining of Composite Materials. Part I: Traditional Methods," *Composites Manufacturing*, **3**, 75-83, 1992.

2. E. Persson, I. Eriksson, and L. Zackrisson, "Effects of Hole Machining Defects on Strength and Fatigue Life of Composite Laminates," *Composites Part A*, **28A**, 141-151, 1997.

3. E. Persson, "Hole Generation Defects in Composite Laminates," Report 95-9, Department of Lightweight Structures, Royal Institute of Technology, Stockholm, Sweden, 1995.

4. S. J. Mander and D. Bhattacharyya, "Drilling of Kevlar Composites," in *Machining of Composite Materials II*, Eds. T. S. Srivatsan, C. T. Lane, and D. M. Bowden, ASM International, Materials Park, OH, USA, 87-95, 1994.

5. L. Zachrisson, I. Eriksson, and J. Bäcklund, "Method and Tool for Machining a Hole in a Fibre-Reinforced Composite Material," Swedish patent 500,933, 1994.

6. E. Persson, I. Eriksson, and P. Hammersberg, "Propagation of Hole Machining Defects in Pin-Loaded Composite Laminates," *Journal of Composite Materials*, **31**, 383-408, 1997.

7. J. Korican, "Water-Jet and Abrasive Water-Jet Cutting," in *Engineered Materials Handbook, Volume 1, Composites*, ASM International, Metals Park, OH, USA, 673-675, 1987.

8. G. Eckold, *Design and Manufacture of Composite Structures*, Woodhead Publishing, Cambridge, UK, 1994.

9. S. Abrate and D. A. Walton, "Machining of Composite Materials. Part II: Non-Traditional Methods," *Composites Manufacturing*, **3**, 85-94, 1992.

10. J. Williams and W. Scardino, "Adhesives Selection," in *Engineered Materials Handbook, Volume 1, Composites*, ASM International, Metals Park, OH, USA, 683-688, 1987.

11. T. J. Reinhart, "Adhesive Bonding Surface Preparation," in *Engineered Materials Handbook, Volume 1, Composites*, ASM International, Metals Park, OH, USA, 681-682, 1987.

12. G. Westerlund, *Armerade Härdplaster*, Swedish Plastics Federation, Stockholm, Sweden, 1992.

13. S. T. Peters, W. D. Humphrey, and R. F. Foral, *Filament Winding, Composite Structure Fabrication*, SAMPE, Covina, CA, USA, 1991.

14. "Bonded Honeycomb Sandwich Construction," TSB124, Hexcel, Pleasanton, CA, USA, 1993.

15. J. Koshorst, "Advanced Composite Components In Airline Service Status and Repair," in *Composites in Manufacturing*, Ed. A. B. Strong, Society of Manufacturing Engineers, Dearborn, MI, USA, 207-217, 1991.

CHAPTER 6

QUALITY CONTROL AND CHARACTERIZATION

From the component manufacturer's point of view the issues of quality control and characterization arise in several different contexts. While the broader concept of quality assurance (QA) is a system of activities to ensure consistent quality of a product or service (which if thoroughly performed, documented, and certified may lead to the much-coveted ISO 9000 certification), quality control (QC) refers to the actual act of ensuring that a product or service is of consistent quality. The quality is usually assessed through quantification of a property of the product or the service, which is then compared to a standard. Characterization is here taken to imply quantification of physical, mechanical, or thermal properties, either for quality control or design purposes.

6.1 Constituent Quality Control and Characterization

In a manufacturing environment there are generally four stages that may require quality control: receiving inspection, inspection prior to use, inspection during manufacturing, and inspection after manufacturing. When raw materials are physically and chemically stable, such as most reinforcements, most fillers, and thermoplastic resins, the second stage may be omitted. If the finished component is stored for a long time before delivery, a fifth stage, inspection prior to shipping, may need to be added [1]. The degree to which a component manufacturer carries out quality control tasks varies tremendously with industry, application, and customer.

Complete quality control of all constituents, i.e. resin components, additives and fillers, reinforcements, cores, adhesives, prepregs, molding compounds, etc., requires Herculean efforts. While there is a vast array of methods available to the polymer scientist to characterize the properties of resin components, most users elect to reduce their quality control efforts to measurement of resin viscosity and determination of crosslinking characteristics. The viscosity is an indicator of molecular weight and may also be used to assess whether the base resin (i.e. without initiator or hardener) has been stored for too long, since resins spontaneously, albeit slowly, crosslink even if not fully formulated; the resin viscosity consequently increases with time and the resin may at some predetermined viscosity be considered as having expended its shelf life. The crosslinking characteristics of the fully formulated resin may be assessed with sophisticated equipment to trace the crosslinking reaction by means of monitoring the heat flow to and from a resin sample to subsequently determine T_g (e.g. using differential scanning calorimetry, DSC). A less sophisticated alternative is to determine the gel time. One way to do this is to maintain a resin sample at a given temperature and then periodically prod the resin to determine when it gels. Alternatively, gel time may be characterized as being when the resin viscosity at a given temperature and shear rate exceeds the viscosity at which gelation is assumed to occur.

Determination of the inherent properties of fillers and reinforcements is normally beyond the capabilities of the average component manufacturer, who therefore has to rely on the raw material supplier delivering material of known and consistent quality. However, the user can easily monitor the quality of the reinforcement in terms of visible defects, such as fractured fibers, yarn non-uniformity, yarn misalignment in fabrics, uneven or incorrect thickness and areal weight, etc. The quality consistency of sandwich cores is often assessed in terms of density only.

While the bottom line is that in many industries the user tends to rely on the raw material suppliers' quality assurance efforts, industries that deal with high-performance components and demanding customers cannot neglect carrying out their own thorough quality control procedures. Such industries essentially include those that deal with aerospace products and military customers, who require adherence to very strict material standards and specifications. In such applications thermoset-based prepregs, particularly carbon-reinforced epoxy, dominate as raw material and a range of specialized test standards consequently have been developed. Prepreg properties of importance include volatile content, resin content, resin flow characteristics, degree of tack, tack time (the time at room temperature before tack is lost due to the progressing crosslinking), gel time, chemical characterization of resin composition, and possibly others. Prepregs are also characterized in terms of properties of basic unidirectionally reinforced composites manufactured from the prepregs; properties include fiber content, void content, thickness, density, transverse and longitudinal tensile modulus and strength, longitudinal compressive modulus and strength, longitudinal flexural modulus and strength, interlaminar shear strength, T_g, and possibly others (see *Section 6.2* for characterization of composites). Just as there are numerous tests dedicated to characterization of properties of thermoset-based prepregs, there are also numerous tests to quantify properties of sandwich cores and adhesives. In aerospace and military applications the raw material supplier and the component manufacturer often both have to perform prescribed tests to verify adherence to extensive property criteria.

Characterization of constituents to determine their inherent properties (such as those quoted in *Sections 3.1, 3.2,* and *3.4*) is important in order to build up data bases in terms of thermal, mechanical, rheological, and several other properties for use in component design and modeling as well as in optimization of the manufacturing process. However, in most cases such characterization is carried out by the raw material supplier and representative properties are generally made available in data sheets. Only in the most specialized circumstances does a component manufacturer carry out such tests; normally the manufacturer focuses on characterization of the resulting composite properties rather than those of the constituents.

6.2 Composite Characterization

The ultimately most interesting properties are after all those of the finished composite. The relevant properties may be physical, mechanical, thermal, optical, electrical, and environmental in nature. Assuming that fibers and matrix have been appropriately selected to ensure that the composite tolerates the general environment it is intended for, the most relevant properties for structural composites are the mechanical ones, which therefore are the focus of this section.

While one ideally might want to perform full-scale tests to determine whether a component fulfills the design criteria or not, this is in reality rarely desirable except as a final verification of the entire process of design, manufacture, and assembly. The main reason for not wanting to perform full-scale tests is naturally the high cost. Instead, it is the norm to first manufacture simple test coupons to characterize a material combination and possibly a layup sequence; the properties thus determined are then used as input to the design process. For relatively simple structures the next step may be to manufacture and test a full-scale prototype. Another approach is to manufacture and test gradually larger scale models before the full-scale product is addressed. When the structure is very complex, say an aircraft, it is common to manufacture and assemble gradually larger and more complex substructures and perform verification tests on each of these before the complete full-scale structure is finally produced and tested.

Most component testing, whether full-scale or not, is unique to an application or possibly field of application and there is therefore little point in delving into such procedures herein. In contrast, methods used to characterize test coupons tend to be the same regardless of final application. It is valuable to become acquainted with the most common and basic test methods, since they are often encountered as direct extensions of manufacturing operations in order to characterize the raw material or manufacturing technique or as part of quality control.

It is important to realize that experimentally determined properties of composites vary widely, mainly due to the complexity and inherent variability of a composite. A single property value, say compressive strength, consequently represents an average of a number of tests and some of the tested specimens obviously failed before this stress was reached. Considering that there are numerous variables involved in raw materials, composite manufacturing, specimen preparation, and testing, it is important that all this very relevant information is recorded together with experimentally determined properties. In general, a proper test report should include the following information about specimens and test conditions (not all items apply to all tests):

- Identification of material tested, including constituent types and fractions, supplier, etc.
- Description of how the composite was manufactured
- Description of how the specimen was prepared
- Description of how the specimen was conditioned prior to the test, e.g. soaked in water or kept at a specific temperature for a week
- Specimen and end tab dimensions
- Test standard designation (and possible deviations from that standard)
- Equipment and fixture types
- Temperature and humidity during test
- Testing rate
- Failure mode
- Number of test specimens (normally a minimum of five)
- Average property and standard deviation
- Date of test

Much effort has been spent by different standardization organizations harmonizing test methods to enable results from different testing facilities to be compared. The standardization organizations of greatest international relevance are ISO (International Organization for Standardization), CEN (Comité Européen de Normalisation, European Committee for Standardization), ASTM (American Society for Testing and Materials), and DIN (Deutsches Institut für Normung, German Institute for Standardization), although most countries have national standardization organizations. In addition to national and international standardization organizations there are many branch organizations and government agencies that issue standards for the testing of composites, but fortunately many test methods specified by different organizations are identical or at least similar. Throughout this section, reference is made to commonly used standards for each type of test (where applicable). The tests discussed in the following by no means represent all tests relevant to composites; only tests to determine the most basic material properties for use in design and some tests commonly used in quality control are discussed. References addressing other test methods are listed at the end of the chapter.

6.2.1 Physical Properties

Apart from obvious physical properties such as dimensions and density, the most commonly determined physical composite properties are constituent contents. There are essentially three different methods to determine constituent contents and the method chosen is usually determined by constituent types.

6.2.1.1 Constituent Fractions through Matrix Burn-off

The first method involves removal of the matrix through burning. A small composite sample is placed in a porcelain crucible of known weight and the crucible containing the composite is weighed to determine the composite weight. The crucible is then placed in a high-temperature oven to ignite the matrix, which is completely reduced to volatiles to leave the bare reinforcement in the crucible. Alternatively, a Bunsen burner (in a fume hood) may be used to ignite the composite and maintain combustion. The burning procedure takes about an hour, but the time required depends greatly on temperature and matrix type. After the crucible has cooled down and its contents have been weighed, the fiber weight fraction may be determined as:

$$W_f = \frac{w_f}{w_c} \qquad (6\text{-}1)$$

where W is weight fraction, w weight, and indices f and c denote fiber and composite, respectively. The fiber volume fraction can then be determined using *Equation 3-2*, assuming that the densities of fiber and matrix are known, or through

$$V_f = \frac{w_f}{\rho_f} \bigg/ \frac{w_c}{\rho_c} \qquad (6\text{-}2)$$

if the densities, ρ, of fiber and composite are known (see also *Section 6.2.1.4*). The matrix weight fraction

$$W_m = 1 - W_f \qquad (6\text{-}3)$$

whereas the matrix volume fraction

$$V_m = 1 - V_f - V_v \qquad (6\text{-}4)$$

where V_v is void volume fraction. Although generally incorrect, it is common to assume that no voids are present in the composite, i.e. the last term in *Equation 6-4* is neglected. Nevertheless, in most cases the matrix content is less important that the fiber content (cf. *Section 3.3.2*).

The described burn-off procedure may only be used as a method to determine fiber content of composites containing reinforcement whose weight is unaffected by oxidation and high temperature and matrices that completely decompose into volatiles when they burn. In practice this reduces the allowable reinforcement types to glass only, whereas most common organic matrices indeed do completely decompose into volatiles. If the composite contains fillers their decomposition behavior must be known. For example, with a filler that just like glass fibers does not lose any weight, such as glass microspheres, only the matrix weight fraction and the fraction of the remainder of the constituents (fibers and filler) may be determined, not the fiber content. With fillers that do decompose completely the fiber weight fraction can be determined, but not the matrix fraction. In an in-between situation no accurate results may be obtained.

Relevant Standards

ISO 1172, "Textile Glass Reinforced Plastics—Determination of Loss on Ignition".
ASTM D 2584, "Standard Test Method for Ignition Loss of Cured Reinforced Resins".

Related Standard

EN 60*, "Glass Reinforced Plastics; Determination of the Loss on Ignition"

6.2.1.2 Constituent Fractions through Matrix Digestion

The second method to determine constituent fractions involves chemical digestion, or dissolution, of the matrix. While the digestion medium varies with the type of matrix the general digestion procedure is the same. A small composite sample and a funnel with a filter are first weighed. The sample is then placed in a glass beaker in a fume hood and the liquid digestion medium is carefully poured on top. The digestion medium is normally heated and stirred occasionally using a glass rod. Once the matrix has been completely dissolved, which may take several hours, the contents of the beaker are emptied into the funnel and the fibers trapped on the filter are rinsed repeatedly with water and possibly a solvent. Filter and fibers are then dried

---

* EN (European Standard; issued by CEN) 60 is identical to ISO 1172.

in an oven at 100°C for 1–2 hours to evaporate remaining liquids. Following drying filter and fibers are weighed and the fiber weight or volume fraction is determined in the same manner as described in *Section 6.2.1.1*.

A parallel experiment should be performed to determine whether the reinforcement does retain its weight during matrix digestion. This may be ascertained by exposing bare fibers to the same digestion environment for the same amount of time as the reinforcement in the composite sample; if there is a weight loss it may be accounted for in the fiber weight fraction calculation. It is not uncommon that the matrix digestion is painstakingly slow and it is therefore sensible to use small samples (large surface area to volume) or perhaps even to grind the composite prior to weighing.

Digestion of most matrices requires strong digestion media, including for example concentrated nitric acid, concentrated sulfuric acid, 50% hydrogen peroxide, meaning that it is of utmost importance that material safety data sheets and safety procedures for the pertinent chemical is carefully reviewed prior to carrying out any experiment. Some successful digestion media are suggested in the standard quoted below.

The matrix digestion method is normally used with carbon and sometimes aramid reinforcement which would not retain their weight during matrix burn-off. Although this method certainly also may be used with glass reinforcement there is little point in doing so since matrix digestion is more time-consuming than matrix burn-off and also involves the use of potentially harmful chemicals.

Relevant Standard

ASTM D 3171, "Standard Test Method for Fiber Content of Resin-Matrix Composites by Matrix Digestion".

6.2.1.3 Constituent Fractions through Microscopy

Microscopy, the third method to determine constituent fractions, is blissfully free of high temperatures and contact with harmful chemicals. To be able to study a composite cross-section under a microscope it must have a finely polished surface. To this end a small sample is normally cast in a thermoset polymer for ease of handling and to reduce the risk of fibers fraying at the edges of the surface during polishing; casting may be omitted for larger samples. The polishing procedure is very important to ensure the good and scratch-free surface required to obtain accurate measurements. Three or four successively finer wet sandpapers should be used followed by two to three polishing cloths and successively finer polishing compounds, which may be in the form of an aqueous solution of aluminum oxide particles or a paste containing diamond particles. Although it is possible to carry out grinding and polishing by hand, it is common to avoid this very time-consuming process by using a rotating polishing table with a device capable of holding several samples for simultaneous preparation and maintaining of a set pressure. Such a sample holder also rotates the samples in relation to sand-paper or cloth to reduce the risk of scratches in any predominant direction. Even with such equipment, preparation of a set of specimens may take an hour or more, not including the time required to cast the specimens in resin. Successful preparation of composite samples for microscopy is an art that takes a long time to learn and polishing methods are very dependent on composite constituents. It is for example difficult to successfully polish composites with ductile matrices, chiefly thermoplastics, and composites reinforced with glass, since glass-fiber ends tend to chip off thus leaving cavities in the polished surface.

Once polishing has been completed the composite cross-section is viewed under an optical microscope. It is now possible to take a photograph, called micrograph (see *Figure 6-1*), of this image and then analyze it through any of a number of arduous methods. However, it is much more convenient to mount a video camera on top of the microscope to monitor the microscopic image in real time and to feed this image into a conventional personal computer. Several different software packages capable of computing area fractions of, for example, fiber ends visible in the image are available. For micrographs such as that in *Figure 6-1*, the area fraction of fibers is often assumed identical to the volume fraction, which is a good assumption only if the fibers are continuous, parallel, and perpendicular to the polished surface. Whereas the human eye can easily distinguish between different constituents of a composite, it is under some circumstances difficult for the software to do so if the contrast between constituents is low. Image-enhancement techniques included in image-analysis software as well as changes in microscope settings can sometimes improve the situation, but it may be difficult or impossible for the software to reliably determine area fractions in for example glass-reinforced composites where the matrix is unpigmented and therefore has a color very similar to that of the glass, see *Figure 6-1*. In contrast, it is relatively straightforward to determine area fractions in carbon-reinforced composites as well as in glass-reinforced composites where the matrix is pigmented.

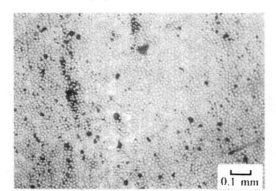

Figure 6-1 Micrograph of cross-section of filament wound glass-reinforced PPS composite perpendicular to parallel fibers. Light areas are matrix pockets and dark areas voids. The average fiber diameter is 16 μm. Photograph reprinted from reference [2] with kind permission from Technomic Publishing Co, Lancaster, PA, USA

A drawback of this method is that very small areas are characterized (cf. the ruler in *Figure 6-1*—the area of this micrograph is approximately 2 mm²) when compared to the relatively large volume of a sample where the matrix is burned off or digested. It is therefore necessary to determine the fiber area fractions of a number of images to arrive at an average that can be said to statistically represent the entire macroscopic cross-section. Important advantages of this method are that it is possible to directly determine void content (assuming the contrast is sufficient) and that one also gets information about fiber and void distribution, matrix-rich areas, size and shape of voids, etc. (cf. *Figure 6-1*).

6.2.1.4 Void Fraction and Density

While the microscopy method may directly yield void fractions, the matrix burn-off and matrix digestion methods may indirectly be used to determine the void fraction if the density of the composite and its constituents are known. The void volume fraction may be calculated as:

$$V_v = 1 - \rho_c \left(\frac{W_m}{\rho_m} + \frac{W_f}{\rho_f} \right) \tag{6-5}$$

where ρ_c is the experimentally determined composite density. ρ_c can be measured through several different methods, including weight and volume measurements of a regular-shaped specimen (if accurate enough), or through immersion of the composite in water to monitor its weight loss upon immersion.

While the density of the reinforcement quoted by the supplier often is accurate enough to determine void fraction, quoted matrix densities may be more questionable (cf. *Table 3-5*). The matrix density is often assumed to be the same as that of cast unreinforced bulk resin, which is not entirely true. With thermosets the bulk resin density is usually lower than that of the matrix, which also would be the case for the semicrystalline thermoplastics that crystallize more readily in the presence of fibers. With semicrystalline thermoplastics the density increases with degree of crystallinity; the PEEK density range quoted in *Table 3-5* illustrates this variance. Due to such uncertainties, as well as several other potential measurement inaccuracies, it is possible to calculate negative void contents or standard deviations that are greater than the average value. It may therefore be necessary to very carefully determine experimentally the constituent and composite densities using methods such as those listed below.

Relevant Test Standards

ISO 7822, "Textile Glass Reinforced Plastics—Determination of Void Content—Loss on Ignition, Mechanical Disintegration and Statistical Counting Methods".
ASTM D 2734, "Standard Test Method for Void Content of Reinforced

Plastics".

ISO 1183, "Plastics—Methods for Determining the Density and Relative Density of Non-Cellular Plastics".

ASTM D 792, "Standard Test Method for Specific Gravity (Relative Density) of Plastics by Displacement".

ASTM D 3800, "Standard Test Method for Density of High-Modulus Fibers".

6.2.2 Mechanical Properties

An immense number of test methods have been developed to characterize every conceivable mechanical property of composites. Many of these tests are dedicated to a specific application or a specific industry, but even when it comes to what may be thought of as common and basic properties, such as those treated in *Section 3.3.5*, there are several alternative test standards and the original intention of test standards—that all property data, regardless of who determined them, should be directly comparable—unfortunately often does not hold true. It is regrettably also common that individual laboratories modify test standards for convenience or some arbitrary reason.

The following sections are limited to tests to quantify tensile, compressive, shear, and flexural properties; several sources for test standards to determine other properties are listed at the end of the chapter. While the tests discussed in the following generally have been developed for unidirectionally reinforced composites, most of them may also be used to characterize fabric-reinforced and cross-plied composites.

All tests to determine mechanical properties that are discussed in the following (and most other tests as well) require a universal testing machine capable of inducing deformation in a specimen under controlled load or rate of displacement; see *Figures 6-2* and *6-3*.

Almost all test methods specify specimen dimensions or relations between dimensions. While the most direct way of obtaining these dimensions is to mold the component to the exact size required, this is seldom possible with continuous-fiber reinforced composites. (In contrast, with unreinforced and short-fiber reinforced polymers, which can be injection molded to size, it is quite common.) In most cases test specimens are therefore machined from larger laminates. It is of paramount importance that specimen machining is very carefully performed to avoid introducing flaws (cf. *Section 5.1.1*), which may cause the specimen to fail prematurely leading to underestimation of the inherent material properties. Satisfactory results may usually be achieved with a water-cooled PCD saw. Some standards even suggest cutting of oversized specimens and use of water-cooled precision milling or grinding to reduce them to nominal dimensions to further limit the risk of inducing flaws during specimen preparation.

Several test standards specify that end tabs be bonded to the specimen, see *Figures 6-2* and *6-3*, to avoid failure initiated at the load-introduction

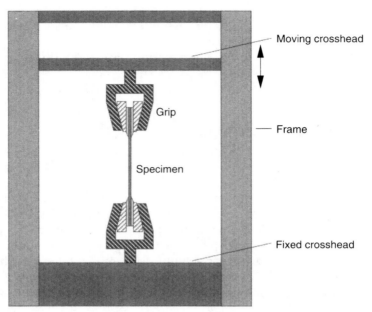

Figure 6-2 Schematic of universal testing machine set up for tensile testing. The standard wedge action friction grips shown are easily replaced to accommodate fixtures for other tests. One of the crossheads that the grips are mounted to is stationary while the other crosshead moves vertically (in this case the upper)

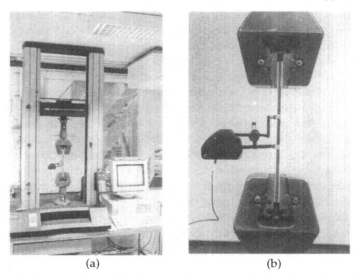

Figure 6-3 Desk-top universal testing machine set up for tensile testing (a) overview, including data-acquisition and control computer; (b) close-up of grips, tensile specimen, and extensometer

points. End tabs are often balanced 0°/90° glass-reinforced epoxy laminates and the procedure of bonding them to the specimen is tedious. Any tough adhesive that will not fail during testing may be used and surface preparation and bonding procedures akin to those described in *Section 5.2.2* should be used. Crosslinking of the adhesive normally takes place in a dedicated fixture to ensure parallelism of the end tabs. If several specimens are to be cut from the same laminate, end tabs in the form of strips may be bonded onto the laminate prior to machining of the specimens to reduce the number of machining and bonding operations. Following crosslinking, specimens are machined to size with end tabs already in place.

A final step in specimen preparation that most tests have in common is bonding of one or more strain gages to the specimen gage section. An alternative to a strain gage is an extensometer, which is attached to the specimen to measure elongation (compression) of the gage section (see *Figure 6-3*); elongation is easily converted to strain. The stress the specimen is subjected to is determined from the known applied load sensed by a load cell divided by the (minimum) cross-sectional area of the gage section, which prior to testing is measured using precision calipers.

6.2.2.1 Tensile Properties

The most basic of all tests surely is to pull on a specimen and then monitor the response. Such a tensile test may yield:

- Tensile (Young's) moduli, E_{ll} and E_{tt}
- Poisson's ratios, v_{lt} and v_{tt}
- Tensile strengths, σ^*_{lt} and σ^*_{tt}
- Ultimate tensile strains, ε^*_{lt} and ε^*_{tt}

where the first subscript denotes reinforcement orientation (in relation to testing direction; *l* is longitudinal and *t* transverse) and the second subscript (*t*) denotes tensile testing. To ensure that a tensile specimen fails where intended, i.e. in the middle, it must normally be made weaker in this region to prevent failure initiated at the grips. One possible way to achieve such a weakness is to locally decrease the specimen cross-section through a (machined or molded) waist in the width direction. This dog-bone shaped specimen is commonly used with plastics and short-fiber reinforced plastics (e.g. ISO 3268, Type I; ASTM D 638). With continuous-fiber reinforced composites it is instead standard procedure to use end tabs, see *Figure 6-4* (e.g. ISO 3268, Type III; ASTM D 3039). While different standards specify slightly different specimen and end tab dimensions, geometries basically remain the same.

Prior to testing, actual thickness and width of the specimen gage section are measured and recorded for at least three locations. A strain gage rosette (two gages on the same carrier film to record strain in two perpendicular directions) is then bonded to the center of the gage section (cf. *Figure 6-4*).

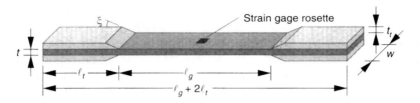

Figure 6-4 Common geometry of tensile test specimen. ℓ_g is the length of the gage section where failure should occur. t is thickness and subscript t refers to end tabs

The specimen is mounted in the grips of the testing machine as shown in *Figures 6-2* and *6-3*, and the crosshead speed (the vertical velocity of the moving crosshead of the testing machine) is set to a constant value in the range 0.5–1 mm/min; the test should proceed until failure. When tested in the fiber direction the failure mode is fiber fracture, fiber pullout, and matrix cracking, while testing transverse to the fibers leads to matrix or fiber–matrix interface failure, which may be brittle or ductile depending on properties of matrix and interface.

During testing a load cell senses the applied load, while the strain gages bonded to the specimen sense the strains. The results are often directly plotted on paper or computer screen, for example as load vs. displacement, while load vs. strain is continuously recorded by a computer. Whatever the original format of the recorded test results, they may be plotted as shown in *Figure 6-5*, since the stress may be calculated from the load divided by the cross-sectional area of the gage section. *Figure 6-5* schematically illustrates results from tensile testing of a unidirectionally reinforced composite parallel and transverse to the reinforcement. If Poisson's ratio is not of interest the strain gages may be omitted and an extensometer used to indirectly record the strain as shown in *Figure 6-3*.

With results such as those illustrated in *Figure 6-5* it is possible to calculate the tensile moduli of elasticity as the initial slopes of the stress–strain relationships; the longitudinal modulus, E_{lt}, is the slope of the longitudinal stress–strain relationship, ε_l, in the figure to the left, while the transverse modulus, E_{tt}, is the slope of the transverse stress-strain relationship, ε_t, in the figure to the right. The major Poisson's ratio is calculated as

$$v_{lt} = \left| \frac{\varepsilon_t}{\varepsilon_l} \right| \qquad (6\text{-}6)$$

determined from the longitudinal test, while the minor Poisson's ratio is similarly calculated as

$$v_{tl} = \left| \frac{\varepsilon_l}{\varepsilon_t} \right| \qquad (6\text{-}7)$$

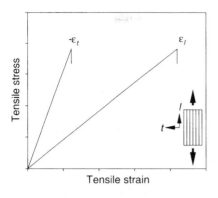 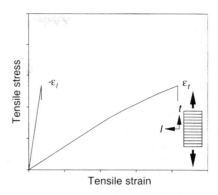

Figure 6-5 Schematic stress-strain relationships for tensile testing of unidirectionally reinforced composites parallel (left) and transverse (right) to the reinforcement

determined from the transverse test. The tensile strengths, σ^*_{lt} and σ^*_{tt}, are calculated from the ultimate load divided by the minimum cross-sectional area, while the ultimate strains, ε^*_{lt} and ε^*_{tt}, are directly obtained from the strain gages.

Relevant Standards

ISO 3268, "Plastics—Glass Reinforced Materials—Determination of Tensile Properties".
ASTM D 3039, "Standard Test Method for Tensile Properties of Fiber-Resin Composites".

Related Standards

ASTM D 638, "Test Method for Tensile Properties of Plastics".
ISO 527, "Plastics—Determination of Tensile Properties".
EN 61*, "Textile Glass-Reinforced Plastics; Determination of Tensile Properties"
EN 2561, "Unidirectional Laminates Carbon-Thermosetting Resin; Tensile Test Parallel to the Fibre Direction".

6.2.2.2 Compressive Properties

Although it is often assumed that elastic properties for metals are the same in tension and compression, this is generally not a good assumption for composites and when it comes to strengths the difference between tension and compression is often significant (cf. *Section 3.3.5*). A compressive test may yield:

---

* EN61 is identical to ISO 3268.

- Compressive (Young's) moduli, E_{lc} and E_{tc}
- Poisson's ratios, v_{lt} and $v_{tt}^{\dagger}$
- Compressive strengths, σ_{lc}^{*} and σ_{tc}^{*}
- Ultimate compressive strains, ε_{lc}^{*} and ε_{tc}^{*}

In compression parallel to the fibers of a unidirectionally reinforced composite, the dominant failure mode is local fiber buckling, see *Figure 6-6*. However, stress concentrations and global instability in the specimen can easily precede the desired fiber instability illustrated in the figure, which may result in determination of properties that are not the inherent properties of the composite. It is the elimination of such undesirable failure modes that make accurate compression testing so difficult. Seemingly insignificant specimen non-uniformities or misalignment of specimen or test fixture result in load eccentricity, which may lead to geometric instability (i.e. buckling) if the gage length is long. If the gage length is too short geometric instability is inhibited, but instead the result is influenced by specimen clamping. Both excessively short and too long gage lengths lead to lower ultimate stresses (i.e. apparent strengths) than those inherent to the composite. There is consequently an optimum gage length where both global instability and influence of clamping are minimized [3] and it is therefore important to closely follow the specimen and test fixture dimensions provided by standards. Compression testing transverse to the fibers of a unidirectionally reinforced composite commonly leads to failure through shearing of the matrix.

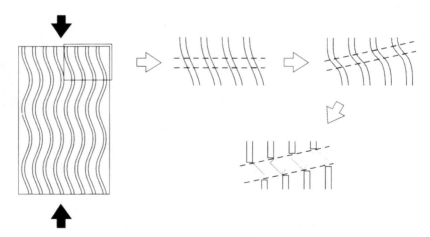

Figure 6-6 Mechanism of kink zone formation of unidirectionally reinforced composite compressively loaded in the fiber direction. Redrawn from reference [3]

---

† The Poisson's ratios are possibly also different in tension and compression, but such a distinction is rarely made.

Compression testing is one of the most difficult tests one can perform on composites and there is considerable disagreement on what method is the most appropriate. The least complicated means of compression testing is to place a short and wide specimen between two parallel surfaces and compress it (e.g. ISO 604, ASTM D 695). With this method it is of utmost importance that both loading and specimen surfaces are indeed perfectly parallel lest stress concentrations be introduced at the ends of the specimen. To this end several different compression fixtures have been developed; *Figure 6-7* illustrates one solution. A compression fixture such as that shown in the figure is intended for thick and wide specimens. To test a thin specimen requires a support jig to prevent global instability, see *Figure 6-8* (e.g. ASTM D 695); this specimen geometry and support jig are intended for specimens thinner than 3 mm.

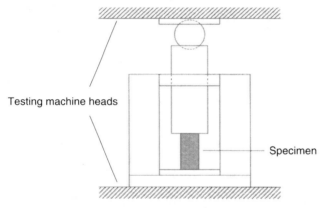

Figure 6-7 Compression fixture for prismatic specimens. Redrawn from ASTM D 695

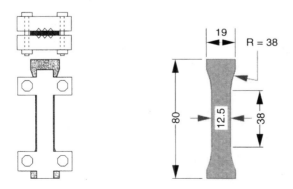

Figure 6-8 Compression specimen geometry and support jig intended for specimens thinner than 3 mm. All dimensions in mm. Redrawn from ASTM D 695

With either type of specimen and whether a support jig is used or not, the specimen dimensions are measured at a minimum of three locations. The specimen is then inserted in jig or fixture and carefully aligned before it is loaded at a crosshead speed of 0.5–1.0 mm/min until failure. In compression testing, deformation is normally measured with a compressometer (an extensometer for compressive testing) attached to the specimen or possibly to the testing machine heads, although strain gages may also be used. The stress–strain curves are then plotted and moduli, Poisson's ratios, strengths, and ultimate strains are determined in the same fashion as for a tensile test. Determination of Poisson's ratios requires strain gages bonded to the specimen (cf. *Section 6.2.2.1*).

The tests illustrated in both *Figures 6-7* and *6-8* are intended for unreinforced or short-fiber reinforced polymers; for testing of continuous-fiber reinforced composites more sophisticated methods are used. One of them is based on the support jig in *Figure 6-8*, which is bolted to an alignment fixture (e.g. SRM 1); the method is consequently often referred to as Modified ASTM D 695. The specimen is in this case not dog-bone shaped, but is instead fitted with end tabs so as to conceptually look like the tensile specimen in *Figure 6-4* (although dimensions vary greatly). Another compression test common with high-performance composites utilizes the test fixture and the specimen illustrated in *Figure 6-9* (ASTM D 3410, Procedure B). This fix-

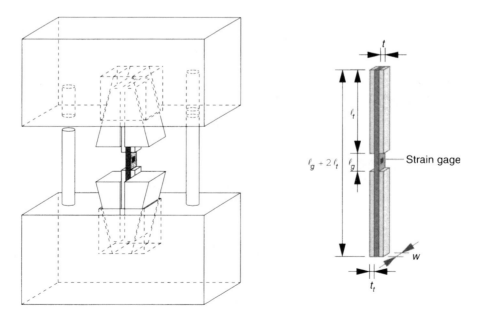

Figure 6-9 IITRI compression fixture and test specimen. Redrawn from ASTM D 3410

ture is often referred to as the IITRI compression fixture after its inventors at the Illinois Institute of Technology Research Institute. Test specimens for both these tests are prepared in much the same way as for the aforementioned tensile tests, but to minimize the risk of geometric instabilities much greater care must be taken in specimen preparation to ensure uniform spec- · imen width and parallel end tabs, which requires a precise end-tab bonding fixture.

IITRI specimens should be equipped with one small (≤ 3 mm) strain gage on each side of the specimen located in the center of the gage section. (If Poisson's ratios are to be determined a strain gage rosette is required on at least one side.) The reason for bonding strain gages to both sides of the specimen is to enable determination of whether the specimen globally buckles prior to failure. *Figure 6-10* illustrates how the strain response may be used to determine whether a test should be discarded due to buckling or not.

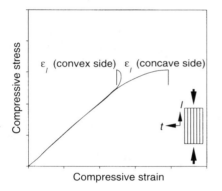 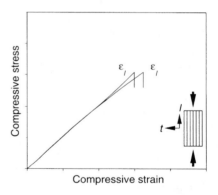

Figure 6-10 Schematic stress-strain relationships for a sample that buckled and thus should be discarded (left) and for a sample that did not buckle (right)

Prior to testing, the specimen dimensions are measured at a minimum of three locations. The specimen is then loaded at a crosshead speed of 0.5–1.0 mm/min until failure. After (or during) each completed test the stress–strain relationships are plotted to determine whether the results should be accepted based on the conformity of the readings from the two strain gages. For a successfully tested specimen (with a response such as that shown to the right in *Figure 6-10*) the strain is normally given as the average from the two gages, see *Figure 6-11*. The compressive properties are then determined according to the previously described method.

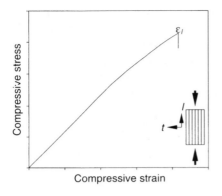 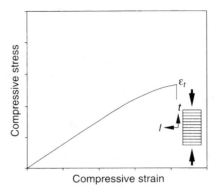

Figure 6-11 Schematic stress-strain relationships for compressive testing of unidirectionally reinforced composites parallel (left) and transverse (right) to the reinforcement

Relevant Standards

ISO 8515, "Textile-Glass-Reinforced Plastics—Determination of Compressive Properties in the Direction Parallel to the Plane of Lamination".
ASTM D 3410, "Standard Test Method for Compressive Properties of Unidirectional or Crossply Fiber-Resin Composites".

Related Standards

ISO 604, "Plastics—Determination of Compressive Properties".
ASTM D 695, "Test Method for Compressive Properties of Rigid Plastics".
SRM 1*, "Compressive Properties of Oriented Fiber-Resin Composites," Suppliers of Advanced Composite Materials Association (SACMA).

6.2.2.3 Shear Properties

Full characterization of composites for design purposes requires determination of shear properties in addition to tensile and compressive properties. Like compression testing, shear testing is difficult and well over a dozen methods, which all tend to provide different estimates of the inherent material properties, have been developed. The main reason for these difficulties is that it has so far proved impossible to develop a test method that induces a uniform shear state in the specimen gage section and tensile and compressive stresses are invariably superimposed on shear stresses [4].

It is commonly assumed that unidirectionally reinforced composites are transversely isotropic. If this truly were the case, in-plane and out-of-plane (or interlaminar) properties should be identical. However, due to imperfections in consolidation and uneven fiber distribution there is often a

---

* SACMA's test methods are recommended methods, not standards.

significant difference between in-plane (*lt*) and interlaminar (*l⊥*) properties (for definitions cf. *Figure 3-4*). Although most shear tests are designed to determine either in-plane or interlaminar properties and not both, shear tests ideally may yield the full set of shear properties:

- Shear moduli, G_{lt}, $G_{l\perp}$, and $G_{t\perp}$
- Shear strengths, τ_{lt}^*, $\tau_{l\perp}^*$, and $\tau_{t\perp}^*$
- Ultimate shear strains, γ_{lt}^*, $\gamma_{l\perp}^*$, and $\gamma_{t\perp}^*$

The easiest way of determining in-plane shear properties of unidirectional composites is to perform a tensile test as described above but on a laminated specimen with ±45° fiber orientation, e.g. $[\pm45]_{2s}$. (For ASTM D 3518 specimen geometry and test procedure are identical to those prescribed for ASTM D 3039, see *Section 6.2.2.1*.) A strain gage rosette in the middle of the gage section aligned with the principal axes of the specimen is also required. With this test, the shear stress is defined as half the longitudinal stress (referring to the direction of the specimen), i.e.

$$\tau_{lt} = \frac{\sigma_x}{2} \tag{6-8}$$

(cf. *Figure 6-12* for coordinate definitions). The shear strain is defined as the sum of the absolute values of the longitudinal and transverse strains (once again referring to specimen directions):

$$\gamma_{lt} = |\varepsilon_x| + |\varepsilon_y| \tag{6-9}$$

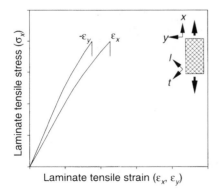 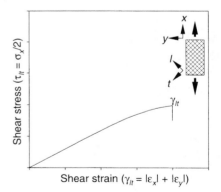

Figure 6-12 Schematic shear test results for tensile testing of ±45° laminate. The left figure shows the laminate stress-strain relationships and the right figure the shear–stress/shear–strain relationship as calculated using *Equations 6-8* and *6-9*

Shear stress is then plotted versus shear strain, see *Figure 6-12*. The modulus is determined as the initial slope of the stress–strain curve, while the shear strength is calculated as

$$\tau_{lt}^* = \frac{\sigma_x^*}{2} \tag{6-10}$$

and the ultimate shear strain as

$$\gamma_{lt}^* = |\varepsilon_x^*| + |\varepsilon_y^*| \tag{6-11}$$

Two related test methods which have been much used in the past but are gradually falling into disrepute use two-rail or three-rail test fixtures, respectively, see *Figure 6-13* (ASTM D 4255). In both these test methods holes are machined in a thin specimen, which is bolted to the test rails as illustrated in the figure. One reason for these test methods becoming less

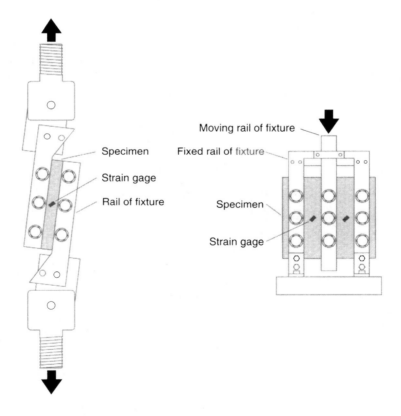

Figure 6-13 Two- and three-rail shear fixtures and test specimens. Redrawn from ASTM D 4255

common are that stress concentrations are introduced at the holes, leading to questionable results [4]. Other reasons are that large specimens are required (76 × 152 mm for the two-rail test and 137 × 152 mm for the three-rail test) and that much time is spent preparing specimens and in assembly and disassembly of the test fixture.

A test method that appears to provide a near-perfect shear stress state in the gage section and that allows determination of shear properties in all three planes is the Iosipescu test method, named after its Romanian inventor (ASTM D 5379). The test specimen is shown in *Figure 6-14* and the test fixture in *Figure 6-15*. Any thickness specimen may be used, but a thickness of 3–4 mm is preferred. In most cases end tabs are not required, but to minimize the risk of crushing failure and twisting of specimens thinner than 2.5 mm, end tabs with a typical thickness of 1.5 mm should be used. In determining properties in directions where the composite has to be 20 and 76 mm thick (cf. *Figure 6-14*), laminates may be bonded from $[0]_n$ or $[90]_n$ laminates to achieve sufficient thickness. Once specimens have been machined to size, a small strain gage rosette, normally with 1.5-mm long gages, is bonded to the middle of the gage section at a 45° angle to the specimen's principal axes, see *Figure 6-15*.

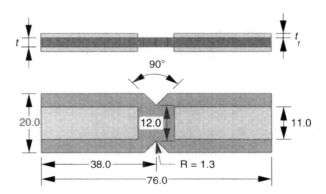

Figure 6-14 Iosipescu test specimen geometry. All dimensions in mm. Redrawn from ASTM D 5379

Following measurements of the gage cross-sectional area (nominally 12.0 × *t* mm), the specimen is inserted in the fixture and is loaded at a crosshead speed of 2 mm/min. The shear stress, determined as the applied load divided by the gage cross-sectional area, is plotted as a function of the sum of the absolute values of the strains measured by the two gages; typical stress–strain relationships are shown in *Figure 6-16*. Modulus, strength, and ultimate strain are calculated with the same methodology as for the previously discussed test methods.

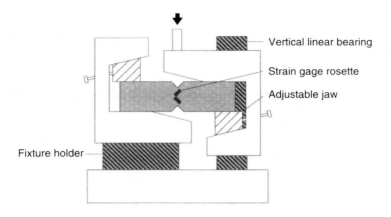

Figure 6-15 Iosipescu shear fixture. Redrawn from ASTM D 5379

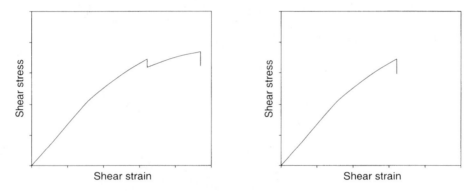

Figure 6-16 Schematic shear test results for Iosipescu shear test on $[0]_n$ and $[90]_n$ laminates. The left curve shows a $[0]_n$ laminate tested in the *lt* plane and the right curve a $[90]_n$ laminate tested in the *tl* plane. In the left curve the first drop in stress corresponds to crack initiation at the notch root. Redrawn from ASTM D 5379

The final test method to determine shear properties that is described herein is the popular short-beam shear test (e.g. ISO 4585, ASTM D 2344, EN 2377) illustrated in *Figure 6-17*. If the support-span/thickness ratio (l_{ss}/t, see figure) is small enough, say 4–5, a laminated specimen normally fails in shear between plies to allow determination of the apparent interlaminar shear strength, often abbreviated ILSS, as:

$$\tau_{/\perp}^{*} = \frac{3P^{*}}{4wt} \qquad (6\text{-}12)$$

where P^{*} is the failure load. Even with a small support-span/thickness ratio it is quite possible to get other failure modes than shear, e.g. tensile or compressive failure of the outermost plies right below the loading cylinder. It is therefore important to monitor and record the actual failure modes and disregard results from specimens that do not exhibit the intended shear failure. Particularly with ductile thermoplastic matrices exhibiting very large ultimate strains (cf. *Table 3-5*) it may be impossible to get a shear failure, thus rendering the short-beam shear test inappropriate.

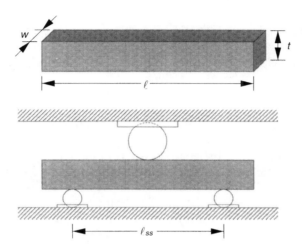

Figure 6-17 Short-beam shear fixture and specimen

There are both notable advantages and disadvantages with the short-beam shear test. Advantages include that the test requires very little material (as an example, ASTM D 2344 prescribes that a 2-mm thick carbon-reinforced specimen should be 12 mm long), that specimen preparation is minimal, and that the test is very simple to perform. It is therefore relatively painless to perform a large number of tests. The two major disadvantages are that the specimen is not loaded in uniform shear and that there is no way to reliably measure strain and hence determine the shear modulus [4]. Due to the non-uniform stress state the test does not provide shear properties meaningful for design purposes (thus *apparent* interlaminar shear strength). However, as long as specimen geometry and material nominally remain the same, the test is a powerful tool for quality control of both raw material and manufacturing technique.

Relevant Standards

ASTM D 3518, "Standard Practice for Inplane Shear Stress-Strain Response of Unidirectional Reinforced Plastics".

ASTM D 5379, "Standard Test Method for Shear Properties of Composite Materials by the V-Notched Beam Method".

ISO 4585, "Textile Glass Reinforced Plastics—Determination of Apparent Interlaminar Shear Properties by Short-Beam Test".

ASTM D 2344, "Standard Test Method for Apparent Interlaminar Shear Strength of Parallel Fiber Composites by Short-Beam Method".

Related Standards

ASTM D 4255, "Standard Guide for Testing Inplane Shear Properties of Composites Laminates".

EN 2377, "Glass Fibre Reinforced Plastics; Test Method; Determination of Apparent Interlaminar Shear Strength".

6.2.2.4 Flexural Properties

The final mechanical test method discussed herein is used to characterize the apparent flexural properties of a composite. A flexural test schematically looks just like the short-beam shear test, cf. *Figure 6-17*. However, the support-span/thickness ratio is significantly larger to ensure flexural failure, which should occur as tensile or compressive failure of the outermost plies right below the loading cylinder. Standards suggest a span-to-thickness ratio of 16, 32, 40, or 60 to minimize interlaminar shear deformation; it has been shown that with increasing support-span/thickness ratio the apparent flexural modulus approaches the tensile modulus of the specimen [5], thus providing an independent check of the tensile modulus. While the test fixture shown in *Figure 6-17* (three-point method) is the most common, it is also possible to use two symmetrically located loading cylinders to half the magnitude of the applied line load and thus reduce the risk of transverse compressive failure (four-point method; e.g. ASTM D 790, Method II).

When the specimen has been machined to size, a strain gage is bonded to the center of the specimen parallel to its principal axis. The specimen is then carefully aligned in the test fixture (with strain gage right below the loading cylinder facing downwards) and loaded until failure. The maximum stress in the specimen, which occurs at the outermost plies right below the loading cylinder, is calculated from:

$$\sigma_f = \frac{3Pl_{ss}}{2wt^2} \qquad (6\text{-}13)$$

where P is applied load, assuming that the moduli are the same in tension and compression. *Figure 6-18* shows this flexural stress plotted vs. the flexural strain measured by the strain gage. The apparent flexural strength is determined from:

$$\sigma_f^* = \frac{3P^*l_{ss}}{2wt^2} \qquad (6\text{-}14)$$

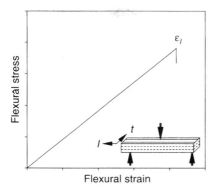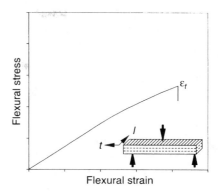

Figure 6-18 Schematic stress-strain relationships for flexural testing of unidirectionally reinforced composites parallel (left) and transverse (right) to the reinforcement

while the apparent flexural modulus is calculated as for previously described methods.

The advantages and disadvantages of flexural testing are similar to those of the short-beam shear test. The test requires small specimens, the only specimen preparation is bonding of a single strain gage, and the test is very simple to perform; large numbers of specimens are therefore readily tested. Just as with the short-beam shear test, experimentally determined properties are not inherent material properties and may consequently not be used in design. Flexural testing is also much used in quality control of both raw material and manufacturing technique.

Relevant Standards

ISO 178, "Plastics—Determination of Flexural Properties of Rigid Plastics". ASTM D 790, "Standard Test Methods for Flexural Properties of Unreinforced and Reinforced Plastics and Electrical Insulating Materials".

Related Standard

EN 63*, "Glass Reinforced Plastics; Determination of Flexural Properties; Three Point Method".

6.2.3 Thermal Properties

A range of thermal properties of the type quoted in *Sections 3.1.1, 3.2.1,* and *3.3.4* may be of relevance in composite design and optimization of the

---

* EN 63 is identical to ISO 178.

manufacturing process. Nevertheless, most of these properties are rarely determined by the component manufacturer, but rather by the raw material supplier. Among the few composite thermal properties of everyday importance to the component designer and manufacturer are coefficient of thermal expansion (CTE) and heat deflection temperature (HDT). While there are several ways to determine the CTE of a laminate, a particularly convenient method which uses temperature-compensated strain gages bonded to the material is described in reference [3]. The HDT is determined in a test where a simply supported rectangular beam specimen is subjected to a constant flexural load (cf. *Figure 6-17*). The specimen is immersed in a heat-transfer medium whose temperature gradually is increased; when a given deflection is obtained the HDT has been reached.

Relevant Standards

ISO 75, "Plastics and Ebonite—Determination of Temperature of Deflection Under Load".
ASTM D 648, "Standard Test Method for Deflection Temperature of Plastics Under Flexural Load".

6.3 Composite Quality Control

One of the perceived disadvantages of composites is the potentially large property variations between nominally identical components. Such variations are not only the result of randomly occurring manufacturing-induced defects (such as those discussed in *Section 4.4*), but also of variations in raw material quality (cf. *Section 6.1*). It is therefore most important to have a quality assurance program in place for the fourth (and possibly fifth) quality control stage mentioned in *Section 6.1* (inspection after manufacturing and inspection prior to shipping).

Just as with constituent quality control (cf. *Section 6.1*), the degree to which quality control is carried out on the finished composite component varies significantly with industry, application, and customer. Quality control is often limited to visual inspection of components and possibly determination of some physical property, such as color, thickness, or planeness, while more demanding applications naturally require more sophisticated quality control procedures. Such methods have been divided below into destructive (or sacrificial) and nondestructive testing. Mechanical testing of randomly selected components—normally to failure—often suffices to ensure that a detrimental change in component quality is detected in time to alleviate the cause of this change. However, some form of nondestructive testing of every single component naturally is preferable (if cost is no objection) and in high-performance applications this is often an explicit requirement from the customer. The most demanding quality control procedures are likely to be found in the aerospace industry, where rigorous chemical and mechanical testing of representative test coupons, as well as nondestructive testing of every single component, is normally required.

6.3.1 Destructive Testing

In many applications standardized mechanical test methods (such as those introduced in *Section 6.2*) are used also in quality control of the finished composite component. The short-beam shear and flexural tests are particularly common due to their simplicity and modest requirements in terms of amount of material. However, quality control testing is for obvious reasons often carried out under conditions as similar to the intended working conditions as possible. This means that a lot of quality control tests are application-specific and for the purposes of this text there is therefore limited point in delving too deeply into such procedures; a few examples are nevertheless briefly mentioned in the following.

A filament wound pressure vessel is naturally pressure tested (but with a liquid instead of a gas to reduce the energy released upon rupture). Randomly selected vessels are usually tested until they burst, whereas every single vessel is probably tested up to a test pressure (well below the failure pressure, but above the intended working pressure) to ensure that no hitherto undetected critical flaw exists in the vessel. (The latter test may be monitored with the quasi-nondestructive test method of acoustic emission described in *Section 6.3.2.6*.)

Similarly, a pultruded I beam intended for use as a structural member is likely to be tested in bending in a setup conceptually like that illustrated in *Figure 6-17*. Since pultruded composites are generally used in very cost-conscious applications, it is most unlikely that every single beam delivered is tested to verify a minimum flexural modulus. However, a limited number of beams may be evaluated in terms of both flexural modulus and flexural strength.

While the aerospace industry certainly carries out very application-specific tests of sub-components, for example gradually larger and more complex wing box sections, eventually whole wings, and finally full-scale tests of entire aircraft, these tests are generally part of verification of the entire design, manufacturing, and assembly process. For quality control, where it would be prohibitively expensive to (destructively) test sub-components, let alone full-scale aircraft, two basic and complementary approaches are often used together. One is to inspect each component using the nondestructive test methods outlined in *Section 6.3.2* before and possibly also after assembly. Another approach is to manufacture components that are a little larger than required, thus allowing coupons to be machined off the edges and separately tested. The idea of the second approach naturally is to obtain test coupons that have experienced the very same storage and processing conditions as the actual component. Such coupons may be characterized in terms of density, fiber content, void content, T_g, and various mechanical properties using the methods described in *Section 6.2.2*. In addition to these basic properties, open-hole tension (tensile testing of a laminate containing a machined hole), open-hole compression, compression after impact (compression of a specimen that has a supposedly well-defined lateral impact damage), and possibly several other properties may be determined.

The concept of machined-off test sections may be used also in less high-performing applications than aerospace. A boat hull is, for example, often laminated without a cut-out for the propeller axis; the sections subsequently cut out to allow for such installation may very well be used for quality control testing.

6.3.2 Nondestructive Testing

Quite often it is not economically feasible to destructively test selected components for quality control purposes and in many applications the customer specifies that quality control procedures must include testing of every single component delivered. A range of nondestructive test (NDT) methods, also called nondestructive evaluation (NDE) and nondestructive inspection (NDI) methods, have been developed to satisfy such requirements. As will be further discussed in *Section 6.4*, NDT methods are extensively used also for in-service inspection to determine if a damaged component must be repaired or replaced.

Among the types of laminate flaws that can be detected in monolithic composites using NDT methods are delaminations, disbonds (partially or completely failed adhesive bonds), voids, incorrect reinforcement orientation, missing reinforcement plies, variations in fiber content, and inclusions of foreign matter. In detecting such anomalies, NDT methods rely on differences in physical and mechanical properties, such as density, modulus, thermal diffusivity, and refractive index, between unflawed and flawed material. Inspection of sandwich components is considerably more difficult due to the huge discontinuity in physical and mechanical properties between faces and core. It is nevertheless generally possible to inspect the near face without problems, while the degree to which the core can be inspected depends on core type. In sandwich components with honeycomb cores the core can be inspected in terms of density changes (buckled and condensed cells); missing or void-containing adhesive layers, including face-to-core disbonds; and water intrusion. With foam cores inspection is more difficult due to the insulative characteristics of foams, but some NDT methods allow detection of face-to-core and core-to-core disbonds, while inspection of the core is generally difficult. Inspection of an inaccessible far face of a sandwich component is normally never possible.

NDT methods are by their very nature comparative, meaning that they only provide qualitative information, which to be of real use must be compared to a reference standard in the form of the response from a component known to be free of critical flaws. In many instances, this reference standard can be the area surrounding a suspected defect. In addition to a reference standard, interpretation of this comparison must be predetermined in terms of whether the inspected component should be accepted as is, repaired, or discarded. In quality control there should be little difficulty in ascertaining nominal component composition (e.g. intended thickness and core-density variations and molded-in metal inserts) in order to use the correct reference standard. In contrast, in in-service inspection (see also *Section 6.4*) where a component, say a fairing on an aircraft, may have been in service for a decade or two the situation is entirely different; the original component composition may not be known since design changes may have been made over the years, nominally identical components have been manufactured by several different companies using different raw materials, and the component has been slowly degraded by the elements over the years. NDT inspection is therefore very dependent on operator skill and experience.

No single method can detect all forms of flaws in all material types and the NDT method consequently has to be carefully selected for each case; typically more than one NDT method may be feasible, but cost generally rises rapidly with accuracy and resolution of the results. While most NDT methods have been developed for aerospace needs and this is still the industry where they are the most common, quality control using NDT methods is becoming increasingly common in many other application areas as well.

6.3.2.1 Visual Inspection

While visual inspection is not considered a fully fledged NDT method, it is nevertheless quite powerful and certainly much used. It was previously mentioned that visual inspection may be in terms of color, thickness, or planeness, but there are several other possibilities as well. In many cases a poor surface, including small "pin-head" porosities, is an indication that there is also a problem (likely with porosities) within the component. Light polishing of the component edges may reveal interlaminar cracks at the edges (where they are the most likely to occur), particularly if the component is viewed under a magnifying glass in good lighting conditions. A glass-reinforced composite with unpigmented matrix is semitransparent if it is not too thick. If viewed with a strong backlight it is generally possible to see whether the reinforcement is correctly oriented and if there are any major inclusions or voids present.

6.3.2.2 Ultrasonic Inspection

Probably the most common NDT method, ultrasonic inspection is widely used in both aerospace and several other industries. When used on composites the method is capable of detecting delaminations, disbonds, voids, and foreign inclusions, as well as changes in fiber fraction, fiber orientation, and thickness, and is best at detecting flaws parallel to the component surface. The concept is quite simple; a sound wave introduced into a component travels at a well-defined speed of sound until it encounters a density change where it is partially or completely reflected. A sound wave that has been reflected by some interior anomaly is compared to a reference wave that has traveled through unflawed material and an interpretation of this comparison is obtained through correlation with reflections from known flaws.

The interrogating sound wave has a frequency in the range 0.5–75 MHz and is generated by a transducer, which is usually a piezoelectric crystal. Two possibilities for sound monitoring are possible: pulse-echo and through-transmission, see *Figure 6-19*. The pulse-echo method uses a transducer that combines the two tasks of sound generation and monitoring to pick up the reflected wave, while the through-transmission method has a dedicated transmitter and a separate receiver that picks up the sound wave penetrating to the back surface. The former method is more common with hand-held equipment and the latter method for stationary facilities. With the through-transmission method the motions of transmitter and receiver naturally have to be coordinated and the method requires full access to the back of the component.

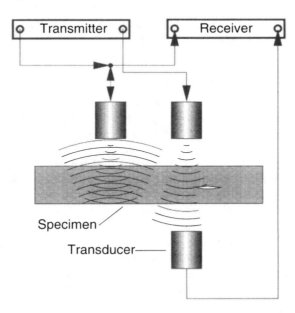

Figure 6-19 Schematics of pulse-echo (left) and through-transmission (right) ultrasonic inspection. In the through-transmission illustration the influence of a flaw is shown. Redrawn from reference [6]

To introduce the interrogating sound wave into the component with small energy losses a coupling medium is required, since only 0.03% of the energy of a sound wave crosses an interface between air and a carbon-reinforced composite. In contrast, 73% of the energy is transmitted across an interface if the coupling medium is water. 99.97% of a sound wave is thus reflected when it encounters an air inclusion (or a dry far surface) and the echo is consequently clear and straightforward to interpret [7]. While water is the most common coupling medium, particularly with automated facilities, oils, glycerin, and gels are often used with hand-held equipment. When water is the coupling medium, the part may be immersed in water as illustrated in *Figure 6-20*. Alternatively, especially for components too large to immerse, water jets may be used to transmit the signals between transducers and component, see *Figure 6-21*.

The results from an ultrasonic inspection may be displayed in several different ways. The most basic is the so-called A-scan, where the reflection from the near surface (when the sound wave enters the component) is related to the reflection from the far surface or possible anomalies within the component, see *Figure 6-22*. With knowledge of the (transverse) speed of sound in the material, the time lag between these two reflections can be directly correlated to distance. If there is no flaw directly beneath the transducer the time

Figure 6-20 Pulse-echo ultrasonic inspection of impact-damaged composite submerged in water tank. The transducer is held by a gantry robot which positions the transducer over the component surface. Photograph courtesy of K. V. Steiner, Center for Composite Materials, University of Delaware, Newark, DE, USA

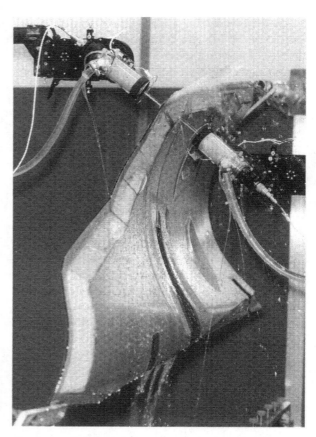

Figure 6-21 Through-transmission ultrasonic inspection of aircraft component using water jets for coupling. The transducer movements are coordinated.
Photograph courtesy of Saab Military Aircraft, Linköping, Sweden

between reflections corresponds to twice the component thickness, but if the time between reflections is less there is some anomaly within the component that is the origin of the reflection. It is important to realize that if there is more than one flaw in the thickness direction, only the one closest to the near surface will be detected if it causes a complete reflection, since this has a shadowing effect. A-scans interrogate the material in the thickness direction at one (in-plane) location only and the transducer therefore has to be moved over the surface to interrogate another location and thus produce a new A-scan. In practice this means that the transducer is moved along the component surface until the known far-surface reflection disappears, meaning that a flaw (or a change in thickness) has been located. Hand-held pulse-echo transducers are often used in in-service inspection and A-scans are readily displayed on a conventional oscilloscope. A B-scan is obtained if the transceiver is moved along a line at a known speed to produce what is essentially a set of combined A-scans. In a B-scan the results are displayed as distance between near surface and reflection as a function of position along the line traveled by the transducer, see *Figure 6-22*.

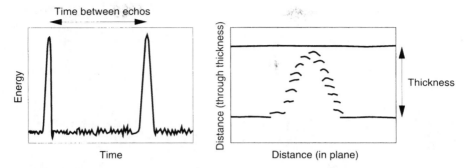

Figure 6-22 Schematics of A- (left) and B-scans (right). In the A-scan the time between echoes corresponds to twice the component thickness or twice the distance to a detected flaw. The B-scan is a schematic representation of impact-induced delaminations. Redrawn from reference [7]

The most common presentation format for ultrasonic inspection in quality control applications is the C-scan, which gives a planar view of the results obtained by moving the transducer(s) line by line over the entire component surface. The conceptual differences between A-, B-, and C-scans lie in the signal-processing software used and in the resulting presentation format and not in the hardware (except for the need for automated positioning of the transducer for B- and C-scans). A conventional C-scan displays the results of an ultrasonic interrogation in a strictly two-dimensional fashion, i.e. information on flaw location in the thickness direction has been lost. With sophisticated signal-processing software, the data may be displayed ply-by-ply in separate C-scans as illustrated in *Figure 6-23*; a conventional C-scan only provides the superimposed information of this ply-by-ply information, which is represented by the last image shown in the figure. Alternatively, the same information may be volumetrically displayed as illustrated in *Figure 6-24*. Such a scan would logically be termed D-scan, but this notation is not (yet) commonplace.

Although not common with composites, it is also possible to locally introduce a sound wave into a component using a laser; both pulse-echo and through-transmission laser ultrasonics have proved feasible. When heated by the laser beam the component surface locally expands to generate a sound wave, which once introduced into the material behaves in the same fashion as discussed above; reflected sound waves are detected by a laser. Laser ultrasonics has advantages in not requiring a coupling medium or a normal angle between laser beam and component, thus facilitating inspection of very complex geometries. Disadvantages include that care must be taken not to damage the component thermally and that, since sound waves are weak, background noise may lead to reduced contrast between flawed and unflawed material.

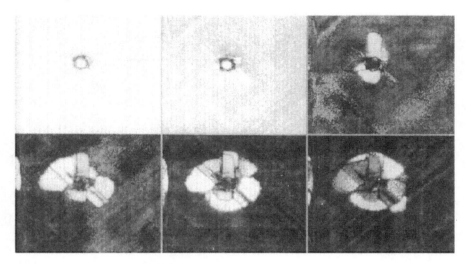

Figure 6-23 C-scans of impact-damaged 32-ply carbon-reinforced PEEK composite. The resulting delaminations are shown for six different thickness locations (one image for each ply may be obtained if desired); the first image shows the indentation on the impact side and subsequent images represent locations further into the laminate. Shadowing effects from delaminations close to the near side prevents effective interrogation behind such a delamination. Note that the reinforcement orientation is visible. C-scans courtesy of K. V. Steiner, Center for Composite Materials, University of Delaware, Newark, DE, USA

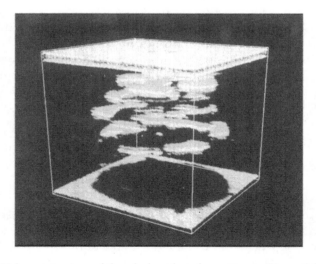

Figure 6-24 Volumetric view of the ply-by-ply information in *Figure 6-23*. Note the shadowing effect, which is particularly clear on the bottom surface of the component image. Volumetric C-scan reprinted from reference [8] with kind permission from Plenum Press, New York, NY, USA

6.3.2.3 *Vibrational Inspection*

Closely related to ultrasonic inspection is vibrational inspection, which also exploits the sonic properties of the component. Natural frequency and damping characteristics of a component can be evaluated either through continuous or pulse excitation, which may be local or global in nature, and the response is assessed using dedicated vibration-analysis equipment. If natural frequency spectrum and damping characteristics are available for a component that is known to be defect-free, the overall structural integrity of any nominally identical component may be assessed, although the location of a possible defect must normally be determined using some other NDT method. It is further possible to determine the elastic moduli of a component with this method since the speed of sound is a direct function of the moduli.

A manual version of vibrational inspection using local pulse excitation is that of "coin tapping", where the component surface is tapped with a metal object (such as a coin). The sound emitted is a function of the local structural integrity of the component and a low-pitched sound in an area of otherwise high-pitched response is likely to be a qualitative indication of a subsurface anomaly, such as a delamination. Coin tapping is widely used in all application areas from boat building to aerospace.

A crude version of vibrational inspection is that if a composite is dropped onto a hard surface (i.e. global pulse excitation) it will ring with an audible frequency proportional to its degree of consolidation; the better the consolidation, the higher the (natural) frequency. This qualitative version of vibrational inspection is particularly effective with continuously carbon-reinforced composites, which if well consolidated emit a high-pitched sound when dropped.

6.3.2.4 *Radiographic Inspection*

Another family of successful and rather common NDT methods rely on the differential in absorption of radiation between flawed and unflawed material. The radiation is normally in the form of x-rays or γ-rays, although other and less common possibilities have been successfully tried. The two most common methods are radiography and computed tomography (CT, also known as computer-aided tomography, CAT), both of which are very similar to the well-known procedures used in hospitals. In radiography the component is exposed to radiation and the unabsorbed radiation passing through the specimen is detected by a photographic film, screen, or video camera, see *Figure 6-25*. In the former case the image is often called a radiograph; *Figures 5-4a* and *5-8* are examples of radiographs. A disadvantage of using a photographic film is that it must be developed, which takes time and consequently does not provide an immediate result of the interrogation. With a viewing screen or video monitor the image can be viewed in real-time and this version of radiographic inspection is sometimes referred to as radioscopy.

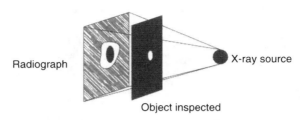

Figure 6-25 Schematic of radiography. Drawing reprinted from reference [9] with kind permission from the author

Radiography is very sensitive to differences in density and may be used to detect delaminations, disbonds, regions of significant voids, foreign inclusions, thickness changes, and sometimes fiber orientation and mold filling pattern in short-fiber reinforced composites. In honeycomb sandwich components density changes, defective adhesive bonds, and water intrusion can be detected. Radiography does not allow detection of very small anomalies and does not easily detect delaminations oriented perpendicular to the interrogating radiation, since they absorb little radiation. The capability to detect small flaws and delaminations is radically enhanced if a radiation-opaque liquid penetrant is introduced into, for example, a delamination. Introduction of a penetrant naturally requires that the defect reaches the surface somewhere to allow entry of the penetrant. The radiographs of *Figures 5-4a* and *5-8* would not have been nearly as clear without the use of penetrant.

Radiography does not provide information on the location of the flaw in the thickness direction unless two or more images of the component taken at different angles are compared. A more refined solution to this is computed tomography, in which a thin collimated beam of radiation interrogates a slice of the component. The component is then rotated slightly and interrogated once again, see *Figure 6-26*. Dedicated software then reconstructs a three-dimensional image of the component from a set of scans representing a 180° rotation, thus retaining information about both size and location of detected anomalies. Also in computed tomography use of a radiation-opaque penetrant enhances flaw detection capability, see *Figure 6-27*. Both radiography and computed tomography require access to the back surface of the component, thus restricting their use.

While radiography is used in high-performance applications where its high cost can be justified, e.g. aerospace, computed tomography, which is even more costly, is usually too expensive for standard quality-control purposes. Both radiography and computed tomography are commonly used in research and development situations.

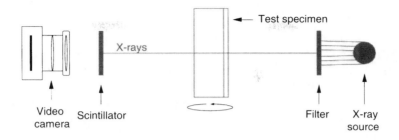

Figure 6-26 Schematic of computed tomography. Drawing reprinted from reference [9] with kind permission from the author

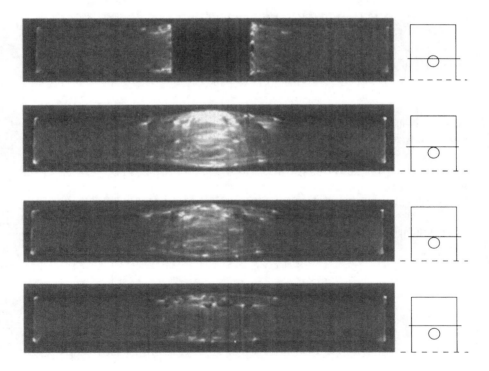

Figure 6-27 CT-scans enhanced through use of radiation-opaque penetrant. The images show four virtual through-the-thickness slices of a laminate with a hole that has been tested in bearing fatigue (cf. *Figure 5-8*). The locations of the virtual cuts are illustrated to the right. CT-scans reprinted from reference [10] with kind permission from Technomic Publishing Co, Lancaster, PA, USA

6.3.2.5 Thermographic Inspection

Thermographic inspection exploits differentials in thermal diffusivity between flawed and unflawed material. A component is actively or passively heated and the temperature on its surface is monitored as a function of time using a video camera sensitive to infra-red light. With active heating the component is subjected to mechanical loads that make the walls of flaws within the component rub together to generate heat through friction; although technically feasible, active heating is not common (although it is used to monitor damage propagation in fatigue-loaded specimens). With passive heating energy is introduced from an external heat source. Regardless of heating method it is important to consider the possibility of overheating the component.

A few different approaches to thermographic inspection using passive heating are possible. The component may be suddenly heated on one side and the temperature on the other side monitored; cooler regions in an otherwise essentially homogenous temperature field indicate defects underneath the surface. This concept, which may be termed through-transmission thermographic inspection, is schematically illustrated in *Figure 6-28*. An alternative is to monitor the temperature on the heated side following a sudden heat pulse (pulse-echo thermographic inspection); since a defect beneath the surface does not conduct heat as well as the bulk material, the surface directly above it will remain warmer than its surroundings. Yet another possibility is to heat the entire component to a homogenous temperature in an oven and then monitor its heat loss upon cooling; a relatively warmer region indicates a subsurface defect.

In comparison to ultrasonic and radiographic inspection, thermographic inspection provides poor resolution. This is particularly true in continuously carbon-fiber reinforced composites where the conductivity in the fiber direction is an order of magnitude greater than through the thickness; differentials in temperature are therefore easily obscured by in-plane heat

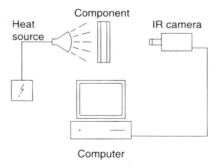

Figure 6-28 Schematic of through-transmission thermographic inspection. Drawing reprinted from reference [9] with kind permission from the author

transfer. Large impact damages and disbonds are nevertheless detectable in carbon-reinforced composites [7]. Thermography has also proved reasonably successful in detecting cavities and core-to-core disbonds and cracks in thick foam-core sandwich components, see *Figures 6-29* and *6-30*, which is difficult or impossible with most other NDT methods. Thermography provides no reliable information on the through-the-thickness location of a defect.

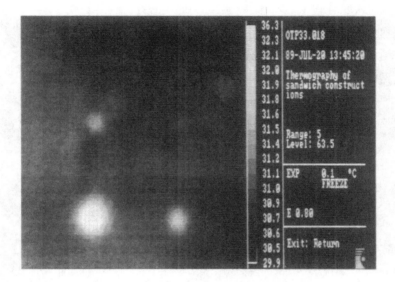

Figure 6-29 Thermal image from through-transmission thermographic inspection of a sandwich panel with 3 mm thick glass-reinforced polyester faces and 30 mm thick low-density PVC foam core (ρ = 100 kg/m^3)containing simulated cylindrical defects. The defect diameters are 4, 8, 12, and 16 mm and their respective heights equal to the diameters. The image was taken 20 s after the end of a 180 s heating period. The image shows that under these conditions the 4 mm defect cannot be detected (the defects are located in the four corners of a square). Thermogram reprinted from reference [11] with kind permission from the author

6.3.2.6 Acoustic Emission Inspection

Acoustic emission is a method that exploits the fact that even a lightly stressed composite will experience microscopic failures long before any significant decrease in macroscopic properties can be detected. Individual fiber fractures and progressing matrix microcracks are sources of sound and stress waves that can be detected using sensitive transducers. If several transducers are located on the component and sophisticated sound-analysis equipment is used, location and sometimes even type of defect can be

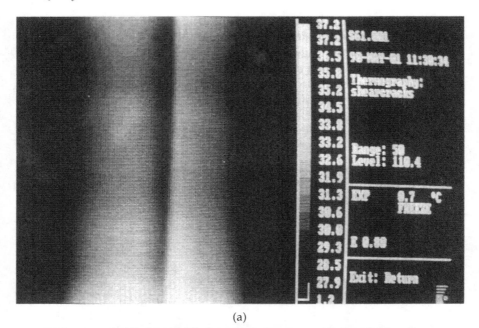

(a)

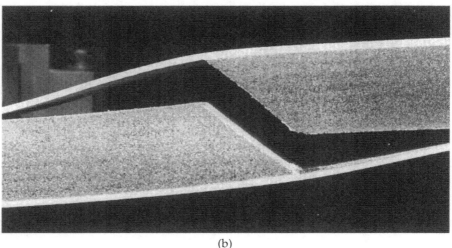

(b)

Figure 6-30 45° shear crack in sandwich panel with glass-reinforced polyester faces and low-density PVC foam core simulated using Teflon film; (a) Thermal image from pulse-echo thermographic inspection using hand-held heater (crack vertical in center of image) (b) Slice of investigated sandwich panel tested in bending to reveal shear crack which further leads to face-core separation. Note that (a) is a planar image of the panel, whereas (b) is a side view. Thermogram courtesy of M. Vikström, Defense Materiel Administration, Stockholm, Sweden; photograph reprinted from reference [12] with kind permission from ASTM, USA

determined, although the latter is very difficult. In glass-reinforced composites an acoustic event (single sound/stress pulse) with an amplitude less than 70 dB corresponds to matrix cracking and events with amplitude in excess of 70 dB to fiber fracture [13]. When compared to other materials (mainly metals), composites are "noisy" and a baseline acoustic response from a component that is known to be without critical defects is necessary to relate the acoustic response to degree of defects; see *Figure 6-31*.

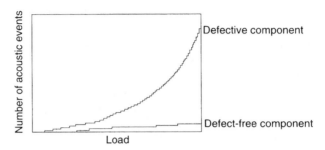

Figure 6-31 Schematic of acoustic response from component containing defects compared to baseline component without critical defects. Redrawn from reference [13]

The Kaiser-effect stipulates that once loaded to a given stress level, a component should not emit any more noise upon reloading to the same stress level. The Felicity ratio, defined as the stress level where noise is emitted again upon reloading divided by the previous maximum stress level experienced, thus should be slightly larger than unity. Failure to obey the Kaiser-effect, i.e. exhibiting a Felicity ratio less than unity, indicates component damage and if noise is continuously emitted at a constant stress level there is reason to suspect ongoing crack propagation [13]. Acoustic emission is used in quality control of, for example, pressure vessels and pipes. To avoid permanent macroscopic damage during loading for acoustic emission inspection it is recommended that [13]:

- Testing is performed at stress levels well below the design load
- The total number of acoustic events is kept below a specified number
- Only few acoustic events above a specific amplitude are permitted
- The Felicity ratio is kept close to unity

6.3.2.7 Eddy-current Inspection

If a current-carrying coil is placed near a conductive material, eddy currents are induced in the material, which in turn affect the impedance of the driving coil. Local variations in resistivity, such as from interior defects, interrupt the eddy currents, thus altering coil impedance. Eddy-current inspection is well-established with metals, but has not been used much with composites due to the difficulty in interpreting the response. Nevertheless, the method has proved capable of detecting impact damage as well as variations in fiber content and component thickness [7]. Since eddy-current inspection requires conductive material, it does not work with glass and organic reinforcement.

6.3.2.8 Optical Inspection

Holography-based inspection methods are capable of detecting minute dimensional changes as a component is stressed. In laser shearography the unstressed component is illuminated with monochromatic laser light and its shape recorded. The component is then stressed to cause deflections and its shape is recorded once more, see *Figure 6-32*. Stressing may be achieved through changes in temperature or ambient pressure (usually through application of partial vacuum), or through vibrational excitation; with heating and decreased pressure the surface above an air inclusion will rise ever so slightly when the entrapped air expands. A photographic double exposure of the two images reveals regions that have deflected as fringe patterns. Surface displacements of order wavelengths of the laser light can be detected, but due to this extreme sensitivity the method is also vulnerable to background noise.

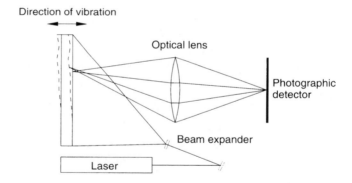

Figure 6-32 Schematic of shearography. Drawing reprinted from reference [9] with kind permission from the author

Electronic speckle pattern interferometry (ESPI) is an enhanced version of shearography. The method still uses a laser, but the excited images are recorded with a video camera and are compared to a prerecorded (and undeformed) reference image to continuously display interferences on a video monitor. Also this form of shearography requires excitation of the component as mentioned above, but a simple hot-air gun has proved to be sufficient in many applications. A range of defects, primarily gaseous inclusions such as voids, delaminations, and disbonds, including ones in the foam core of sandwich panels, can be detected, but no information on the through-the-thickness location of a defect is obtained. The method is a relative newcomer to NDT and is not yet in widespread use.

6.4 In-service Inspection

In a range of applications it is necessary to have the capability to assess the structural integrity of components or entire structures while in service to prevent unpredictable and potentially catastrophic failures. This may either be carried out at predetermined service intervals, following repair, or may be called upon by an incident where a component is known to have been exposed to a load that may have been in excess of the design load. The most common example of the former case is probably the routine inspections carried out in both civilian and military aerospace applications. While the latter cause for inspection—often in the form of unexpected impact loads—is most common in aerospace applications due to bird strikes, collisions during taxiing, dropped wrenches, etc., there are analogies in other application areas. One example is a sailing yacht that may have been dropped or dented during launching, or it may have hit the bottom or a dock a little too hard. High-speed powerboats and passenger ships are subject to fatigue loading from the incessant slamming contact between bow and waves, and sandwich hulls are particularly vulnerable to face–core disbonds stemming from such conditions.

An in-service inspection method should preferably be nondestructive, portable, accurate, provide real-time results, and not require access to the back of components. The most powerful tool for in-service inspection is the human eye, although the extent of many types of damage, particularly if caused by impact, are not necessarily visible on the component surface. Airlines nevertheless largely depend on visual inspection of composite components during routine checks. In this application it is assumed that serious damage, e.g. from impact, overheating, lightning strike, etc., apart from causing interior laminate damage also leads to easily detectable penetration or chipping of the paint layer. However, if damage to the paint has been detected, closer inspection is needed and is usually in the form of coin tapping and closer visual inspection using magnifying glass and good lighting. This close visual inspection can be facilitated by application of a pigmented penetrant, which through capillary action penetrates the finest cracks and thus makes them easier to see. Visual inspection and coin tapping are normally sufficient to determine whether the damaged component does not need repair, if repair can wait, if repair can be carried out without removing the component (cf. *Figure 5-40*), or if the component must be replaced. If visual inspection and coin tapping provide inconclusive results, radiography is the preferred NDT method to inspect both composites and sandwich components [14].

Another rather common NDT method for in-service applications employs a hand-held transducer for pulse-echo ultrasonic inspection. While the coupling medium in this case is usually oil or gel, transducers with rubber rollers that provide sufficient acoustic contact offer an alternative. Thermographic inspection and shearography have also proved capable of fulfilling most of the aforementioned requirements on in-service inspection methods but are nowhere near as common. A couple of these requirements may be omitted if a component can be replaced by a spare and evaluated separately; in this case all of the previously described NDT methods can be used. One attractive possibility arises if the NDT response of the very same component when it was new (and at possible previous in-service inspection occasions) is available. Such an earlier "portrait" offers an excellent possibility to assess the decay from aging or impact.

6.5 Summary

Quality control is of utmost importance throughout manufacturing to ensure uniform and predictable properties in the finished component. The degree to which quality control is carried out depends significantly on industry, application, and customer. The raw material quality is often merely assessed in terms of resin reactivity (for thermosets) and only in specialized application areas, e.g. aerospace, is the raw material extensively screened. In the composite industry as a whole, the bulk of quality control efforts concentrate on characterization of the finished component.

Composites are most commonly characterized in terms of mechanical properties, whereas physical and thermal characterization is carried out to a lesser degree. Mechanical characterization is very common in terms of tensile, compressive, shear, and flexural properties, i.e. stiffness, strength, Poisson's ratio, and strain to failure. Tensile testing is straightforward and there is a general consensus on how testing should be performed. While conceptually simple, compressive testing is very difficult, since failure may be initiated at the load introduction points and the specimen may buckle; end tabs are used to minimize the risk of these unintended failure modes. There is considerable disagreement within the composite industry on how compressive properties should be determined. Shear testing is also difficult with little consensus on what test methods to use. A range of different test methods have been developed to determine in-plane and interlaminar shear properties. Flexural testing is as straightforward as tensile testing and there is consensus on how testing should be carried out. While tensile, compressive, and most shear tests are intended to provide inherent material properties that may be used for design purposes, flexural tests, as well as the conceptually similar short-beam shear test, only provide qualitatively useful properties. The most commonly assessed physical property is fiber content, which may be determined through burning of the matrix (only with glass fibers), acid digestion of the matrix, and microscopy of a polished cross-section. Thermal composite properties of relevance are the coefficient of thermal expansion and the heat deflection temperature, but these are seldom determined in routine experimental characterization.

Destructive quality control testing employs a range of application-specific tests as well as the aforementioned mechanical test methods and is usually carried out on randomly selected components. In particular the flexural and short-beam shear tests are commonly used since they are simple to perform and require little material. Nondestructive testing may be carried out through a range of methods that all rely on differences in physical and mechanical properties between flawed and unflawed material. Detectable flaws in laminates include delaminations, disbonds, voids, incorrect reinforcement orientation, missing plies, variations in fiber content, and inclusions of foreign matter. The most common nondestructive quality control means, visual inspection and coin tapping, are rarely considered fully fledged NDT means. The most common instrumented NDT method is ultrasonics, where a sound wave is introduced into the component and the wave reflected by the component's back surface or an internal defect is monitored. In radiography and computed tomography radiation passed through the component is absorbed to a greater degree by defects; the penetrating radiation is recorded with photographic film or video camera. Thermography is similar in concept to radiography, except that in this case a heat pulse is used to interrogate the component and a video camera monitors the temperature development on the component surface. Acoustic emission exploits the fact that microscopic failures occur in a component even at stress levels far below levels leading to a decrease in macroscopic properties. Vibrational, eddy current, and shearography inspection are less common NDT methods.

A component may need to be inspected for damages during service either due to prescribed service inspection, repair, or due to an incident where a component may have been exposed to an unusually large load. Apart from visual inspection and coin tapping, which are the most common methods for in-service inspection, radiography and ultrasonics and to a lesser degree also thermography and shearography are used on components still part of a larger assembly, such as an aircraft. If the component may be removed from the assembly any NDT method may be used.

Suggested Further Reading

For most of the topics in *Sections 6.1* and *6.2* the actual test standards offer some of the best information available. A good and thorough overview of all types of test standards for both composite raw materials and finished components (primarily ISO, EN, and ASTM standards) is:

R. M. Mayer, *Design with Reinforced Plastics*, The Design Council, London, UK, 1993.

While ISO and EN standards are available on a standard-by-standard basis only, ASTM and some DIN standards are conveniently available in book form. Standards are generally available from national standardization organizations. Most ASTM standards relevant to testing of composites have been collected in the first reference below, which is a compilation of pertinent standards from the three annually published volumes following it:

ASTM Standards and Literature References for Composite Materials, 2nd edn, ASTM, Philadelphia, PA, USA, 1990.

Plastics (I), **8.01**, ASTM, Philadelphia, PA, USA.

Plastics (II), **8.02**, ASTM, Philadelphia, PA, USA.

Space Simulation; Aerospace and Aircraft; High Modulus Fibers and Composites, **15.03**, ASTM, Philadelphia, PA, USA.

Twenty-six highly relevant recommended methods for testing of continuous-fiber reinforced composites are provided in:

SACMA Recommended Methods (SRM), SACMA, Arlington, VA, USA, 1994.

Critical evaluations of test methods to determine tensile, compressive, and shear properties are available in:

S. Chatterjee, D. Adams, and D. W. Oplinger, "Test Methods for Composites a Status Report, Volume I: Tension Test Methods," DOT/FAA/CT-93/17, United States Department of Transportation, Federal Aviation Administration, Atlantic City International Airport, NJ, USA, 1993.

S. Chatterjee, D. Adams, and D. W. Oplinger, "Test Methods for Composites a Status Report, Volume II: Compression Test Methods," ibid.

S. Chatterjee, D. Adams, and D. W. Oplinger, "Test Methods for Composites a Status Report, Volume III: Shear Test Methods," ibid.

These three reports are available from the National Technical Information Service, Springfield, VA 22161, USA.

An excellent undergraduate-level introduction to mechanical testing of continuous-fiber reinforced composites is:

L. A. Carlsson and R. B. Pipes, *Experimental Characterization of Advanced Composite Materials*, 2nd edn., Technomic, Lancaster, PA, USA, 1996.

Nondestructive testing of composites is the topic of:

Nondestructive Testing of Fibre-Reinforced Plastics Composites, Volume 1, Ed. J. Summerscales, Elsevier, Barking, Essex, UK, 1987.

Nondestructive Testing of Fibre-Reinforced Plastics Composites, Volume 2, Ed. J. Summerscales, Elsevier, Barking, Essex, UK, 1990.

References

1. G. D. Mayorga, "Quality Assurance and Quality Control," in *International Encyclopedia of Composites*, Ed. S. M. Lee, VCH Publishers, New York, NY, USA, **6**, 1-7, 1991.

2. B. T. Åström and R. B. Pipes, "Thermoplastic Filament Winding with On-Line Impregnation," *Journal of Thermoplastic Composite Materials*, **3**, 314-324, 1990.

3. L. A. Carlsson and R. B. Pipes, *Experimental Characterization of Advanced Composite Materials*, 2nd edn., Technomic, Lancaster, PA, USA, 1996.

4. D. F. Adams and E. Q. Lewis, "Current Status of Composite Material Shear Test Methods," *SAMPE Journal*, **31**(6), 32-41, 1994.

5. C. Zweben, "Static Strength and Elastic Properties" in *Delaware Composites Design Encyclopedia*, Technomic, Lancaster, PA, USA, **1**, 49-70, 1989.

6. K. V. Steiner, "Einsatz einer robotergestützten Anlage zum Bandablegen von thermoplastischen Verbundwerkstoffen," Fortschritt-Berichte VDI, Reihe 2, Nr. 369, VDI Verlag, Düsseldorf, Germany, 1996.

7. D. E. W. Stone and B. Clarke, "Nondestructive Evaluation of Composites," in *Concise Encyclopedia of Composite Materials*, Ed. A. Kelly, Pergamon/Elsevier, Oxford, UK, 208-214, 1994.

8. K. V. Steiner and T. C. Lindsay, "Correlation of Full-Waveform Ultrasonic NDE Data and Low-Velocity Impact Damage to Composite Panels," in *Review of Progress in Quantitative Nondestructive Evaluation*, Eds. D. O. Thompson and D. E. Chimenti, **13B**, Plenum Press, New York, NY, USA, 1283-1290, 1994.

9. E. Persson, "Hole Generation Defects in Composite Laminates," Report 95-9, Department of Lightweight Structures, Royal Institute of Technology, Stockholm, Sweden, 1995.

10. E. Persson, I. Eriksson, and P. Hammersberg, "Propagation of Hole Machining Defects in Pin-Loaded Composite Laminates," *Journal of Composite Materials*, **31**, 383-408, 1997.

11. M. Vikström, "Thermographic Nondestructive Testing of Sandwich Constructions," Report 89-14, Department of Aeronautical Structures and Materials, Royal Institute of Technology, Stockholm, Sweden, 1989.

12. D. Zenkert and M. Vikström, "Shear Cracks in Foam Core Sandwich Panels: Nondestructive Testing and Damage Assessment," *Journal of Composites Technology & Research*, **14**, 95-103, 1992.

13. G. Eckold, *Design and Manufacture of Composite Structures*, Woodhead Publishing, Cambridge, UK, 1994.

14. A. L. Seidl, "Inspection of Composite Structures," *SAMPE Journal*, **30**(4), 38-44, 1994.

CHAPTER 7

RECYCLING

While by no means a new concept (mankind was certainly much better at reuse and recycling in the past), recycling has over the past couple of decades once again become a natural part of our lives. In most of the world it is now common practice to recycle or reuse newspapers, glass bottles, plastic bottles, aluminum cans, automobile parts, clothes, and sometimes even entire houses. The concept of recycling polymers is really as old as the plastic industry itself, but it was not until the end of the 1960s and the beginning of the 1970s that polymer recycling finally gained momentum, not least fueled by the oil shocks in 1974 and 1978–79, which resulted in drastic increases in raw material cost [1]. Recycling of composites is an even more recent occurrence with significant work generally not starting until the latter half of the 1980s and technical publications starting to appear in the early 1990s. Compared to the widespread and widely documented activities on polymer recycling, the work on recycling of composites is still modest. However, as composites are more widely used in an increasing number of commodity products, not least automobiles which consume up to a quarter of all composites manufactured (cf. *Figures 1-4* and *1-5*), the issue of composite recycling is becoming ever more important. In broad terms, successful composite recycling requires incentives, infrastructure, recycling techniques, and market commitment.

7.1 Incentives

While the public perception seems to be that plastics make up a significant proportion of municipal solid waste (MSW) deposited into landfills, this is not entirely accurate. *Figure 7-1* gives the relative content in United States' landfills in 1991 [2]. Slightly different figures may be found in other studies from the beginning of the 1990s, which show that of the MSW in the United States approximately 7 weight percent (equivalent to 10–20 volume percent) is plastics [1]. Of a total of 180 million tons of MSW in the United States, 70 weight percent ends up in landfills, while the rest is recycled or incinerated (1991 figures) [2]. The volume percent plastics in European MSW appears to be in excess of 25 percent [3]. However, in both the United States and Europe the current recycling rate for plastics is around 1 percent, which is much lower than for paper, metals, and glass [1,3]. Figures for the amount of composites in MSW are few and are generally rough estimates; in Germany the amount of composites has been estimated at a maximum of 5 percent of the plastics content [4] and in the United States the equivalent figure has been estimated at a maximum of 10 percent [5]. In relation to the overall waste volume, composites waste clearly is not yet a major concern. It should also be remembered that a significant portion of plastic products that end up in MSW, e.g. packaging, has a life span in the order of days or possibly weeks, whereas most composite components have life spans on the order of decades.

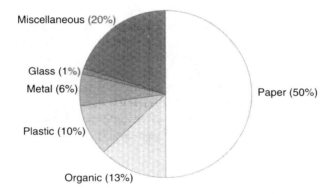

Figure 7-1 Relative volume content of landfilled MSW in the United States in 1991. Organic includes wood, yard waste, and food, while Miscellaneous includes construction and demolition debris, tires, textiles, rubber, and diapers. Data from reference [2]

Incentives behind recycling initiatives can broadly be divided into environmental, legal, and economical categories, although many incentives belong to more than one category. The former category includes a general desire to have a clean (and green) conscience (mainly for individuals), to have an image of environmental consciousness and political correctness (mainly for industry), to utilize resources efficiently, and to reduce the overall amount of waste. The legal category comprises legal impediments (mainly applicable to industry), such as required recycling and permitted waste amount and content. Finally, if there is money to be made from recycling, it certainly will be carried out by individuals as well as industry; this category also includes the effective incentive of ever-increasing costs for waste deposition in fewer and fewer landfills. As an example, the United States had 6,000 landfills in 1990, 3,000 landfills in 1995, and is expected to have 2,000 landfills in 2006, according to the Environmental Protection Agency (EPA) [6].

The dominating incentive varies strongly with waste category and country. For unreinforced plastics the strongest ones are legislation (spearheaded by Germany), rapidly escalating costs for waste deposition and incineration (mainly in the United States, Japan, and some European countries), and the general desire of companies to have a green image (everywhere). While plastic recycling is certainly technically feasible, it is unfortunately so far often difficult to make money from recycling of post-consumer waste (unless it is through avoidance of fines or high deposition fees), since virgin plastics are relatively inexpensive.

The incentives for composite recycling are similar to those for plastics, but since the amount of composite waste is comparatively small the incentives are so far much weaker than for plastics. On the economical side it is presently nearly impossible to make composite recycling profitable based solely on material value. However, there is little doubt that legal and environmental (or political-correctness) incentives to recycle composites will grow stronger. One such example is the German initiatives to keep the weight percentage of an old automobile that must be deposited (i.e. not reused or recycled) at the end of its service life equal to or less than 15 percent by the year 2002. For cars sold after 2002 the equivalent figure is 10 percent and for cars sold after 2015, 5 percent [7].

7.2 Infrastructure

Once there are sufficient incentives for recycling, the lack of infrastructure for waste collection is the next hurdle. An example of a first attempt at creating such an infrastructure on a notable scale is in Germany, where a mobile shredder travels around the country to collect composite waste, which it shreds to reduce waste volume and then brings to a recycling center. This modest but important initiative and others that have followed it in other parts of the world basically only deal with industrial scrap, which is production waste that is clean and of known and uniform content. The real challenge will be to collect post-consumer waste that is well-dispersed geographically. Since post-consumer waste is also dirty, aged, commingled (i.e. a mixture of different plastics; not to be confused with the commingled yarns discussed in *Section 2.3.4*), and of unknown composition it must following collection be identified, separated, cleaned and dried prior to recycling. With post-consumer waste the issue of how aging influences recyclability is an unresolved issue. To date, all semi-commercial pilot programs for composite recycling have used industrial scrap and there are so far no initiatives to recycle post-consumer composite waste on a commercial scale.

7.3 Recycling Techniques

The bulk of the remainder of this chapter discusses technically successful techniques to recycle composites. The techniques are divided into four classes which apply to polymer-based waste in general [1]:

- *Primary recycling* Conversion of waste into material having properties equivalent to those of the original material
- *Secondary recycling* Conversion of waste into material having properties inferior to those of the original material
- *Tertiary recycling* Conversion of waste into chemicals and fuel
- *Quaternary recycling* Conversion of waste into energy

Most tried techniques for primary and secondary recycling essentially involve mixing of some waste material with virgin raw material, which is then processed as if it were all virgin material. Whether the proper designation for the procedure is primary or secondary recycling is a matter of how successful it is. Tertiary recycling refers to chemical decomposition of the polymer, i.e. depolymerization, into useful chemical substances and fuel, while quaternary recycling is synonymous with incineration with utilization of the energy released. The technique of incineration without energy utilization is of course not recycling at all, merely a means of reducing waste volume.

Most composite recycling efforts to date have centered around primary and secondary recycling; such techniques are treated in some detail in *Section 7.3.1*. Tertiary and quaternary recycling, which have been thoroughly investigated for plastic waste but rarely tried with composites, are discussed in *Sections 7.3.2* and *7.3.3*, respectively. From the point of view of material utilization it is generally preferable to succeed with the highest possible level of recycling, e.g. secondary rather than quaternary recycling. However, from economical and overall resource utilization perspectives this need not necessarily be the case, since for example secondary recycling may require excessive amounts of energy and other resources (facilities, manpower, chemical additives, etc.), while quaternary recycling is straightforward and often does not require energy or specialized resources.

7.3.1 Primary and Secondary Recycling

The most attractive recycling alternative certainly is to recycle waste into material with properties equivalent or at least comparable to those of the original material (as common with metals and glass, for example), since it implies a possibility for an infinite number of recycling circuits. The division between primary and secondary recycling is relative and a question of what extent of property degeneration that one is willing to allow while still saying that the properties are equivalent. A common alternate notation for primary and secondary recycling is material recovery.

7.3.1.1 Thermoset Composites

Although it may be conceptually less straightforward to recycle thermoset composites than their thermoplastic brethren, the common misconception that thermoset composites due to the presence of crosslinks cannot be recycled is a myth. Proven techniques for material recovery of thermoset composites are all based on shredded and then ground or milled composites used as filler or possibly as replacement for some of the reinforcement in composites that generally end up having properties inferior to those of the original material.

To enable material recovery of thermoset composites, the component to be recycled must first be reduced in size. With a large composite the first step is generally to saw or break it into manageable pieces that can then be fed into a shredder to further reduce their size to approximately tens of centimeters. Prior to shredding, metal components, such as inserts and fasteners, should be removed for the sake of shredder longevity. The shredded material is then ground to small pieces or milled to a powder; the final particle size may be anywhere from several millimeters to fractions of a millimeter, see *Figure 7-2*. This so-called regrind is separated by size using gradually coarser meshes. In coarse regrind the reinforcement remains partially intact and the regrind thus may be used for reinforcing purposes, while the fine powder is only useful as filler.

Figure 7-2 Composite regrind. From left to right: coarse SMC regrind (fiber length 6–20 mm), fine SMC regrind (fiber length < 0.25 mm), and coarse GMT regrind

Several studies have investigated the feasibility of replacing filler and reinforcement in molding compounds with regrind. These studies have shown that it is indeed possible to recycle composites in this fashion, although component properties are usually inferior to those achievable with virgin material. *Figure 7-3* illustrates properties of composites compression molded from SMC containing varying amounts of the regrind in the center of *Figure 7-2*, while maintaining a constant glass content of 25 weight percent. In the first two formulations following the "Control SMC", the regrind merely replaces filler, while in the last formulation 16 weight percent extra resin had to be added to reduce viscosity to manageable levels. A benefit of using regrind instead of virgin filler is that while conventional SMC fillers have densities of about 2,800 kg/m³, regrind has a density of about 1,800 kg/m³, resulting in overall weight savings (cf. *Figure 7-3*).

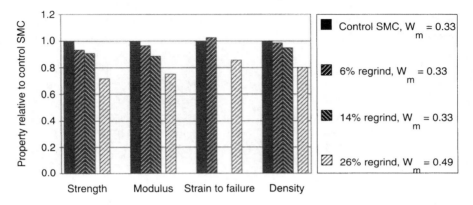

Figure 7-3 Tensile properties and density of composites compression molded from SMC containing varying amounts of fine SMC regrind normalized with respect to properties of composites molded from virgin "Control SMC". Percentages regrind refer to weight percent. Data from reference [8]

Although the data presented in *Figure 7-3* are specific to one particular study, they are largely representative of the property trends that one should expect when replacing virgin filler with fine thermoset regrind. While several studies confirm this type of results, other studies have shown that properties need not suffer at all and also that coarse regrind may be used to replace part of the virgin glass reinforcement. It would appear reasonable to assume that further refinements in compound formulations and techniques will result in primary recycling becoming the norm for use of regrind in SMC. SMC containing SMC regrind is used in series production of selected automobile components in Germany, the United States, and Japan, although widespread market acceptance appears to be quite some way off.

Results similar to those discussed above for SMC have been achieved with both fine and coarse regrind replacing filler and part of the reinforcement in BMC as well as to reinforce thermoplastics. While it is quite straightforward to include regrind in the manufacture of BMC and short-fiber reinforced injection molding compounds using standard equipment, it is somewhat more difficult to do so with SMC. Inclusion of coarse regrind, which still retains some glass yarn integrity, in SMC manufacture requires specialized solutions to avoid penetration of the carrier film (cf. *Figure 2-53*). Successful techniques have nevertheless been developed in recent years, such as first depositing half the virgin glass onto the lower resin layer, then the coarse regrind, and finally the remainder of the virgin glass so as to sandwich the regrind in the center. In contrast, inclusion of fine regrind poses no problems since it can be mixed into the resin in the same manner as virgin filler.

A method to use regrind in sprayup has also been developed. Through use of specially designed equipment, coarse regrind up to several millimeters in size may be mixed with virgin resin and sprayed in the fashion described in *Section 4.2.1.2*; this method has been used to manufacture the boat shown in *Figure 7-4*.

Another potential use for coarse regrind is as filler bound together by a resin, e.g. a foaming PUR or polyester, to produce board for use in construction applications. The technical feasibility of manufacturing such board from commingled plastics waste, including sandwich components consisting of glass-reinforced polyester faces and crosslinked PVC core, has been

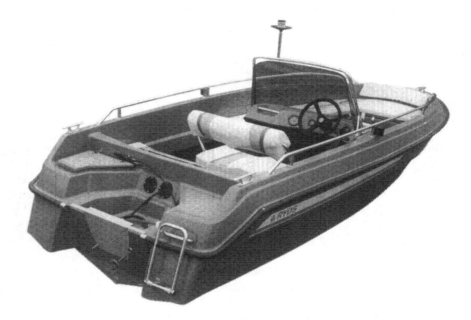

Figure 7-4 Sprayed up boat containing 20 weight percent regrind. The conventional single-skin hull has been replaced by a regrind sandwich structure. First virgin material is sprayed onto the mold to build up a laminate thickness equivalent to a quarter of the conventional single skin, then a resin mixture containing 40 weight percent regrind is sprayed to build up an areal weight equivalent to half the conventional single skin, and finally another layer of virgin material is sprayed on. Due to the success of the development work resulting in this prototype boat as well as it solving the costly disposal problem of overspray, this recycling concept is now commercially applied. Photograph courtesy of Ryds Båtindustri, Ryd, Sweden

demonstrated. Although the mechanical property retention of such board under high humidity conditions appears to be far superior to plywood and chipboard, it is too expensive to currently allow it to be competitive with wood-based products.

So far the treatment of thermoset composite recycling has been limited to glass-reinforced unsaturated polyesters, which is essentially the only material system used in a sufficient number of commodity applications to warrant any major concern about waste volumes. However, due to the high value of carbon fibers it may be economically feasible to recycle fibers from not-crosslinked prepreg scrap. Before the epoxy in the prepreg has crosslinked it may be dissolved in industrial solvents and thus completely removed. Retrieved carbon fibers may be chopped to uniform lengths and used in manufacture of nonwovens or as reinforcement in molding compounds. While it is technically possible to extract the reinforcement also from crosslinked epoxy composites through acid digestion (cf. *Section 6.2.1.2*), it is likely not to be economically feasible. An alternative to solvent dissolution is to burn off the not-yet-crosslinked resin (or, for that matter, the crosslinked resin), but such a procedure changes the surface properties of the fibers enough to deteriorate their interfacial properties [9]. Since carbon fibers can be recycled from prepregs, it is naturally also feasible to recycle unimpregnated carbon fabric cutoffs in the same manner. The concept of recycling unimpregnated glass reinforcement scrap as random reinforcement in molding compounds and through remelting into glass raw material (although not for fiber manufacture) has proved technically feasible. On the same note, not-yet-crosslinked resin in SMC scrapped for whatever reason may be dissolved in industrial solvents to retrieve all constituent [6,10].

7.3.1.2 Thermoplastic Composites

Since thermoplastics can in theory be melted and cooled to solidify an infinite number of times, recycling of thermoplastic composites through material recovery should be easier than for thermosets. While this is generally the case, each remelting unfortunately causes the matrix to gradually degrade through chain scission and/or (unintentional) crosslinking (cf. *Section 2.1.2.3*). The tolerance to such degradation varies significantly from polymer to polymer. Just as with thermoset composites, the main focus of thermoplastic composite recycling efforts is on commodity materials and products, in this case GMT and the automobile industry. Several studies have proved the feasibility of introducing GMT regrind in manufacture of GMT blanks. *Figure 7-5* illustrates that a property degeneration similar to that with SMC regrind in SMC is to be expected (cf. *Figure 7-4*), unless fiber content is increased.

A commercially viable option for material recovery of both GMT-based and injection molded thermoplastic composites is to use the shredded material, possibly together with virgin raw material, in injection molding

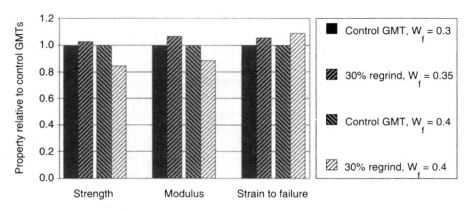

Figure 7-5 Tensile properties of composites compression molded from GMT containing GMT regrind normalized with respect to the properties of composites molded from two virgin "Control GMTs". Data from reference [11]

(this is the most common way of recycling unreinforced industrial scrap from thermoplastic injection molding, such as sprues and runners). An intermediate alternative is to extrude GMT scrap—with or without virgin raw material—and use the extrudate in compression molding. From a general composite recycling point of view it is encouraging to note that while material recovery of thermoset composites is still economically difficult to justify in most cases, the reverse holds true with industrial thermoplastic composite scrap. It is for example not uncommon that compression molders manage to sell their GMT scrap to injection molders.

Thermoplastic composites also offer the possibility of reshaping one component into another. Thus, a scrapped component of moderate curvature may be used as blank in a compression molding operation (possibly with intermediate reconsolidation into a flat blank). Such a procedure may very well amount to true primary recycling without any property degeneration [12]; properties may indeed even be improved due to enhanced impregnation resulting from repeated melting and consolidation [13].

A most relevant concern is what happens to material properties in recycling of material that contains previously recycled material. *Figure 7-6* shows normalized tensile properties of repeatedly recycled injection molded glass-reinforced PP components. The change in properties can be attributed to fiber attrition (the screw in the injection molding machine (cf. *Figure 4-32*) mechanically reduces fiber length) and polymer degradation (as evidenced by discoloration). While such strength and modulus degradation is hardly unexpected, it can be eliminated at least for one recycling cycle through addition of chemical additives that improve the fiber–matrix interface and through addition of virgin glass fibers [10]. While these results also are specific to one material system and one particular study, they are largely representative of the property trends that one should expect in repeated recycling of thermo-plastic composites. In contrast, it appears as if fine thermoset regrind can be reused as filler numerous times without property deterioration since it is essentially chemically inert.

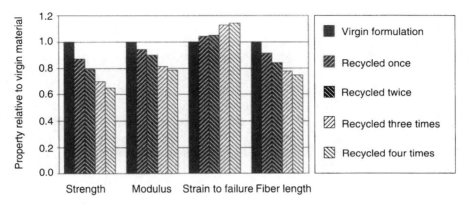

Figure 7-6 Tensile properties and fiber length of composites injection molded from recycled PP reinforced with 30 weight percent short glass fibers normalized with respect to the properties of the (unrecycled) "Virgin formulation". Data from reference [10]

7.3.2 Tertiary Recycling

There are several feasible techniques for tertiary, also called chemical, recy-cling of unreinforced plastics. Among these techniques are hydrolysis, alcoholysis (e.g. methanolysis and glycolysis), hydrogenation, and pyroly-sis. In hydrolysis the ground plastic scrap is exposed to steam at an elevated temperature causing polymer chains to depolymerize into substances of low molecular weight, which can be used in the synthesis of new polymers. In concept methanolysis and glycolysis are similar to hydrolysis, but instead of water the decomposing media are methanol and glycol, respectively. For example, both PET and PUR are amenable to hydrolysis and methanolysis. While hydrolysis of PET merely reverses the polymerization reaction to retrieve the original monomer compounds, most investigated tertiary recy-cling techniques result in low molecular-weight compounds other than those used in the original polymerization [14,15]. Glycolysis of PET may result in compounds useful in the production of unsaturated polyesters and PUR foams [14]. Also unsaturated polyesters can be hydrolyzed [15]. In hydrogenation milled plastic waste is disperged in used oil and subjected to high pressure and temperature. Due to the prevailing hydrogen the polymer molecules are split into gas, oil, and coke [16]. In pyrolysis the shredded material is subjected to high temperatures in the absence of oxygen resulting in controlled thermal decomposition while preserving the energy content of the material. Pyrolysis may be used to reduce any polymer, including contaminated scrap (i.e. commingled plastics containing paints, adhesives, etc.), to chemical compounds of lower molecular weight, but the resulting products are dependent on waste composition and pyrolyzation temperature. There are also efforts underway to develop thermoset resins that have reversible crosslinks allowing them to be completely dissolved and reused following completion of their service life, but such resins are not yet commercially available.

Pyrolysis of unsaturated polyester-based SMC has been successfully evaluated. The process, run at temperatures between 700 and 1,000°C, produces gas (14 weight percent), oil (14 weight percent), and a solid byproduct (72 weight percent) consisting of carbon, filler, and glass fibers. The gas produced has sufficient fuel value to sustain the pyrolysis process once equilibrium has been reached and the oil is suitable stock for fuel blenders. While the mechanical properties of the fibers are degraded by the high temperature, the solid byproduct, once milled to a powder, may be used as filler in the manufacture of SMC, BMC, and thermoplastic molding compounds [17].

While the reinforcement is a desired ingredient in primary and secondary recycling, it is only the organic parts of the composite (normally only the matrix, but potentially also aramid and PE reinforcement) that are useful in tertiary recycling. Indeed, the reinforcement may very well make tertiary recycling more difficult, particularly since the organic content often is as low as 20–30 weight percent in commodity composites (cf. *Table 2-3*). Partially for this reason, but also due to unfavorable process economics, tertiary recycling is currently of little commercial significance in composite recycling.

7.3.3 Quaternary Recycling

Incineration may be used merely as a means to reduce the amount of waste that must go into landfills (volume reductions are up to 90 percent) or the waste may be burned for its calorific value (and, of course, also to eliminate the amount of waste). When the energy content is recovered one talks of quaternary recycling. *Figure 7-7* illustrates that the calorific value of most unreinforced plastics is comparable to conventional fuels, suggesting that incineration of plastic waste clearly is not a waste of resources when pure oil and natural gas are burned for heating purposes (without an intermediate life as plastic). It is also noteworthy that the polymer content in MSW raises the calorific value to such a level (approximately 8.5 MJ/kg) that less or no fuel is needed to sustain incineration [16].

In contrast, incineration of composites generally yields much less energy than unreinforced plastics since only the organic fractions burn, see *Figure 7-8*; the energy value is roughly proportional to the amount of matrix. The

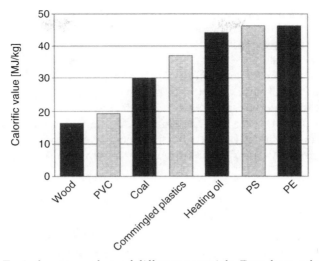

Figure 7-7 Typical energy values of different materials. Data from reference [18]

non-combustible matter, i.e. reinforcement and fillers, in some cases amounts to 70 weight percent of a composite (cf. *Figure 7-8*); these are normally unaffected by incineration and are retrieved as bulky ash. The ash creates a problem in incineration since it may clog up the furnace, but small amounts of composites may be incinerated together with general plastic or municipal waste without problems. The ash resulting from composite incineration has potential as cement filler, but due to chemical conversion of the SMC filler it cannot be used as filler in manufacture of new SMC [17]. With

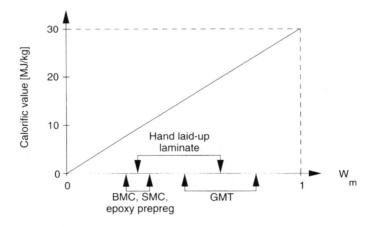

Figure 7-8 Typical energy values of composites as function of matrix weight fraction. Data from reference [19]. Approximate ranges of volume fraction matrix for common material forms indicated for reference

appropriate facilities, plastics and composites can be incinerated without risk of unacceptable air-borne pollution; in fact, they often produce less pollution than conventional fossil fuels, partially due to low or no sulfur content [15,19].

In general, the most sensible use of oil and natural gas probably is to convert it into a polymer, manufacture a molding compound, mold a component, recycle the material once or twice in the primary or secondary sense, and finally recycle it in the quaternary sense. However, quaternary recycling is not politically correct in many countries despite it being technically, economically, and environmentally advantageous if all facts are carefully examined.

7.4 Market Commitment

It is all well and good that it is technically feasible to recycle composites through any of the techniques outlined above, but if no one is interested in buying the recycled material the recycling scheme obviously cannot be economically justified. As discussed above, it is currently essentially impossible to economically justify composite recycling based solely on material value, since so much effort goes into most recycling techniques and virgin raw materials are relatively inexpensive. Some recycling schemes are nevertheless limping along in anticipation of future legal requirements and other incentives that would change process economics in favor of recycled material. Case in point, it has proved to be very difficult to convince molders to use SMC containing recycled instead of virgin filler despite similar pricing and equivalent properties. This general lack of market commitment is partly blamed on worries of possible processing difficulties and partly on worries of lack of material consistency. A seemingly successful Canadian–American SMC recycling venture appears to have failed for lack of such market commitment, despite several SMC components containing recycled SMC having been commercialized.

7.5 Green Thinking in Design

As part of the design process, the designer has the opportunity and indeed the responsibility to take environmental considerations into account. To promote optimum resource utilization, the Commission of the European Union (CEU) has adopted a set of directives that should form the basis of all product design [18]:

1. Prevent and reduce waste at source
2. Increase reuse and recycling
3. Recover energy safely
4. Dispose of unavoidable waste safely

7.5.1 Design for Minimization of Waste

While the title of this chapter would appear to indicate that only the subjects of the CEU directives 2 and 3 are treated herein, it is important that the subjects of directives 1 and 4 are not forgotten in the design process. In order to design for reduced waste, i.e. to reduce the amount of material that will need to be recycled in the future already at the source, the designer should strive to [18]:

- use less raw material through structural optimization
- use less virgin raw material through introduction of recycled material
- select the manufacturing technique that produces the least amount of scrap
- design for equal lifetimes of components within an assembly for applications where component exchange is not feasible (so that when one component fails the others are also near the end of their respective lives)
- design for low weight to save energy in transportation
- design for damage tolerance and repair to extend component lifetime
- design to enhance afterlife potential in terms of both reuse and recycling

In this context it deserves to be pointed out that biodegradable polymers are generally considered to contribute little in terms of reduction of waste, since biodegradation requires oxygen and sunlight and neither is available in landfills. Excavation of landfills have proved that degradation is painstakingly slow; while 20 to 50 percent of organic waste degrades in the first 15 years, newspapers may still be fully legible after several decades in a landfill [2,20]. Nevertheless, biodegradable plastics are most useful in, for example, certain farming applications.

7.5.2 Design for Recycling

While methodologies for design for assembly (DFA) have become commonplace in recent years—not least in the automobile industry—the concept of design for recycling (DFR) is a considerably more recent introduction. To design for recycling can mean several things, but the overall intent is of course to minimize cost and maximize yield in recycling of a product following completion of its service life. Guidelines that may aid in recycling-oriented design of polymer-based materials are discussed in the following.

Design for Disassembly (DFD)

Components of an automobile, for example, should be easily removable and disassembly routines documented. If disassembly is to be carried out through reverse assembly such manuals are more important than in disassembly through brute force. In the latter case possibilities for grasping the component and consideration of where it should break during disassembly must be taken into account in design. (However, one should keep in mind that if it is too easy to remove components, one may invite theft and vandalism.)

Label Components

Components should be clearly labeled in an easy-to-see location for rapid and accurate identification of component contents, see also *Section 7.5.3*.

Use Few Materials

Complex and integrated components in for example an automobile often contain many different materials, such as metals, polymers, paper, etc., which makes cost-effective recycling nearly impossible. To eliminate or at least reduce the cost of material separation of a polymer-based component and thus enhance the chance of economically successful recycling, the monomaterial concept, i.e. use of a single polymer throughout the component, is advantageous. Combining the polymer with reinforcement and filler makes recycling more difficult, but their inclusion is often a necessity and certainly does not prevent recycling. Polymers that are compatible (e.g. PET and polycarbonate, PC) may be recycled together and thus may be used in the same component without impeding recycling. With the monomaterial concept it is generally acceptable to use different versions and grades of the same base polymer within a component. There are several initiatives evaluating the feasibility of producing various automobile components, such as complete bumpers and dashboards which are normally anything but monomaterial components, using one polymer only. In the case of the bumper, the bumper beam (the structural part) may be a combination of glass-reinforced PP GMT and unidirectionally glass-reinforced PP prepregs, an expanded PP foam core, and an elastomer-modified PP fascia. A similar combination of a glass-reinforced PUR carrier, an expanded PUR foam, and a PUR skin is in use in dashboards.

Avoid Painting and Adhesive Joining

Paints and adhesives should be avoided since they are pollutants from a recycling point of view and cannot easily be removed. While, for example, (unreinforced) plastic panels for use as exterior vehicle body panels may have class-A finish and be self-colored, composite body panels unfortunately normally require painting if they are to be visible.

Minimize Number of Metal Fasteners

Although undesirable from a recycling point of view, metal fasteners often cannot be avoided. Material separation may be conceptually straightforward, but adds cost and remaining metal parts may severely damage shredder knives; removal of metal fasteners can be facilitated through thoughtful design.

These DFR guidelines above are in many cases in conflict with structural, aesthetical, and economical requirements, so some kind of compromise always has to be sought. Use of the monomaterial concept often results in use of more material and thus also may end up a more costly and heavy solution, since optimum materials (in structural and aesthetical terms) are not used throughout the component. An example of this is a slalom ski, which through use of a number of materials (e.g. thermoplastics for exterior surfaces; thermoset matrix for laminates reinforced with glass, aramid, carbon, and/or PE fibers; wood or polymer foam core; steel edges) is virtually impossible to recycle. However, a monomaterial slalom ski would be unacceptable from performance and economic points of view.

There is no doubt that an all-metal automobile is a whole lot easier to recycle than one combining metals, plastics, composites, etc. and in recent years some composite components have been replaced in later models in favor of steel or aluminum to simplify recycling. (Similar and somewhat misinformed reasoning has also led to GMT gaining ground at the expense of SMC and BMC, since conventional wisdom incorrectly has it that thermoplastics are recyclable and thermoset are not.) However, if the intention of facilitating recycling is to reduce the overall environmental impact of a component, this kind of reasoning is strongly flawed since composite (and plastic) components reduce vehicle weight and thus fuel consumption. While estimates differ depending on the assumptions made, each kilogram vehicle weight saved on an automobile reduces gasoline consumption on the order of 10 kg over its lifetime; each saved kilogram thus eliminates emissions from 10 kg gasoline and saves the same amount of fuel for use by future generations. Furthermore, each kilogram saved on body and chassis allows half a kilogram to be saved in drivetrain, brakes, suspension, etc., thus further reducing weight and fuel consumption [17].

A life-cycle assessment (LCA) may be used to estimate more rigorously the overall environmental impact of for example an automobile component in a cradle-to-grave perspective. An LCA calculates the impact of raw material extraction and manufacture, component manufacture, service life, and recycling or deposition. Several LCA studies have shown that if one compares the overall environmental impact of a composite automobile component to steel and aluminum alternatives, one finds that the composite often causes less impact over its entire life due to its lower weight. However, if a component is not part of a vehicle, i.e. no savings in fuel consumption and emissions result from reduced weight, the situation is often in favor of conventional materials. In vehicles it is thus not necessarily the component that is the easiest to recycle that causes the least environmental impact.

7.5.3 Component Labeling

One of the prerequisites for successful recycling of any material is knowledge of what material one is attempting to recycle. While there are techniques to recycle commingled plastics, the result of most recycling techniques is drastically improved if only one polymer is present.

Identification of plastic and composite components may be achieved using explicit labeling (if any) or through exploitation of differences in physical properties, such as density (does it float or not?) or static electricity build-up. Commercially available techniques and equipment to identify and separate polymers are available. Several schemes have been introduced to label plastic products to facilitate separation of commingled waste. The universally most common scheme, illustrated to the left in *Figure 7-9* and in *Table 7-1*, is mainly used by the packaging industry and is only intended for a few common polymers. A more all-encompassing and standardized labeling system that covers all common plastics and commercial blends is shown to the right in *Figure 7-9* and in *Table 7-1* [21,22]. There are labeling standards also for composites, but none of them is universally accepted, resulting in different systems being concurrently used, see *Figure 7-10*.

Figure 7-9 The universally accepted labeling system of SPI in the United States to the left and the internationally standardized and more all-encompassing system to the right. Notations for some other polymers given in *Table 7-1*

Table 7-1 Examples of labeling system notations. "—" denotes lack of symbol

Polymer	SPI system symbols		ISO 1043 symbols
PET	1	PETE	PET or PETE
HDPE	2	HDPE	HDPE
PVC	3	V	PVC or V
LDPE	4	LDPE	LDPE
PP	5	PP	PP
PS	6	PS	PS
Other	7	Appropriate abbreviation	—
Unsaturated polyester	—	—	UP
Epoxy	—	—	EP

>UP.GF(30)<   >UP-GF 30<   >GM40PP<   >GM30PP-R<

Figure 7-10 Examples of composite labeling. The two notations to the left both denote glass-fiber reinforced unsaturated polyester with $W_f = 0.30$, the third notation glass-mat reinforced PP with $W_f = 0.40$, and the fourth glass-mat reinforced PP with $W_f = 0.30$ of which at least 10 weight percent is recycled fibers

7.6 Summary

Composite waste is not yet a major problem, but environmental, legal, and economical incentives are likely to change the situation in the near future. While it is currently difficult to economically justify composite recycling, factors such as ever scarcer and more expensive landfill space, increasing environmental awareness, and emerging legal requirements are already creating niche markets. Apart from incentives, successful recycling requires infrastructure, recycling techniques, and market commitment. Lack of infrastructure for waste collection and market commitment to using recycled raw material are currently among the greatest hurdles to composite recycling.

In contrast, a number of technically feasible techniques to recycle all kinds of industrial composite waste have been developed. Secondary recycling is possible with both thermoset and thermoplastic composites and is usually accomplished through grinding of the waste, which is then used as filler or partial replacement of the reinforcement in molding compounds. Some techniques for secondary recycling even approach and in some cases succeed in the ultimately desirable primary recycling. Thermoplastic-based industrial scrap is somewhat more straightforward to recycle in the primary sense than thermoset-based waste, since the matrix can be melted, but there is no doubt that there are few technical obstacles to the recycling of thermosets as well. Tertiary recycling of composite waste has proven technically feasible in trials, but is not commercially practiced to any degree. This lack of acceptance is largely due to the comparatively low organic content of most composites as well as the high inorganic content, which creates processing difficulties. In contrast, tertiary recycling of plastics is carried out. While quaternary recycling is not always politically correct, it is nevertheless a sensible end-of-life solution for commingled and contaminated composite waste. However, also in this case the low organic content is a disadvantage and the high inorganic content a processing complication. Experience has shown that techniques to primarily and secondarily recycle thermoset-based waste are relatively insensitive to mixing of related materials, whereas with thermoplastic composites the waste usually has to be uniform and chemically unaltered to permit primary and secondary recycling.

While it is certainly very important to minimize industrial scrap, the real challenge from both technical and economic perspectives, is to recycle post-consumer composite waste. With the exception of quaternary recycling, few studies have so far seriously tackled this issue. The first difficulty in recycling of post-consumer waste is in the huge logistics problem of collection. If the intent is anything more sophisticated than incineration, the waste must then be identified, separated, cleaned, and dried, before recycling. The present state of the art of composite recycling seems to suggest that secondary and sometimes primary recycling is feasible with industrial composite scrap of all types. Although there are some technically feasible secondary and tertiary recycling techniques for post-consumer composite waste, the best solution for the time being is likely to be quaternary recycling. The challenge for the future is to make the aforementioned technically feasible recycling techniques economically attractive.

Through thoughtful component design, the designer has both opportunity and responsibility to greatly reduce future recycling tasks. First, the inherent need for recycling may be minimized through use of less material. Second, what recycling eventually is necessary can be facilitated through design for disassembly, component labeling, and use of few materials, as well as through minimization of paints, adhesives, and metal fasteners.

Suggested Further Reading

Comprehensive treatments of plastic recycling include:

Plastics Recycling: Products and Processes, Ed. R. J. Ehrig, Hanser, Munich, Germany, 1992.

Recycling von Kunststoffen, Eds. G. Menges, W. Michaeli, and M. Bittner, Hanser, Munich, Germany, 1992.

W. Michaeli, M. Bittner, and L. Wolters, *Stoffliches Kunststoff-Recycling*, Hanser, Munich, Germany, 1993.

A great deal of recent work on plastic recycling has been carried out in Germany, which explains why much of it is published in German. Published work on recycling of composites is limited and scattered throughout conference proceedings and trade journals. Annual conferences where work on composite recycling has been presented in the past include:

SPI Composites Institute Annual Conference, USA

Internationale AVK-Tagung, Germany (organized by Arbeitsgemeinshaft Verstärkte Kunststoffe, Frankfurt, Germany)

References

1. R. J. Ehrig and M. J. Curry, "Introduction and History," in *Plastics Recycling: Products and Processes*, Ed. R. J. Ehrig, Hanser, Munich, Germany, 1–16, 1992.

2. W. L. Rathje, "Once and Future Landfills," *National Geographic*, **179**(5), 116–134, May 1991.

3. W. Freiesleben, "Role and Importance of Material Recycling in Plastic Waste Management During the Next Decade in Western Europe," in *Recycling von Kunststoffen*, Eds. G. Menges, W. Michaeli, and M. Bittner, Hanser, Munich, Germany, 464–471, 1992.

4. P. Stachel, "Entsorgung von GFK-Teilen in der Automobil- und Elektroindustrie," Technische Vereinigung, Internationale 6. Duroplasttagung, **6**, 137–147, 1992.

5. J. McDermott, "Fiberglass Recycling Earning Respect," *Composites Fabrication*, **10**(5), 9–13, May 1996.

6. J. C. Bradley, W. D. Graham, and R. Forster, "Solvent Separation: A Method for Recycling of Uncured SMC," SPI Composites Institute Annual Conference, **49**, 15-C, 1994.

7. "BMW Recycling", BMW AG, 1994.

8. J. Schiebisch and G. W. Ehrenstein, "Effect of Adding Particle and Fiber Recyclate to SMC on Mechanical Properties and Structure," SPI Composites Institute Annual Conference, **49**, 5-D, 1994.

9. H. Richter and J. Brandt, "Recycling Carbon Fiber Scrap," in *Engineered Materials Handbook, Volume 1, Composites*, ASM International, Metals Park, OH, USA, 153–156, 1987.

10. W. D. Graham, R. B. Jutte, and D. L. Shipp, "Recyclability of Glass Fiber Reinforcements in Thermoset and Thermoplastic Applications," Technical Paper, European Owens-Corning Fiberglas, Brussels, Belgium, 1993.

11. "GMT Recycling," Symalit, Lenzburg, Switzerland.

12. G. Schinner, J. Brandt, and H. Richter, "Recycling Carbon-Fiber-Reinforced Thermoplastic Composites," *Journal of Thermoplastic Composite Materials*, **9**, 239–245, 1996.

13. H. Hamada, S. Yamaguchi, G. O. Shonaike, T. Kimura, and Z. Maekawa, "Recyclability of Long Glass Mat Reinforced Thermoplastic Composite," SPI Composites Institute Annual Conference, **49**, 15-B, 1994.

14. J. Milgrom, "Polyethylene Terephtalate (PET)," in *Plastics Recycling: Products and Processes*, Ed. R. J. Ehrig, Hanser, Munich, Germany, 45–72, 1992.

15. W. J. Farrissey, "Thermosets," in *Plastics Recycling: Products and Processes*, Ed. R. J. Ehrig, Hanser, Munich, Germany, 231-262, 1992.

16. H. Emminger and G. Menges, "Recycling, eine Existenzfrage für Kunststoff," in *Recycling von Kunststoffen*, Eds. G. Menges, W. Michaeli, and M. Bittner, Hanser, Munich, Germany, 51–73, 1992.

17. "Recycling of SMC—The Energy/Environment Picture," SMC Automotive Alliance, Bloomfield Hills, MI, USA, 1992.

18. "Designing Green—A Guide," Dow Europe, Horgen, Switzerland.

19. S. J. Pickering, "New Recycling Technology for Processing Scrap Composites," Verbundwerk '92, **4**, M3, 1992.

20. E. G. Wogrolly, "Deponie," in *Recycling von Kunststoffen*, Eds. G. Menges, W. Michaeli, and M. Bittner, Hanser, Munich, Germany, 158–180, 1992.

21. ISO 1043, "Plastics—Symbols".

22. ASTM D 1972, "Standard Practice for Generic Marking of Plastic Products".

CHAPTER 8

HEALTH AND SAFETY

In addition to more general concerns shared with all industrial environments, manufacturing of polymer composites brings on health and safety concerns mainly due to the organic materials involved. The intention of this chapter is to raise the reader's awareness of potential dangers, since a prerequisite for a safe work environment is knowledge of the potential dangers involved so they can be minimized. It is important to realize that with proper routines and facilities, the dangers for the operator can be kept within limits that are no higher than in any normal industrial environment.

8.1 Terminology

The toxicity of a chemical is an inherent property that indicates the harmful effect it may have on the body. Whether a chemical causes a hazard or not depends on whether exposure is significant enough to cause a harmful effect. Therefore, minor exposure to a very toxic substance need not necessarily pose a greater hazard than extensive exposure to a slightly toxic substance. A risk, finally, is the probability that a hazard results in harmful effect. While discussing health and safety issues it is important to remember the differences between toxicity, hazard, and risk [1].

Acute toxicity is the ability of a substance to cause harmful effect following a single or short-term exposure. Acutely toxic substances, which cause localized damage, may be classified as irritating or corrosive. Irritating substances may or may not cause cell death, whereas corrosive materials destroy cells. A chronically toxic substance may cause harmful effect from repeated or prolonged exposure (occurring over a significant portion of the life of an individual) or from long-term effects following a single or short-term exposure. Some chronically toxic substances may cause cancer (carcinogenicity) or changes in cells' genetic material (mutagenicity) [1].

Sensitization is an allergic reaction that may develop from repeated exposure to a substance, although a single exposure in rare cases may suffice.

Even exposure to non-irritating concentrations of a substance may over time cause sensitization. Sensitization is not lost with time. Once sensitized, very low concentrations may be sufficient to cause an allergic reaction and the extent of the reaction need not correlate to the level of exposure. Exposure to one substance may also lead to sensitization towards other substances; this is referred to as cross-sensitization [1].

Legally acceptable exposure levels for toxic substances vary greatly between countries and in many cases there may not even be any applicable legislation. The permissible exposure level (PEL), or occupational exposure level (OEL), may be defined in several different ways. The time-weighted average (TWA) refers to an eight-hour working day and a forty-hour working week. The short-term exposure limit (STEL) denotes the acceptable exposure level for a short (and specified) period of time, while the ceiling PEL (PEL-C) is the concentration that never should be exceeded under any circumstances. For volatile materials, PELs are generally expressed in (volume) parts per million (ppm), i.e. 1 ppm equals 0.0001 volume percent. PELs for solid particulates are expressed as amount of solids per air volume; units vary widely and include mg/m^3 and number of fibers per cm^3 (f/cc). The significance of PELs is that no seriously harmful effect is believed to occur due to exposure at or below the limit PELs. PELs have been determined for normal and healthy individuals, but the tolerance to a substance is individual, meaning that some persons may not be able to tolerate exposure even below the PEL.

8.2 Routes of Exposure and Harmful Effects

Exposure to a toxic substance may take place through:

- inhalation
- direct skin or eye contact (absorption)
- ingestion
- injection

Exposure through inhalation is generally relevant at all times when dealing with volatile organic substances as well as during machining of composites. Exposed skin and unprotected eyes are at risk of absorption at any time that toxic substances are handled. Accidental ingestion is for example possible if eating or drinking in manufacturing areas. Injection may take place if impregnated fibers or fiber yarns protrude along the edge of a crosslinked component; such yarns are easily stiff and sharp enough to penetrate skin. By far the most common exposure route is inhalation followed by direct contact. Although possible, ingestion and injection are rather uncommon and therefore are not further discussed herein.

The following paragraphs outline potential effects of exposure to materials commonly used in the composite industry. Many of these effects, particularly the graver ones, are the *possible* results of lifelong exposure, often at very high concentrations. Thus, while hazards in composite manufacturing should not be marginalized, they should not be exaggerated either. The symptoms common to specific materials are further discussed in *Section 8.4*.

The acute symptoms of inhaled toxic organic substances are irritated mucous membranes of the respiratory tract and possibly coughs, as well as central nervous system (CNS) depression. CNS depression manifests itself as headache, nausea, dizziness, fatigue, increased reaction time, impaired coordination and balance, etc. Long-term inhalation of toxic organic substances may lead to sensitization, normally with asthma-like symptoms, and in some cases to chronic CNS depression and chromosome changes. Inhalation of airborne composite particles may be acutely irritating, but generally is not (with commonly used material systems) considered harmful nor as having any long-term effects as long as the matrix is fully crosslinked.

Direct skin contact with toxic organic substances may lead to dermatitis, or skin inflammation, with acute symptoms such as skin becoming red, swollen, tender, and itchy; in severe cases blistering and weeping may occur. Most solvents degrease, or defat, skin, i.e. remove the natural body oils on the skin, leaving it vulnerable to absorption not only to the solvent, but also to substances dissolved within it. Manual handling of reinforcement may lead to irritation from mechanical abrasion; following such irritation the skin has reduced tolerance to absorption of for example toxic substances. Direct skin contact with both organic substances and reinforcement may result in sensitization, commonly in the form of dermatitis. Direct eye exposure to organic substances may lead to acute irritation and in some cases permanent damage.

In addition to direct effects on the body, the possibility of fire and explosion must not be forgotten. All organic compounds are flammable to some degree and most solvents and reactive dilutents are highly flammable. Since solvents and reactive dilutents are also very volatile, the risk of explosion of solvent vapor may be significant.

8.3 Protective Measures

Most exposure possibilities may be eliminated or reduced with appropriate administrative controls, engineering controls, and use of personal protective equipment (PPE). While PPE may be quite effective in reducing exposure, the most effective approach is to first minimize the inherent need for PPE in the workplace. In order of importance, one should therefore strive to reduce the use of toxic substances, reduce emissions, contain remaining emissions, and finally minimize operator exposure. PPE should only be used as a last resort when all other measures to limit exposure have been exhausted.

8.3.1 Administrative Controls

The first step in reducing hazards and risks is knowledge of the potential dangers involved with each specific material system and manufacturing technique. While this chapter is a general introduction to health and safety issues, it should under no circumstances replace the much more detailed and material-specific information in the material safety data sheets (MSDS) available from raw material suppliers. Amongst the information to be found in an MSDS is physical properties, reactivity, toxicity and other hazards, PELs, appropriate protection, spill and disposal routines, fire-fighting measures, etc. It is the responsibility of the employer to ensure that personnel receives training in handling of toxic materials, that training is fully understood, and that safety routines are followed. MSDSs must naturally be reviewed and personnel given proper training *prior to* handling the material in question.

If use of a specific toxic substance cannot be avoided, exposure may be reduced through administrative routines such as:

- Isolation of operations, e.g.:
  - only allowing toxic materials in dedicated areas and eating and drinking in others
  - physically isolating welding, ignition sources, and hot surfaces (ovens, boilers, etc.) from areas where solvent vapor may be present and where (dust-creating) composite machining is carried out
  - not allowing smoking where solvents are used
- Limiting the number of workers present during tasks that result in significant emissions
- Scheduling worker rotation between tasks
- Scheduling tasks that result in significant emissions to take place at the end of the workday
- Promoting personal hygiene, e.g. washing of hands before eating, going to the bathroom, and at the end of the workday
- Labeling of containers, including clearly visible hazard labels
- Continuous cleaning of the workplace, including immediate cleanup of spills and vacuuming of dust (rather than sweeping or blowing, which merely spreads dust)
- Keeping discarded but soiled tools, resin containers, cleaning rags, etc. in closed containers or under local ventilation
- Preparation of and training in emergency procedures should an accident occur
- Undertaking regular maintenance of equipment
- Enforcing established routines and procedures for material handling, cleaning, manufacture, use of PPE, etc.

The list above is not exhaustive and merely provides examples of possible administrative controls to reduce exposure. It also deserves to be pointed out that many sensible and exposure-reducing routines are specific to a manufacturing technique; some of these are discussed in *Section 8.5*.

8.3.2 Engineering Controls

Engineering controls to reduce exposure may be implemented in areas such as resin formulation, process control, automation, and ventilation. In terms of resin formulation it is for example possible to limit evaporation through faster crosslinking and reduction in crosslinking temperature (though these desires are unfortunately generally in conflict). Resin-specific means to reduce exposure through changes in resin formulation are discussed in *Section 8.4.1*.

In terms of worker safety, process control is often synonymous with temperature management. Dangers may arise from excessive temperatures during crosslinking, which apart from increased evaporation may lead to thermal decomposition of the matrix into potentially harmful airborne substances. While such excessive temperatures may be caused by overheating, another very real danger is heat from the crosslinking exotherm; both overheating and out-of-control exotherm may be limited or eliminated with closed-loop temperature control. In most manufacturing techniques molds are heated creating a burning hazard, which can be minimized through use of insulation.

Process automation of most sorts lead to decreased or even eliminated worker exposure, since workers then can be physically separated from toxic substances. Examples include enclosure of laminating, painting, or machining robots in individually ventilated rooms which humans only rarely enter, but also manufacturing tehniques such as filament winding and pultrusion.

Arguably the most effective of all engineering controls is ventilation, which may be accomplished on several different levels, including dilution ventilation, booth ventilation, and local ventilation. Dilution ventilation refers to continuous replacement of the contaminated air of the entire workplace with fresh air, see *Figure 8-1*. As illustrated in the figure, air is sucked out through one end of the workplace, while replacement air is fed in through another. To be effective, the air flow must be evenly distributed, i.e. sucked out through an entire wall and fed in through the entire opposite wall (or, alternatively, from ceiling to floor). The air velocity must be low to ensure that volatile evaporation is not enhanced and settled dust does not become airborne again. Dilution ventilation is quite limited in its effectiveness, partially because it is difficult to achieve an even air flow throughout a workplace, thus creating still spots as well as areas with too high air velocity, and partially because it easily spreads contaminated air to parts of the workplace that would not otherwise become contaminated. A means of limiting the spread of contaminated air to other rooms of a building than those where it is generated is to keep these rooms under slight negative pressure. Despite its inadequacies, dilution ventilation is widely used, albeit with limited success.

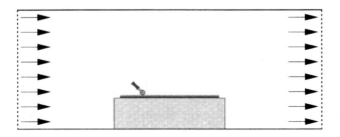

Figure 8-1 Schematic of dilution ventilation

A variant of dilution ventilation is booth ventilation, which is common with spray-up. In this case it is mainly the booth, i.e. a partially enclosed section of a room, that is ventilated and the intention is to suck the contaminated air away from the operator towards the opposite wall where it is sucked out, see *Figure 8-2*. Adequate results require that the operator stands in a narrow opening on the inlet side of the booth so as to create a locally high flow rate around himself. This may be achieved with heavy drapes that can be moved by the operator (or automatically) as he moves along the mold. If drapes (or some other means of locally increasing the air flow around the operator) are not used, booth ventilation loses most of its effectiveness and worker exposure increases greatly [2].

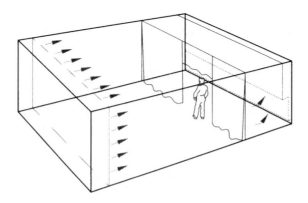

Figure 8-2 Schematic of booth ventilation. Redrawn from reference [2]

Local ventilation is the most effective form of ventilation, since it removes contaminated air at the source. Since much smaller air volumes get contaminated and thus must be exhausted, local ventilation is much more effective than dilution ventilation. While local ventilation is characterized by low flow rate and high velocity, dilution and booth ventilation in contrast involve high flow rates and relatively low velocities. A given ventilation capacity (volume flow rate) is therefore much more effectively used for local ventilation. Local ventilation is generally in the form of fume hoods and hoses, also known as (elephant) trunks which are manually adjusted over a work or impregnation area, see *Figure 8-3*.

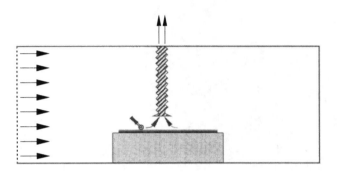

Figure 8-3 Schematic of local ventilation with adjustable exhaust trunk

Local ventilation is often permanently installed over a frequently used work area, see *Figure 8-4*. Another quite effective version of local ventilation is push-pull ventilation, where air gently sweeps the laminate surface away from the operator towards the extraction device, see *Figure 8-5* [3].

For a local ventilation system to function efficiently, several requirements must be met [2]:

- Extraction devices must be located near the generating source (not crosslinked resin, grinding disk, etc.)
- Efficiency is improved if the end of the trunk is rounded or funnel-shaped
- If it is not possible to place extraction devices right above the surface of a laminate, it is better to place extraction slots around the workpiece rather than at floor level
- Flow rate must be relatively high, while velocity is of secondary importance
- Efficiency is improved if a given ventilation capacity is divided onto several trunks rather than one or a few trunks

Figure 8-4 Local ventilation permanently installed over frequently used work bench

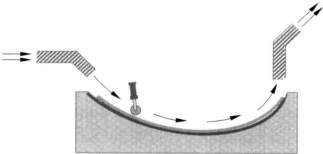

Figure 8-5 Schematic of push-pull ventilation

- A given ventilation capacity should be used for local ventilation at the expense of dilution ventilation
- Efficiency is improved if replacement air is diffuse (cf. *Figures 8-1* to *8-3*)
- Extraction devices must not restrict work, or they will not be used as intended
- The operator's head should never be in between workpiece and extraction device

- The operator should be made aware that he can improve his own work environment through proper use of extraction devices

Since composite machining easily creates intolerable amounts of airborne dust, much work has been invested in on-tool extraction, as illustrated in *Figure 8-6*; also the tools in *Figures 5-9* and *5-10* are equipped with on-tool extraction. To be truly effective such ventilated machines must have custom-fit manifolds and should be hooked up to a high-vacuum dust-collecting system, as illustrated in *Figure 8-7*. In such a system, a vacuum pump creates a negative pressure significantly lower than can be created with a fan in a normal ventilation system, while a dust collector removes dust from the extracted air. Due to the powerful vacuum created by such a system, ventilation efficiency is extremely good and no respiratory protection is therefore necessary. The powerful vacuum also allows use of small-diameter extraction hoses that do not restrict work. Less powerful vacuum requires larger diameter hoses to give the same flow rate and since such a system becomes bulkier, there is a risk that it will not be used. Dust-extraction systems are generally permanently installed, but mobile units (high-powered "vacuum cleaners") are available as well.

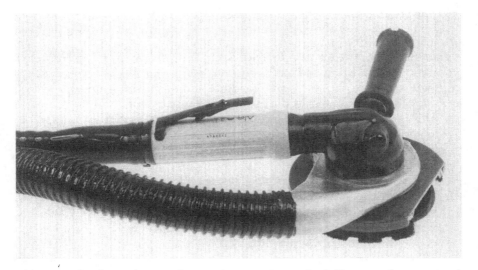

Figure 8-6 Sander with on-tool extraction system attached. Photograph courtesy of Nederman High Vacuum Systems, Sweden

All types of ventilation systems, but particularly local ventilation, are sensitive to disturbances and locally high air velocities, for example from open windows and doors. The main reason for this lies in the fact that disturbances caused by a pushing air flow reach much further than those of a

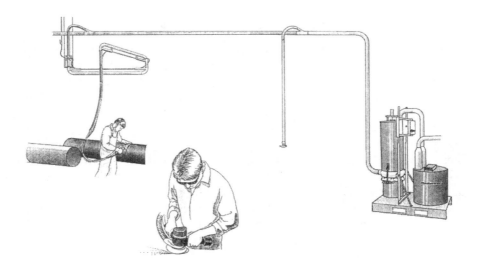

Figure 8-7 Schematic of on-tool ventilation system for dust extraction. Figure courtesy of Nederman High Vacuum Systems, Sweden

pulling air flow. Already at a distance equal to the trunk diameter, the velocity in a pulling air flow is less than 10 percent of that at the opening of the trunk. In contrast, with the same flow rate in a pushing air flow, the velocity is still 10 percent of that at the opening at a distance of 30 outlet diameters [2]. This is the reason why replacement air should be as diffuse as possible. While replacement air may be supplied through perforated metal sheets, it is generally more effective to pass it through fabric or felt to achieve a diffuse air flow.

Although requirements vary widely depending on numerous variables, air flow rates may have to be several thousand m³/h to achieve PELs in for example wet layup processes. The need for such flow rates may translate into a significant cost, since replacement air must generally be controlled in terms of temperature and humidity. While it is possible to recycle extracted air following cleaning, it is more common to use heat exchangers where contaminated air and replacement air never come into direct contact.

The contaminated air that is exhausted from the workplace sooner or later must end up in the atmosphere. While it is technically possible to clean the air, the current reality is often that no cleaning is carried out except by larger industries, which are more likely targets of government watchdogs. This situation is bound to change as legislators show increasing interest in industrial emissions of all kinds and economically feasible alternatives become available. Filtration of air containing volatiles may be accomplished through incineration with energy recovery (quaternary recycling) if concentrations are high enough and through adsorption and biofiltration when concentrations are low to intermediate. In biofiltration bacteria oxidize volatiles into carbonic gas and water vapor [3].

All ventilation systems have in common the need for professional installation and regular maintenance. An incorrectly designed or inadequately maintained ventilation system, particularly a push-pull system, can do more harm than good. Filtration installations similarly require regular maintenance and exchange or cleaning of filters. It is nevertheless not uncommon that maintenance of ventilation and filtration systems is carried out far too seldom or not at all and the system consequently does not deliver the expected results.

8.3.3 Personal Protective Equipment

While PPE can be very effective in reducing exposure of all kinds, it should always be used as a last resort or a temporary solution. The correct approach is to reduce or preferably completely eliminate the inherent need for PPE through administrative and engineering controls. After all such measures that are practically feasible have been implemented, a need may nevertheless still remain for use of PPE to further reduce exposure or merely as an added safety feature.

Respirators may be used to reduce inhalation exposure to vapors or dust. The most common respirators merely purify inhaled air and must be chosen for the specific vapor or particle size in question; a nonwoven dust mask, for example, is useless against solvent vapor. Respirators, or separate filter cartridges in multiple-use respirators, should be replaced periodically, since they must be clean and unclogged to provide adequate protection. Air-supplying respirators in principle work the same way as scuba (diving) equipment in that air is supplied from a totally separate and uncontaminated source and not from the immediate environment.

The most obvious personal protection for bare skin is gloves and clothes to eliminate direct skin contact. When gloves are used as barriers to organic substances, it is important that the right kind of glove is used to ensure that it is not destroyed by the substance. Sometimes it may be acceptable to use a skin-protecting barrier cream instead of (or together with) gloves. Particularly when handling prepregs it is of paramount importance that the barrier cream or whatever substance the gloves may be coated with does not interfere with ply consolidation; indeed, a barrier cream may very well have the same effect as a mold release, thus possibly ruining the component. Whether the risk of manually handling an organic material is considered high or not, one should always wear gloves, since the toxicity of all materials used in composite applications is probably not yet known. In many composite manufacturing-related activities a need for thermal protection of hands arises, e.g. when placing charges in heated molds or demolding of still hot components. Gloves intended to provide thermal protection are by their very nature bulky and are often made from aramid fabric (instead of asbestos, which formerly was used). Another alternative is offered by welder's gloves. While gloves for thermal protection may be used many times, latex-type gloves used to provide protection against toxic substances should be discarded after each use, since it is virtually impossible to put a glove back on without touching its possibly contaminated outside surface with the other hand.

Safety glasses protect against impact, but not chemical splash, which requires special goggles or safety glasses with face shield. Due to the possibly catastrophic consequences of an eye injury, it is good common sense to always wear eye protection in industrial environments. Contact lenses should not be worn if there is a potential for splash from corrosive substances. The reason is that if the substance penetrates between lens and eye it will have ample time to do damage to the eye before the lens is taken out to properly rinse the eye and remove the substance.

One reason why PPE, with the exception of clothing, eye protection, and the often inevitable glove, is not a good long-term solution is that wearing protection often is a nuisance and there is consequently a significant risk that it will not be used (just as with some forms of local ventilation). It is therefore most important that the inevitable discomfort is minimized. Respirators thus must allow unrestricted breathing without causing sweating, gloves must not restrict dexterity and must allow feel, while glasses and goggles must not restrict vision or fog up.

8.3.4 First Aid

Despite rigorous administrative and engineering controls as well as use of PPE, accidents inevitably happen. It is for this reason very important that safety procedures are periodically reviewed and that safety equipment is readily available and functioning. Apart from normal first-aid kits, access to eye wash and showers should be provided in the workplace; in many countries these are legal requirements. First-aid procedures for a specific material are given in the MSDS; these procedures should be reviewed prior to handling the material.

While by no means intended to provide complete first-aid instructions, the following procedures may serve as general guidance should no other and more material-specific instructions be available:

- *Inhalation of toxic vapors* Remove individual from source or location of high vapor concentration and provide clean air. Give artificial respiration and oxygen if necessary.
- *Skin contact with toxic material* Remove contaminated clothing. Carefully clean skin with soap and water. Do *not* use solvent.

- *Skin contact with molten thermoplastic* A molten thermoplastic will adhere to skin and since its heat transfer properties are poor it remains hot for quite some time. Despite this, molten thermoplastic should *not* be removed from the skin since severe tissue damage may result. The appropriate approach is to cool the thermoplastic under running water.
- *Eye contact with toxic material* Rinse eye with copious amounts of running water for 10–15 minutes.
- *Ingestion of toxic material* Rinse mouth thoroughly and drink lots of liquids, but do *not* induce vomiting.

In all cases except the most trivial, medical assistance should be sought immediately.

8.4 Constituent Materials

A fully crosslinked composite is basically inert and considered harmless, whereas the constituents on their own in some cases are not. The following sections discuss the most common constituents with reference to the terminology defined in *Sections 8.1* and *8.2*. The toxicities and hazards discussed herein do not constitute a complete listing and are only intended as guidance; MSDSs provide detailed and material-specific information.

Some PELs are stated in the following but it must be understood that such data inevitable become outdated, since for many substances the trends are that PELs are gradually lowered as knowledge and awareness of the ill effects of these substances increase. However, since it is unlikely that all toxic effects of all materials used in composite manufacturing are known, it is good practice to always try to minimize exposure.

8.4.1 Matrices, Adhesives, and Solvents

Most of the potentially serious hazards in composite manufacturing and related activities are brought on by the toxicity of the organic constituents and in particular unreacted thermoset resins, reactive dilutents, crosslinking additives, adhesives, and solvents.

8.4.1.1 Thermoplastics

One of the major advantages of thermoplastic matrices over their thermoset brethren is that no active chemistry is involved in composite manufacturing. The only matrix-specific danger during normal processing of thermoplastics is associated with the high temperature of melt and manufacturing equipment. However, if a thermoplastic is erroneously heated to a temperature in excess of the intended processing temperature, it may decompose into airborne and potentially toxic substances and an inhalation hazard may arise. Most thermoplastics, with the odd exception such as PPS, decompose into relatively harmless substances that at worst are temporary irritants to eyes and respiratory tract. The above applies to the pure polymer, but all poly-mers contain additives (cf. *Section 2.1.8*) that potentially may decompose into toxic substances. Adequate local ventilation should therefore be avail-able also where thermoplastics are processed.

The aforementioned minor health issues associated with thermoplastics only apply to fully polymerized thermoplastics. As briefly discussed in *Section 2.1.5*, thermoplastic composites may be manufactured using a low molecular-weight prepolymer, which following reinforcement impregna-tion is fully polymerized. With this approach, which includes reactive and possibly toxic substances, there are substance-specific hazards akin to those discussed for thermosets in following sections.

8.4.1.2 Unsaturated Polyesters and Vinylesters

Health concerns with unsaturated polyesters and vinylesters are generally considered synonymous with the most common crosslinking agent, i.e. styrene, and not with the polymers themselves. Since styrene is very volatile it easily evaporates and becomes an inhalation hazard. The reported levels that cause a specific acute reaction vary widely, partly because tolerance is individual and built up and partly because reactions are subjective. At con-centrations in the range 20–100 ppm, styrene is a mild, temporary irritant to eyes and respiratory tract. Above 200 ppm styrene is a definite irritant caus-ing CNS depression, and above 500 ppm it is a severe irritant [1,3,4]. The risk of acute styrene poisoning through inhalation is quite low since the human nose is extremely sensitive to the very characteristic styrene smell; the odor threshold is typically 0.1 ppm [3]. Styrene is said to have excellent warning properties, since the odor threshold is orders of magnitude below PELs. *Table 8-1* gives PELs for styrene in a range of countries and this illus-trates the ambiguity in what styrene concentrations that are considered critical. Long-term occupational exposure to styrene increases the frequency of chromosome changes in one type of blood cells (lymphocytes) and may possibly also cause brain damage at concentrations as low as 10 ppm, but sensitization is uncommon [4]. There is no evidence to prove that styrene is a carcinogen, but since there is insufficient information to draw a definite conclusion the International Agency for Research on Cancer (IARC) classi-fies styrene as a possible carcinogen [1,4,5].

A range of resin-specific solutions have been tried to limit styrene emis-sions. An obvious solution is to reduce the styrene content in the resin so there is less that can evaporate. A conventional unsaturated polyester resin contains 40–50 weight percent styrene, but if the molecular weight of the polyester molecules is reduced the same viscosity may be achieved with styrene contents as low as 25 weight percent without significantly affecting the properties of the crosslinked resin. Another possibility is offered by the LSE resins mentioned in *Section 2.1.6.1*, which contain a substance that migrates to the surface of the resin to create a thin film impenetrable to

Table 8-1 PELs for styrene in different countries [3,5]

Country	TWA ppm	STEL ppm
Australia	50	100 (15 min)
Austria	40	80 (30 min)
Belgium	50	100 (15 min)
Canada	50[a]	100[a] (15 min)
Denmark	25[b]	40 (30 min)
Finland	20	100 (15 min)
France	50	—
Germany	20	40 (30 min)
Israel	50	100 (15 min)
Italy	50	100 (15 min)
Japan	50	100 (15 min)
Luxembourg	20	40 (30 min)
Netherlands	25	—
Norway	25	37.5 (15 min)
South Africa	50	—
Spain	50	100 (15 min)
Sweden	20[c]	50 (15 min)
Switzerland	50	100 (4 × 10 min)
UK	100[d]	250 (10 min)
USA	50[e]	100 (15 min)

[a] Regulations are in effect for each province
[b] PEL-C
[c] 10 ppm for new installations
[d] Maximum exposure limit, obligation to reduce as low as possible
[e] Average limit, varying from state to state

styrene. LSE resins do not reduce evaporation during spray-up, layup, and rolling, since it takes some time for the film to form, but during crosslinking evaporation may be reduced by more than 50 percent. Both reduction of styrene content and use of LSE resins are effective means of reducing the styrene concentration in the air of the workplace.

Styrene is a mild to severe irritant to both skin and eyes upon direct contact. In terms of PPE it is important to realize that no glove material is good for long-term exposure to styrene.

Since styrene is highly flammable, high vapor concentrations may cause explosions. Styrene vapor has a higher density than air and there is a common misconception that styrene flows along the floor. In theory this is true, but the density difference between uncontaminated and styrene-containing air is minute, particularly at low to moderate concentrations, and temperature differences and normal air movements are much more dominant and thus obscure any settling tendencies [2,3].

While styrene is generally considered the main health concern with unsaturated polyesters and vinylesters, the organic peroxide initiators used (cf. *Section 2.1.6.1*) are toxic and may be severe irritants and sensitizers to skin and eyes and may in high concentrations be corrosive [4]. Organic peroxides are also highly flammable and may decompose with explosive violence if not handled correctly. Such a situation may occur if initiator and accelerator are directly mixed. It is therefore absolutely imperative that initiator *or* accelerator is first thoroughly mixed with the resin before the other component is added.

8.4.1.3 Epoxies

With epoxies several substances are of relevance from a toxicological point of view. The high molecular-weight epoxy resin is first mixed with a low molecular-weight reactive epoxy dilutent to achieve low viscosity. Prior to processing, the resin is mixed with a low molecular-weight hardener and in some cases also with an accelerator.

While different types of unsaturated polyester and vinylester resins have very similar toxicities, the opposite holds true for epoxies since so many resin types and crosslinking possibilities are available. Also in contrast to unsaturated polyesters and vinylesters, epoxy systems are with the exception of some hardeners, generally not particularly toxic through inhalation. Most hazards with epoxies stem from direct contact with resin constituents, which may lead to dermatitis and sensitization.

As a rule of thumb it can be said that the toxicity of an epoxy decreases with increasing molecular weight. Reactive dilutents, which are low molecular-weight epoxies, are thus often highly toxic, whereas the polymer itself generally has a low level of toxicity. Taking as an example the common epoxy DGEBPA (cf. *Section 2.1.6.2*) one finds that it has a low order of toxicity, is slightly to moderately irritating, and is a possible sensitizer. Due to lack of sufficient evidence, the IARC classifies DGEBPA as a possible carcinogen, although most evidence appears to suggest the opposite. However, DGEBPA is not necessarily representative of epoxies, since some epoxies have higher levels of toxicity [1].

Most hardeners have high levels of toxicity and are severe irritants and sensitizers through both direct contact and inhalation. Some hardeners are severe corrosives, while others are suspected of causing damage to liver and possibly kidneys following repeated exposure; several hardeners are suspected carcinogens. Hardeners should consequently be treated with utmost care following strict safety procedures. Not surprisingly, suggested handling procedures and safety measures tend to be the strictest for the hardener, but it is good practice to apply the same safeguards to the resin. Hardeners with reduced toxicity are under development [1,4]. Epoxy systems have poor or no warning properties against inhalation exposure.

Dust from incompletely crosslinked epoxies can cause reactions in already sensitized individuals. Epoxy dust or vapor that is heated to a temperature in excess of 300°C can produce skin damage which may be permanent [4].

8.4.1.4 Other Thermosets

Phenolics and PURs, both of which contain toxic substances, are discussed in the following paragraphs, while the toxicities of most other thermosets used as composite matrices, including imide-based resins such as BMI and PI, have not been thoroughly investigated [1].

The main health concern with phenolics is traces of formaldehyde in the not-crosslinked resin. Formaldehyde is a colorless gas with a distinct smell and is an irritant through all exposure routes; it is also a strong sensitizer and a suspected carcinogen. Exposure to not-yet-crosslinked phenolics should consequently be avoided. Formaldehyde begins to irritate at concentrations of 0.5–2 ppm and causes tearing of eyes at 3–5 ppm. Another concern is traces of phenol, which is also an irritant. With an odor threshold of 0.3–3 ppm phenol has good warning properties, since TWAs tend to be an order of magnitude higher [1,4].

PURs are the result of a reaction between an isocyanate and a polyol, where the latter has a low level of toxicity. In contrast, most isocyanates and particularly the two most common ones, toluene diisocyanate (TDI) and methylene diphenyl diisocyanate (MDI), are highly toxic and irritating; TDI is a suspected carcinogen. The main danger with isocyanates is inhalation and very low concentrations may cause dryness of throat, tightness of chest, and headache. Repeated and sometimes even a single exposure may lead to sensitization and extremely low concentrations may cause reactions following sensitization. TDI may give rise to cross-sensitization where individuals not only become sensitive to extremely low concentrations of TDI, but also to smells such as perfume, cigarette smoke, automobile exhaust, cold air, etc. Poor warning properties and delayed symptoms make isocyanates particularly deceptive. Although both TDI and MDI have characteristic smells their odor thresholds, 0.1–0.4 ppm and 0.4 ppm respectively, are much higher than concentrations that may cause sensitization. Symptoms from inhalation of isocyanates may in some cases be delayed for several hours, thus easily being mistaken for reaction to something else. Isocyanates are also irritants to skin and may cause permanent eye damage upon direct contact. Isocyanates should clearly be handled with great care using PPE [1,4].

8.4.1.5 Solvents

Solvents are encountered in a wide range of composite-related applications, all the way from fiber, resin, and prepreg manufacture to cleaning of tools and molds. Solvents constitute health hazards both through inhalation and skin contact, but are usually not very toxic through ingestion. In direct skin contact solvents readily degrease the skin and may cause dermatitis; already irritated skin is more susceptible to irritation than healthy skin. The all too common practice of using solvents to clean off skin is consequently not recommended. Inhalation of solvent vapors may cause irritation and result in CNS depression. Most solvents are also highly flammable [1,6].

The much-used solvents of the ketone family, which include acetone, methyl ethyl ketone (MEK), and methyl isobutyl ketone (MIBK), have low levels of acute toxicity, are mild to moderate skin irritants, and moderate to severe eye irritants. Although inhalation leads to CNS depression, ketone solvents possess good warning properties [1]. There is some evidence that exposure to acetone and MEK may cause liver damage [6].

8.4.2 Reinforcements and Fillers

Airborne fibers and general machining dust may irritate eyes and respiratory tract. Airborne particles easily penetrate gaps in clothing and may thus cause considerable discomfort to the skin. While discussing inhalation of airborne dust and fibers, a distinction must be made between total dust and respirable dust. Only respirable dust, which has dimensions of less than 3.5–5 µm, is a potential health concern since such small particles are fine enough to penetrate all the way to the alveolar surfaces deep in the lungs where oxygen is taken up and carbon dioxide expelled. Dust with particle dimensions in excess of 7–10 µm is not considered respirable, since such particles are trapped by the normal defense mechanisms and are eventually expelled from the body [1].

Glass fibers used in composite applications normally have diameters in excess of 10 µm and when they fracture they generally retain their diameter; consequently, airborne glass fiber particles are normally too large to be respirable. Carbon fibers, which typically have diameters of 7 µm, may splinter lengthwise when fractured and limited fractions of carbon fiber dust thus may be respirable. Carbon fiber dust may also severely damage electrical and electronic equipment due to its conductivity and its ability to penetrate into the most inconceivable places. With typical diameters of 12 µm, aramid fibers are themselves too large to be respirable. However, when aramid fibers fracture fibrils (splinters), which have dimensions that potentially make them respirable, are created.

Manual handling of unimpregnated glass and carbon reinforcement may in some cases cause mechanical irritation, known as glass itch in the former case, and may lead to dermatitis. Such irritation makes the skin more susceptible to absorption of for example toxic substances. At the edges of a crosslinked composite, glass and carbon fibers or yarns sometimes protrude and may penetrate the skin and break off to cause discomfort. In contrast, aramid reinforcement generally causes no discomfort to exposed skin.

The reinforcement types commonly used in composite applications are not acutely toxic and appear not to have any lasting effects, although the IARC classifies glass fibers as a possible carcinogen from lack of sufficient evidence to the contrary [1]. A fiber type that no longer is in significant use as composite reinforcement is asbestos, which is a well-known carcinogen. Although it is beyond doubt that asbestos fibers may cause cancer, it is not positively established why. Evidence suggests that the carcinogenicity of asbestos fibers lies in their needle-like shape and very small diameters (on the order of single μm). Since the mechanism behind the carcinogenicity of asbestos is not positively established, it is good practice to limit exposure to all reinforcement types despite the lack of evidence of carcinogenicity in the currently common reinforcement types [6]. While the common reinforcement types are not toxic in themselves, fibers are generally coated with organic substances for handling purposes and to enhance properties of the fiber–matrix interface (cf. *Section 2.2.6*). Particularly with reinforcement intended for use with epoxy matrices, where fibers are often coated with an epoxy, there is a risk of sensitization to this epoxy. Irritation or sensitization to this coating or the composite matrix is sometimes mistaken for reaction to the reinforcement.

Most fillers are inert and not toxic, although filler dust may be a temporary irritant. An exception is silica, which may be an inhalation hazard [1].

8.4.3 Preimpregnated Reinforcement

When using prepregs and molding compounds as raw material instead of impregnating the reinforcement with a liquid resin, health hazards for the operator are significantly reduced. One reason is naturally that the raw material supplier has taken care of resin formulation and impregnation, which provide the greatest exposure hazards. Another reason is that the resin is already partially crosslinked, meaning that lower levels of volatiles and reactive groups are present and the inhalation hazard is therefore essentially eliminated. While the hazard to direct skin is significantly reduced, gloves specifically resistant to the matrix in question should be worn during handling of prepregs and molding compounds since sensitization possibly may occur. Another reason for wearing gloves is that the protective oils on the skin may act as a release agent preventing proper consolidation between plies.

8.4.4 Core Materials

In sandwich manufacturing the cores used are generally delivered as blocks that can be assumed inert. The only potential danger common to all cores is from machining dust, which may lead to temporary irritation. With polymer-based cores there is a potential for overheating causing thermal decomposition. As an example, overheating of expanded PS may produce styrene, but other core materials may decompose into substances that are potentially more toxic. While premanufactured cores are inert during normal sandwich manufacture, in-situ foaming of thermoset cores—usually PURs—involves reactive compounds, thus bringing on the same type of health concerns as for the resin in question. The other commonly used sandwich cores, balsa, metal honeycomb, and Nomex honeycomb, should present few if any health concerns.

8.5 Manufacturing Techniques

While the toxicity of the composite constituents are highly relevant, the manufacturing technique largely determines exposure and thus the resulting hazard. *Table 8-2* lists and briefly comments on some of the steps that may give rise to health and safety concerns during manufacturing.

Table 8-2 Examples of exposure possibilities during different manufacturing steps. For all steps the inhalation hazard has been implicitly assumed

Manufacturing step	Exposure possibility
Cleaning of mold	Exposure to toxic solvent
	Direct contact possible
Formulation of gelcoat	Mixing of toxic chemicals
	Stirring promotes evaporation
	Direct contact possible
Application of gelcoat	Application, particularly spraying, promotes evaporation
	Direct contact possible
Cleaning of tools	Exposure to toxic solvent
	Movement promotes evaporation
	Direct contact possible
Crosslinking of gelcoat	Large surface area promotes evaporation
Cutting of reinforcement (possibly preimpregnated)	Reinforcement particles may become airborne
	Direct contact possible
Formulation of matrix	See Formulation of gelcoat
Layup of laminate	See Application of gelcoat
Rolling of laminate	Rolling promotes evaporation
	Skin contact possible
Cleaning or disposal of tools and protective equipment	See Cleaning of tools
Crosslinking of laminate	See Crosslinking of gelcoat
Formulation of topcoat	See Formulation of gelcoat
Application of topcoat	See Application of gelcoat
Cleaning of tools	See Cleaning of tools
Crosslinking of topcoat	See Crosslinking of gelcoat

Although the list primarily applies to open mold techniques, several of the steps are generic and also apply to other techniques.

8.5.1 Mold and Tool Cleaning

Cleaning of molds, resin containers, and tools has traditionally been carried out using solvents, but the current trend is towards compounds that are better for both worker and environment. Since mold cleaning inevitably involves a fair degree of manual work, PPE should be worn. The common practice of removing loose debris from molds with compressed air is generally not a good idea, since it makes dust airborne instead of collecting it; vacuuming is a sounder approach. In the name of worker health, it may be defensible to use disposable tools and resin containers to the greatest degree possible since cleaning leads to significant exposure to both resin and solvents. In all cleaning operations, local ventilation should be used to limit worker exposure and contamination of the workplace air.

8.5.2 Resin Formulation

In most manufacturing situations it is the resin formulation, or mixing (both of gelcoat, matrix, and topcoat), that constitutes the first exposure possibility. Toxic chemicals are handled and the possibility of direct eye and skin contact can and should not be neglected. Appropriate gloves and splash-safe goggles should be used to minimize the possibility of direct contact. While respiratory protection is rarely used during mixing (particularly since exposure time is comparatively short), local ventilation, such as fume hood or trunk, should be used. As mentioned in *Section 8.4.1.2*, it is highly important that resin mixing instructions are carefully followed since there is otherwise a risk of explosive reactions.

8.5.3 Wet Layup

It is symptomatic that the manufacturing techniques most common in small companies with limited financial resources, spray-up and wet hand layup, are the ones associated with the gravest health concerns. In most cases, a gelcoat is first sprayed onto the mold resulting in significant styrene evaporation (cf. *Figure 4-11*). Following crosslinking, the laminate is sprayed or laidup and compacted using a roller (cf. *Figure 4-13*); particularly with spray-up evaporation is considerable (cf. *Figure 4-18*). During both hand layup and rolling the worker usually has his head at arm's length from the laminate, thus inhaling air with high volatile concentrations. The topcoat is generally applied in the same fashion as the gelcoat and brings on the same kind of health concerns. Until the resin in the top layer gels, volatiles are free to evaporate—generally from a large surface area.

With changes in administrative and engineering routines as well as use of improved resins (cf. *Section 8.4.1.2*), the risk of exposure may be notably reduced in all steps of wet layup. The most severe exposure problem is through inhalation and the most effective means to improve the work environment consequently is ventilation (cf. *Section 8.3.2*), since respirators should only be used as a last resort. Booth ventilation to create a high flow rate around the operator is the best solution in spray-up (cf. *Figure 8-2*). If correctly designed and used, booth ventilation may reduce exposure to below TWAs. The degree of evaporation during spray-up may be reduced using airless spray guns and lower pressures, although too low pressures may force the operator to stand closer to the mold, thus increasing exposure. Due to the very nature of hand layup the operator has the laminate at arm's length, which inevitably leads to significant exposure unless ventilation is adequate. A way to reduce exposure during layup is to use a resin-dispensing roller (cf. *Figure 4-15*). In terms of ventilation, push-pull (cf. *Figure 8-4*) and local (cf. *Figure 8-3*) ventilation are the most effective solutions.

In most kinds of wet layup the laminate has to be rolled to complete impregnation and work out voids and rolling promotes evaporation, partially due to the creating of airborne droplets with large surface area-to-volume ratio, see *Figure 8-8*. Since rolling often takes place using rollers with short handles, rolling also may lead to significant exposure. Although long handles may halve exposure during rolling, there will always be a need for short handles in confined spaces and critical areas (cf. *Figure 4-13*). Use of a roller splash guard further reduces exposure, since the portion of the fine resin mist thrown off the roller due to centrifugal forces reaching the operator's breathing zone is reduced, see *Figure 8-8*. In many countries it is for good reason prohibited for one worker to roll a laminate that is simultaneously sprayed up at the other end of the mold.

Figure 8-8 Schematic of effect of splash guard on roller (right) to be compared to spread of resin mist without splash guard (left)

Approximately 80 percent of volatile emissions originate from the top layer of a laminate, so it is good practice to finish a laminate once lamination has started and not use intermediate crosslinking unless it is absolutely necessary [3]. Once layup and rolling has been completed the laminate should be left to crosslink in an enclosed space or under adequate local ventilation; a "tent" completely covering the laminate attached to local ventilation is an effective yet inexpensive solution. LSE resins provide another effective means of significantly reducing styrene emissions from a laidup laminate left to crosslink.

The risk of direct eye and skin contact is considerable throughout all steps of wet layup and covering clothing, gloves, and eye protection should consequently be used at all times. In wet layup of for example (storage) tanks and boats it is not uncommon that work has to be performed both upside down and in confined spaces. Under such conditions, protection to completely eliminate the possibility for direct contact as well as the use of air-supplying respirators are generally necessary, since the risk of exposure is much greater than under most other conditions.

8.5.4 Prepreg Layup

There are few health hazards associated with prepreg layup. Avoiding direct skin contact with the prepreg through use of gloves is in general sufficient, although respiratory protection may be required in rare cases. If cutting is performed manually the hand opposite the knife hand should be protected by both a cut-resistant and a resin-resistant glove (often meaning double gloves). Most composites manufactured from prepregs are consolidated in autoclaves, which may bring on some additional safety issues. During crosslinking gaseous products that may be both a source of discomfort and a health hazard are created. To avoid exposure to these byproducts, the autoclave should be purged prior to opening, i.e. the internal atmosphere should be replaced. The temperature in an autoclave may be high enough to cause fires should oxygen be present and inert pressurization gases (normally nitrogen) are therefore used to eliminate this possibility. Due to the use of inert gases, there is then a risk of an oxygen-deficient atmosphere in and near a newly opened autoclave.

8.5.5 Liquid Molding

One of the major advantages of all liquid molding processes is that they use closed molds, i.e. at no point in time, except in some cases during formulation, is the unreacted resin free to evaporate to any significant degree. While there is naturally a resin–air interface at the advancing resin front, the interface area is negligible compared to the exposed areas in techniques using open molds, and the air escaping, or being drawn out of, the mold, is often ventilated to the external atmosphere. Consequently, inhalation concerns are minimal. Molds are often heated and temperature-resistant gloves may be needed for demolding (cf. *Figure 4-39*).

8.5.6 Compression Molding

While preparing the charge, the SMC is normally handled manually and gloves should therefore be worn to eliminate direct contact. Just as with manual cutting of prepregs, the hand opposite the knife hand should be protected against both knife and resin (cf. *Figure 4-43*). Since the mold is heated and the components emerge hot from molding, temperature-resistant gloves should be worn during charging (cf. *Figure 4-44*) and demolding (cf. *Figure 4-45*).

8.5.7 Filament Winding

The health hazards associated with filament winding stem from evaporation at the impregnation station and the exposed external composite surface. Both problems may be alleviated with partial or complete enclosure of these emission sources and ventilation of volatiles to the outside, see *Figure 8-9*. Since the technique is highly automated the risk of exposure is essentially limited to removal of wound components and rethreading of the machine upon commencement of winding of the next component. Due to the short exposure times involved, inhalation hazards are small and the possibility of skin contact is easily minimized using disposable gloves. Following completion of winding, components are normally crosslinked in an oven that also should be ventilated to the outside since emissions may be considerable. The crosslinked components exit the oven hot and temperature-resistant gloves are thus likely needed.

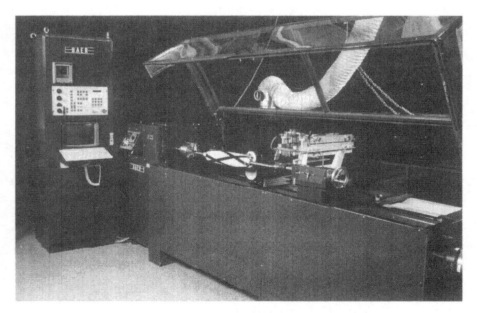

Figure 8-9 Illustration of completely enclosed filament winding facility attached to an exhaust trunk. Photograph courtesy of Swedish Institute of Composites, Piteå, Sweden

8.5.8 Pultrusion

Inhalation hazards in pultrusion are highly dependent on the impregnation technique. With the injection–pultrusion solution illustrated at the bottom of *Figure 4-57* the exposure possibility is minimal. In contrast, the other impregnation options shown in *Figures 4-53, 4-57,* and *4-55,* which have open resin baths, necessitate local ventilation to reach the volatile levels required in most industrialized nations. It is hardly surprising that injection pultrusion is most common in countries where PELs for styrene are the lowest (and vice versa), cf. *Table 8-1*. In this context it is a problem that many pultruders are small or very small companies with resources so limited that they feel they cannot afford adequate ventilation or a change to injection pultrusion. Local ventilation of an open impregnation bath may conceptually be accomplished in the same fashion as shown in *Figure 8-9*. Pultrusion is essentially completely automated and the operator rarely works near unreacted resin and normally only handles the fully crosslinked component. The continuously emerging composite is cut to specified lengths and there is a potential dust problem associated with this cutting, although local ventilation easily solves this problem. The pultrusion die is quite hot and there is a theoretical possibility of burns.

8.5.9 Thermoplastic–Matrix Techniques

As already discussed in previous sections, there are few health and safety issues relevant to the processing of thermoplastic composites other than the risk of direct contact with molten matrix, the possibility of thermal decomposition of the matrix into toxic by-products upon overheating, and the fact that molds are significantly hotter with thermoplastics than with thermosets. Molds should consequently be thermally insulated (not least because insulation also lowers the electricity bill) and even greater care taken to wear temperature-resistant protective gloves.

8.6 Machining

Probably the second greatest health concern in composite-manufacturing related activities after exposure to toxic substances is exposure to dust and decomposition products from machining of crosslinked or solidified components. Part of the problem stems from the fact that the chips produced with conventional cutting techniques are significantly finer than when machining traditional construction materials. Although it is likely that few particles small enough to be respirable are produced, dust from machining can be a serious nuisance, primarily to eyes and upper respiratory tract, but also to skin. The most common and by far the most effective way to reduce exposure to machining dust is high-capacity local ventilation at or on the cutting tool (see *Section 8.3.2*). If such ventilation is not available, dust masks and covering clothing provide adequate protection unless concentrations are excessive. Another possibility for dust control during machining is offered by water dousing of the cutting zone, although not all composites fare well from excessive water exposure. With the exception of the hazards of epoxy dust mentioned in *Section 8.4.1.3*, machining dust containing conventional reinforcement and matrices appears to be a temporary irritant only.

With AWJs machining dust is contained in the water and is washed away from the cutting zone thus eliminating any inhalation hazard. There is, however, a danger associated with the jet itself and it should therefore be impossible to get in its way. AWJs are also quite noisy and ear protection may be required.

In laser machining the material is thermally decomposed and toxic airborne substances are possibly created. While laser machining facilities generally include a coaxial gas jet to remove and contain waste products, good local ventilation is a necessity, since dust particles smaller than 1 μm are created [1]. The laser beam itself constitutes a danger, so safety glasses should be worn and procedures adopted to eliminate accidental exposure.

8.7 Summary

The two main health concerns associated with composite-manufacturing related activities are exposure to toxic substances and airborne particulates. Exposure may be in the form of inhalation, absorption, ingestion, and injection, and the former two are the most common and most dangerous exposure routes. Inhalation of toxic substances irritates the mucous membranes of the respiratory tract and may lead to CNS depression. Skin and eye absorption of toxic substances irritates the tissue and may lead to dermatitis. Both inhalation and absorption may lead to sensitization.

Exposure hazards may be drastically reduced through administrative controls, engineering controls, and use of PPE. Administrative controls include enforcement of sensible and safe working practices as well as limiting each individual's exposure through worker rotation, for example. The most effective engineering control means is ventilation, which should strive to extract contaminants as close to the generating source as possible. A ventilation system needs to be professionally designed, installed, and maintained to achieve proper functionality. While clothing and hand and eye protection may always be required to reduce exposure hazards, respiratory protection should only be seen as a temporary solution until administrative and engineering controls are implemented to eliminate the need for such protection.

Thermoplastic resins are not toxic under normal processing conditions. With unsaturated polyesters and vinylesters the main health concerns are due to the styrene. Styrene has low acute toxicity and excellent warning properties, as well as low chronic toxicity; the main hazard is inhalation. Sweeping statements about the toxicity of epoxy systems are difficult to make since variation possibilities are vast. Most high molecular-weight epoxies have low acute toxicity, whereas reactive dilutents generally have higher toxicities courtesy of their lower molecular weight. The greatest potential health hazard with epoxies is absorption of the hardeners which are often highly toxic. Some components of phenolics and PURs are also highly toxic. In contrast to matrices, common reinforcement types are not toxic and at worst cause temporary irritation, although this irritation may make skin more susceptible to absorption of organic substances. Prepregs and molding compounds effectively eliminate most concerns of inhalation of toxic substances and absorption.

Throughout manufacturing the main exposure routes are inhalation and absorption. The gravest concerns are with open-mold techniques and in-process impregnation, since intense manual labor is involved. Implementation of appropriate administrative controls, efficient ventilation, use of LSE resins, elimination of bare skin, and use of suitable gloves are effective means of significantly reducing exposure hazards. In prepreg layup onto open molds the health concerns are practically eliminated as long as appropriate gloves are worn. The closed molds in liquid molding effectively eliminate most exposure possibilities and thus offer attractive alternatives to open-mold techniques. In compression molding of molding compounds exposure hazards are low as long as gloves are worn. In their most common incarnations, filament winding and pultrusion provide ample opportunities for volatile emissions and direct contact. However, both techniques are automated and easily enclosed to eliminate accidental contact and provide efficient ventilation. All techniques employing heated molds bring on a possibility for burns from mold and demolded component. Since thermoplastic resins are not toxic under normal processing conditions the concerns in manufacturing are limited to contact with heated mold, molten resin, and inhalation of potentially toxic decomposition products from overheating.

Dust from machining of composites using conventional cutting techniques may be a considerable nuisance both in terms of inhalation and direct contact, but since such dust generally is not respirable it is a temporary irritant. In most machining operations the workpiece is heated and organic matter thermally decomposed. Decomposition products are potentially toxic and adequate ventilation is necessary.

This chapter probably does not provide pleasant reading. Although the hazards described herein should not be underestimated ("better safe than sorry"), it must be understood that most of what has been said concerning adverse health effects is based on studies on laboratory animals under conditions of extreme exposure and on experience from occupationally exposed workers (i.e. in general 40 hours a week, year after year). The bottom line is that with appropriate safety procedures and equipment, composites can be manufactured and machined without creating significant health and safety hazards to the operator. The key is knowledge and understanding of the potential hazards and enforcement of established safety routines.

Disclaimer

The contents of this chapter are only intended as an introduction to health and safety issues in activities related to manufacturing of polymer composites; it does not constitute a complete, nor necessarily accurate, listing of toxicities, hazards, risks, and appropriate precautions. MSDSs provide the most detailed information available in terms of toxicity and health and safety precautions, while government regulations put forward local legal requirements and PELs in handling of toxic materials. While each individual should familiarize himself with the toxicity of materials to be used prior to handling them, it is the responsibility of the employer to ensure that personnel receive proper training and that established routines are followed.

Suggested Further Reading

Since the potential health and safety issues in composite-manufacturing related activities encompass a wide range of disciplines, it is impossible to find any one good text covering all areas. MSDSs, which for many substances are available on the Internet, provide detailed information on substance toxicity and health and safety precautions. Governments of most industrialized nations determine legal requirements in terms of, for example, PELs and working conditions and generally also publish guides suggesting safe working practices and provide general guidance. Some trade organizations also issue publications which provide guidance in terms of handling of toxic organic substances; examples of the latter are:

UP-Resin Handling Guide, European Organisation of Reinforced Plastics/Composite Materials (GPRMC), Brussels, Belgium.

Safe Handling of Advanced Composite Materials, 3rd edn., SACMA, Arlington, VA, USA, 1996.

References

1. *Safe Handling of Advanced Composite Materials*, 3rd edn., SACMA, Arlington, VA, USA, 1996.
2. G. Isaksson, "Metoder att reducera styrenexponering vid AP-produktion," Swedish Plastics Federation, Stockholm, Sweden, 1976.
3. *UP-Resin Handling Guide*, GPRMC, Brussels, Belgium.
4. *Härdplaster*, The Swedish Occupational Safety and Health Administration,

AFS 1993:4, Publikationsservice, Solna, Sweden, 1993.

5. Anon., "Running a Green Production Plant," *Reinforced Plastics*, 24–36, March 1995.

6. J. W. Braddock, "Safety and Health," in *International Encyclopedia of Composites*, Ed. S. M. Lee, VCH Publishers, New York, NY, USA, **5**, 85–93, 1991.

INDEX